KNOTS OF TIME
n. 1

Book Series directed by
Alessandra Papa

Scientific Board
Roberto Dell'Oro, Roberto Diodato, Alessio Musio, Alessandra Papa,
Adriano Pessina, Elena Postigo Solana

Editorial Board
Luca Alici, Luigi Alici, Antonio Allegra, Ingrid Basso, Roberta Corvi,
Carla Danani, Giuseppe D'Anna, Antonio Da Re, Francesco Fistetti,
Alessandra Gerolin, Francesco Miano, Donatella Pagliacci, Enrico Peroli,
Silvia Pierosara, Paolo Ponzio, Maria Vita Romeo

The text is peer-reviewed in a double-blind refereeing process

Quaderni di ricerche filosofiche del CrifipAB
Centro di ricerca sulla filosofia della persona Adriano Bausola
Università Cattolica del Sacro Cuore di Milano

Roberto Dell'Oro

BIOETHICS AT THE BOUNDARY

Explorations between Philosophy and Theology

Cover Image: Freepik - Midjourney

© 2024—Mimesis International
www.mimesisinternational.com
e-mail: info@mimesisinternational.com

Isbn: 9788869774850
Book series: *Knots of Time*, n. 1

© MIM Edizioni Srl
P.I. C.F. 0241937030

TABLE OF CONTENTS

ACKNOWLEDGEMENTS 9

INTRODUCTION 13

PART I
BIOETHICAL OPENINGS

CHAPTER 1: THE ETHOS OF AMERICAN BIOETHICS:
BETWEEN LIBERAL INDIVIDUALISM AND SOCIAL SOLIDARITY 19

CHAPTER 2: METHOD AND MEANING: IN SEARCH OF A
RICHER BIOETHICS 31

CHAPTER 3: ON THE ULTIMATE, WHICH IS THE FIRST:
THINKING BEYOND BIO (ETHICS) 55

CHAPTER 4: REPRODUCTIVE CLONING, MARRIAGE, AND FAMILY:
IDEOLOGICAL PREMISES AND FORGOTTEN ISSUES 87

CHAPTER 5: HUMAN GENOME EDITING: A *PROLEGOMENON* 107

CHAPTER 6: ON THE ETHICS OF TWINS SEPARATION:
THE FORGOTTEN CASE OF BABIES JODY AND MARY 123

CHAPTER 7: BETWEEN RESISTANCE AND SURRENDER: IN SEARCH OF
AN ANTHROPOLOGICAL HORIZON FOR AN ETHICS OF DYING WELL 143

CHAPTER 8: CONFRONTING THE DOBBS DECISION:
ON THE *LEGALITY* OF ABORTION 159

PART II
PHILOSOPHICAL PROBINGS

CHAPTER 9: INTERPRETING CLINICAL JUDGMENT:
EPISTEMOLOGICAL NOTES ON THE PRAXIS OF MEDICINE   169

CHAPTER 10: WHY CLINICAL ETHICS? EXPERIENCE, DISCERNMENT,
AND THE ANAMNESIS OF MEANING AT THE BEDSIDE   189

CHAPTER 11: THE PHYSICIAN PATIENT RELATIONSHIP: A PHILOSOPHICAL
PERSPECTIVE ON MEDICAL PROFESSIONALISM AND VIRTUES   205

CHAPTER 12: THE MARKET ETHOS AND THE INTEGRITY OF
HEALTH CARE   219

CHAPTER 13: EMBODIMENT AS SATURATED PHENOMENON: MEDICINE,
THEOLOGY AND SOME METAPHYSICAL PREMISES OF MODERNITY   227

CHAPTER 14: CAN A ROBOT BE A PERSON?
DE-*FACING* PERSONHOOD AND FINDING IT AGAIN WITH LEVINAS   245

CHAPTER 15: *PASSIO ESSENDI*: ON SUFFERING   275

PART III
THEOLOGICAL EXPANSIONS

CHAPTER 16: THEOLOGICAL DISCOURSE AND THE POSTMODERN
CONDITION: THE CASE OF BIOETHICS   283

CHAPTER 17: RECASTING FUNDAMENTAL MORAL THEOLOGY:
NOTES ON THE CHRISTOLOGICAL ANTHROPOLOGY OF KLAUS DEMMER   305

CHAPTER 18: CONSCIENCE AFTER VATICAN II: THEOLOGICAL PREMISES
FOR A DISCUSSION ON CATHOLIC HEALTH CARE   329

CHAPTER 19: INSIGHTS FOR A MORAL THEOLOGICAL REFLECTION
ON CONSCIENCE: POPE FRANCIS' APOSTOLIC EXHORTATION
AMORIS LAETITIA   349

CHAPTER 20: SHOULD WE HOPE FOR A CURE?
RAISING A CHILD WITH AUTISM   379

CHAPTER 21: LIFE TAKEN, LIFE GIVEN:
META-ETHICAL REFLECTIONS ON THE COVID-19 PANDEMIC   393

We have looked too low
The ground beneath us
Falls away
& joy leaps up in us
Out of nothing
Leaps out of itself
Leaps out of more than itself
& the elemental world is there
Again.

William Desmond, *God and the Between*

ACKNOWLEDGMENTS

Versions of many of the chapters in this book appeared in earlier publications. Some of them have been revisited and revised, others appear here for the first time. I thank the publishers for permission to use my work.

Chapter 2 (with Warren Reich), as «A New Era for Bioethics: The Source of Meaning in Moral Experience,» in (ed.) Allen Verhey, *Religion and Medical Ethics: Looking Back, Looking Forward* (Grand Rapids, MI: Eerdmans, 1996): 96-119.

Chapter 3, as «On the Ultimate That Is the First: Thinking Beyond Bioethics,» *Gregorianum* (2019): 621-647.

Chapter 4, as Cloning, Marriage, and Family: Ideological Premises and Forgotten Issues,» *INTAMS review* 11(2005): 67-79.

Chapter 6, as «On the Ethics of Twins Separation: Analysis of a Recent Case,» *Itinerarium* 10 (2002): 89-106.

Chapter 7, as «Between Resistance and Surrender: In Search of an Anthropological Horizon for the Ethics of Dying Well,» *Itinerarium* 30 (2022) 80-81, 145-155.

Chapter 9, as «Interpreting Clinical Judgment: Epistemological Notes on the Praxis of Medicine,» in (ed.) Corrado Viafora, *Clinical Bioethics: A Search for the Foundations* (Dordrecht: Springer Academic Publisher, 2005), 155-168.

Chapter 10, as «Why Clinical Ethics? Experience, Discernment, and the Anamnesis of Meaning at the Bedside,» *Persona y Bioetica* vol. 20 (2016)1: 86-99.

Chapter 11, as «The Physician Patient Relationship: A Philosophical Perspective on Medical Professionalism and Virtues,» Marco Curcovic and Anna Borovechi, ed., *Medical Professionalism: Bridging Bioethics and Medical Practice* (Dordrecht: Springer Verlag, 2022), 217-229.

Chapter 12, as «Market Ethos and the Integrity of Health Care,» *Journal of Contemporary Health Law and Policy*, 18 (2002) 3: 641-647

Chapter 13, as «Embodiment as Saturated Phenomenon: Medicine, Theology, and Some Metaphysical Premises of Modernity,» in *International Journal of Philosophy and Theology*, vol. 2, n. 4 (2014): 69-84

Chapter 14, as «Can a Robot Be a Person? Defacing Personhood and Finding It Again with Levinas,» *Journal of Moral Theology*, XI/1 (2022): 132-156

Chapter 15, as «Suffering and the Joy of Life Put to the Test,» in *Etica teologica della vita: scrittura, tradizione, sfide pratiche*, ed. Vincenzo Paglia (Citta' del Vaticano: Libreria Editrice Vaticana, 2022), 345-350. Chapter 16, as «Theological Discourse and the Postmodern Condition: The Case of Bioethics,» *Medicine, Health Care and Philosophy*, 18 (2002) 3: 127-136.

Chapter 17, as «Recasting Fundamental Moral Theology: Notes on Klaus Demmer's Christological Anthropology,» in *Gregorianum* 93/3 (2012): 463-483.

Chapter 18, as «Conscience after Vatican II: Theological Premises for a Discussion on Catholic Health Care,» in David De Cosse and Thomas Nairn, ed. *Conscience and Catholicism: From Clinical Context to Government Mandates* (Maryknoll: NY, Orbis Books, 2017), 33-48.

Chapter 19, as «Insights for a Moral Theological Reflection on Conscience: Pope Francis's Apostolic Exhortation *Amoris Laetitia*,» *Marriage, Families &Spirituality* 26/2 (2020): 208-231.

Chapter 20, as «Should We Hope for a Cure? Raising a Child with Autism,» in Aldegonde Brenninkmeijer-Werhahn, ed., *Marriage – Constancy and Change in Togetherness* (Berlin: LIT Verlag, 2017), 199-208.

Chapter 21, as «Life Taken, Life Given: *Meta*-Ethical Reflections on the COVID-19 Pandemic,» in Adriano Pessina, ed., *Vulnus: Persone nella Pandemia* (Milano: Mimesis Edizioni, 2022), 153-171.

Acknowledgments

I owe a debt of gratitude to my research assistant at the Bioethics Institute of Loyola Marymount University, Trevor McCarthy, MA for the extraordinary competence in editing the original manuscript.

I also would like to thank Prof. Alessandra Papa and Prof. Adriano Pessina, from the Catholic University of Milan, for their encouragement and support in producing this book and suggesting Mimesis International as the venue for its publication.

I dedicate this book to Regina, Marco and Sandro, my family. Their love has informed these pages, sustaining my efforts, and calling me back to gratefulness for the life we share.

INTRODUCTION

The book I am introducing has a long history. It represents a kind of synthesis, both conceptual and existential, of my engagement with the field of bioethics over the span of, now, almost thirty years. The results of my journey, evidenced in the chapters that follow, do not entirely fit the *orthodox* definition of bioethics, when considering the latter's method, preoccupations, and theoretical developments in the Anglo-American context. It is almost like a journey «at the boundary,» between philosophy and theology. Thus, the title of this book, whose liminality and, perhaps, fragmentation, almost expresses a search for more systematic completion. The latter will come soon in a monograph.

I pursued studies in the field of bioethics following undergraduate and graduate work in philosophy and theology in Milan, Rome, and Munich. This led to a doctorate, in 1992, at the Pontifical Gregorian University under the direction of Klaus Demmer. Afterwards, I did bioethics at Georgetown University's Kennedy Institute of Ethics, with the late Edmund Pellegrino as a mentor. Georgetown came to me as an offer of scholarly engagement in the form of a post-doctoral fellowship, made possible by the generosity of the *International Study Group on Bioethics* (ISGB) led, at the time, by a visionary Catholic bishop, Cardinal Carlo Maria Martini, from Milan. Since the early '80s, Martini had brought together, on a regular basis, scholars from various Catholic universities, philosophers and theologians, to address, in interdisciplinary fashion, some of the emerging issues of the then burgeoning field of bioethics.[1]

---

[1] Klaus Demmer began documenting the work of the ISGB in 1986, with an article for *Herder Korrespondenz*, followed by another one in 1988. See Klaus Demmer, «Das bioetische Gespräch: Initiativen katholischer Universitäten,» in *Herder Korrespondenz* 40 (1986): 489-93, and «Orientierungsversuche auf schwierigem Feld. Ein Symposon über Fragen der Bioethik,» in *Herder Korrespondenz* 42 (1988): 438-41. Two volumes from the series *Philoso-*

The invitation of Martini to pursue a specialization in bioethics at Georgetown came with the additional support of my *Doktorvater*, Klaus Demmer, who was one of the most revered scholars in the group. Thus, began for me a new intellectual adventure in America.

I speak of «new» adventure in several respects. First, in terms of its existential challenges. I moved to the States, an unfamiliar context, much different from Europe, leaving my country, family, and friends. Secondly, I knew little of American culture and, more specifically, of the Anglo-American philosophical tradition. My training had been squarely in the European stream of «Continental philosophy,» which, I learned, had to be distinguished from the «Analytic» proclivity of English-speaking philosophers.

I quickly realized that the field of American bioethics, though exciting for the questions it addressed, especially those in the areas of reproductive technologies and genetics, had been barely nourished by the theoretical traditions with which I was most familiar. I had to confess my humble surrender to a different mindset, learn its theoretical grammar and developments, all the while trying to create connections, measuring differences, and envisioning potential overlapping. Mine has been a difficult hermeneutical exercise, in search of something like a «fusion of horizons.»

The reader will find traces of this perhaps ambitious attempt in the chapters that follow. The topics I address are all rooted in the American discourse but follow a theoretical development that might appear surprising to the average American scholar of bioethics. Broadly speaking, I feel that the usual reduction of bioethical discourse to the orchestration of four ethical principles, autonomy,

---

phy and Medicine, vol. 34 and vol. 41, published by Kluwer, speak to the initiative: Edmund Pellegrino, John Langan, and John Collins Harvey, eds., *Catholic Perspectives on Medical Morals: Foundational Issues* (Dordrecht: Kluwer Academic Publishers, 1989); Kevin Wm. Wildes, Francesc Abel, and John Collins Harvey, eds., *Birth, Suffering, and Death: Catholic Perspectives at the Edges of Life* (Dordrecht: Kluwer Academic Publishers, 1992). Another book, resulting from a conference at Georgetown, with a preface by Francesc Abel, references the initiatives of the International Study Group in Bioethics (ISGB), in the context of the International Federation of Catholic Universities. See International Federation of Catholic Universities, *Human Life: Its Beginnings and Development. Bioethical Reflections by Catholic Scholars* (Louvain-la-Neuve: CIACO, 1988). The book is dedicated to the memory of Andre' Hellegers.

beneficence, non-maleficence, and justice, remains inadequate to the deeper requirements of theoretical sophistication called for by the enormous challenges posed by today's developments in science and technology.

I put the matter with reference to a kind of thought experiment: I imagine the pioneers of bioethics looking on to the field, today, many of them gone by now. I wonder what they see. I mean, how would they judge our methodological instabilities, the spectacle of widespread conceptual shallowness, too thin to provide meaningful signposts to our ethical quandaries? Our scholarly fore fathers (and mothers) might ask us to step back, to retrieve a more *critical* stance: without fearing the disapproval of the scientific community; without subservience to the allures of public visibility, short lived in the show of media-fed self-importance. Chocked by the impelling demands of technological advancements, we have become unable to breathe properly. I mean, breathe *deeply* in a philosophical sense. What would taking a deep breath entail? I submit that, at a minimum, it might entail retrieving a broader horizon of meaning for our own reasoning. It might also call for a different attunement to what, later on, I will refer to as the *primal ethos of life*. The first calls for an epistemological recharging. The second for renewed metaphysical sensibility.

The long statement might help the reader make sense of the various parts of the book. Its various chapters can be read separately, since each of them has its own integrity, and stands rather independently of the others. Thus, the reader should not expect a progressive unfolding of the content, even if, in the end, one can identify something like a unified *Gestalt*.

Part I, *Bioethical Openings*, offers a number of essays dealing with questions of method, together with issues currently under scrutiny in the conversation of bioethics scholars; among others, cloning, human genome editing, euthanasia, assisted suicide, and abortion.

Part II, *Philosophical Probings*, takes up topics of a more obvious philosophical nature, spanning epistemological, anthropological, and social-ethical questions.

In Part III, *Theological Expansions*, I offer a broader vista, pushing philosophical mindfulness to the point where it merges with the preoccupations of theology. The work of earlier practitioners of bioethics, many of them theologians, was precisely of such nature. In fact, what made their contributions so substantive might have something to do with their background premises, informed by metaphysical confidence and thick anthropological notions, rooted in theological categories, yet open to public argumentation.[2]

In the end, I am not pleading for a *confessional* exercise, driven by missionary intention. Nor is the quality of my presuppositions lost in the «rational» accessibility of the normative solutions I am suggesting. For sure, a bioethics theologically informed entails a kind of reversal of perspectives with respect to the prevailing methodological orthodoxy, which, on the other hand, tends to see any hospitality to religious content by philosophy as a dangerous promiscuity, unfaithful to the secular purity of its premises. Such «heterodox bioethics» stands between two opposite extremes: the claim to neutrality of moral discourse, articulating a purely secular account of ethics, and shunning hospitality to theological contributions, on the one hand; the colonizing pretense of theological discourse to entirely subsume secular reason under its religious premises, on the other.

I hope this book will invite to a different dialogue between philosophy and theology in bioethics, one defined by dialogical confidence more than dialectical antagonism. The testimonial character of theological discourse requires, for sure, witness to the good in addition to mindfulness of ethical quandaries. But the poverty of theology, like that of philosophy, will not castigate the dynamism of the spirit into critical submission, in Kantian fashion; nor lead theology to reject its own limits in a dialectical movement of progressive self-surpassing, as in Hegel. It will rather open any rationalistic self-imposed limitation or, conversely, self-proclaimed pretention, to a different recognition: that the adventure of thought finds its completion beyond itself, in a transcending movement of self-giving, into the vast, yet promising, ocean of life itself: the *bio*s of bio-ethics.

---

[2] I am thinking of Hans Jonas, James Gustafson, Paul Ramsey, Karen Lebacqz, Edmund Pellegrino, David Thomasma, Leon Kass, Albert Jonsen, and Warren Reich. But this is still a short list.

PART I
BIOETHICAL OPENINGS

CHAPTER 1
THE ETHOS OF AMERICAN BIOETHICS
Between Individual Liberalism and Social Solidarity

Bioethics is a relatively new field of study characterized by an interdisciplinary approach to ethical questions raised by advances in biotechnologies and the life sciences. Its definition is well known, at least in the version that Professor Warren Reich, a scholar at Georgetown University, crafted for the *Encyclopedia of Bioethics* in the late '70s: «Bioethics is a composite term derived from the Greek word *bios* (life) and *ethike* (ethics). It can be defined as the systematic study of human conduct in the area of life sciences and health care, insofar as this conduct is examined in light of moral values and principles.»

For the past forty years, the field of bioethics has not only enriched scholarly reflection, but also contributed to a richer understanding of the implications of medical research for society at large. However, although an established academic field, bioethics has struggled to find a clear methodology, to articulate a common philosophical language for all the disciplines involved, and to define its specific location in the intellectual discourse of our pluralistic moral landscape.[1]

Bioethical questions can never be settled on the basis of one perspective alone: interdisciplinary enterprises are, by definition, continuous efforts. Yet the lack of a sense of finality in bioethics can hardly be understood as the result of methodological instability only.

I suggest to look at the complexity of American bioethics as the necessary result of the larger framework within which it operates. More specifically, the difficulty in coming to conclusive convictions about specific ethical issues, most recently whether to pass health care reform legislations, allocate public funding for research

---

[1] The methodological question is addressed by Albert Jonsen together with the context of bioethics' history. See his *The Birth of Bioethics* (New York: Oxford University Press, 1998).

on embryonic stem cells, or enact provisions at a state level that legalize physician assisted suicide, depends upon larger notions of a social and, ultimately, political nature.[2]

I call such a framework, the *ethos* of American bioethics. The presupposition here is that we never think about the morality of our actions or about criteria for conduct in a vacuum. Terms that circulate within ordinary discourse, such as «justice» and «freedom,» are also central themes within social and political thinking. This has practical implications for bioethics. No matter how strenuously bioethicists may hope to isolate their perspective from wider civic imperatives, social and political theory frames and penetrates all bioethical considerations. Indeed, to reiterate a point deeply rooted in the philosophical tradition, to separate politics from ethics is to fail to understand both.

This chapter comprises three basic considerations. In the first, I make general claims on the relation between ethics and political theory. In the second, I look at contemporary debate in social and political theory as reflected by the bioethical field. In particular, I will deal with the debate between liberal and communitarian thinkers, with a small foray into feminist theories in bioethics. Finally, I consider the importance of the debate just mentioned for understanding the relationship between the individual, the family, and the state.

1. *The Normativity of Social and Political Theory*

Terms of ordinary moral discourse function like a conceptual prism through which we view different human relationships, activities, and forms of life. Most of the time we take such terms for granted. In the West, notions like rights, individual freedom, autonomy, and justice, have become part of ordinary language, yet the interpretation of their meaning may generate different understandings

---

[2] «Some would want to separate ethics and politics, but while the two might not be identical, a metaxological philosophy of the between cannot be blind to a certain porosity between them. All forms of community ultimately bear on what we love, as Augustine stresses, and how are loves are expressed and organized in shared ways of life,» William Desmond, *The Intimate Universal: The Hidden Porosity Among Religion, Art, Philosophy, and Politics* (New York: Colombia University Press, 2016), 173.

of human capacities for purposive activity and, ultimately, different normative conceptions of the society in which we live. Therefore, one task of the ethicist is to examine critically the resources of ordinary language and to reveal their latent meanings and nuances.

Consider, for an example, our understanding of the terms «public» and «private».[3] *Publicus* is that which belongs to, or pertains to, «the public,» the people. As such, the public is «open to scrutiny.» «Private,» on the other hand, defines a sphere of intimacy not subjected to the persistent gaze of publicity. In light of this understanding of the distinction between public and private, defenders of constitutional democracy have long insisted on the protection of privacy as the condition for preventing government from becoming intrusive. The distinction also helps articulating the difference in the relationships that define the social engagement of the self: to be a mother or a father, for an example, is different from being a citizen; to be a friend, different from being a public official. Of course, it is inescapable for us to be involved in a number of competing ethical or normative perspectives. The way to solve the possible conflict of opposing claims will be influenced by what we take to be the appropriate relationship between «private» and «public» life. This, in turn, will define our understanding of what politics should or should not attempt to define, regulate, and even control.

There is, however, widespread disagreement over the respective meaning of public and private within societies. The boundaries between the public and the private help to create a moral environment for individuals. They establish norms for what is appropriate or worthy actions, and raise barriers to actions in different areas, particularly the taking of human life, regulation of sexual relations, promulgation of familial duties and obligations, and the arena of political responsibility.

In the history of Western political thought, public and private imperatives, concepts, and symbols have been ordered in a number of ways. They include the demand that the private world be fully integrated within the public arena; the insistence that the public sphere be «privatized,» with politics controlled by the standards,

---

3   For a brilliant reconstruction of the meaning of the distinction, see William Desmond, *The Intimate Universal*, 156-197: «Politics and the Intimate Universal.»

ideals, and purposes emerging from a particular vision of the private sphere; or, finally, a continued differentiation and bifurcation between the two spheres.

Bioethics is deeply implicated in each of these broad, general theoretical tendencies that often touch on the private and the public. Consider the example of a couple who decides to conceive a child through artificial insemination by donor (AID), and to establish surrogacy agreements with a woman to carry the couple's child to gestation. In spite of the relative frequency of the practice, which, for good or for ill, counts California as its pioneers, one cannot fail to wonder about the impact of this particular practice upon society's view of the family and of inter-generational ties: What is the effect on the psycho-social development of donor children? What are the responsibilities, if any, of the donor father beyond the point of sperm donation for a fee? What those of the surrogate mother? Do contractual agreements suffice to «cover» not just the legal but also the ethical implications of such arrangements?

All these questions could simply be solved by an appeal to *privacy*, justified on the basis of a presumed right to «procreative liberty». In this view, such questions are considered the exclusive business of the individuals involved in the contractual transaction at stake. Yet, other questions loom large: does society have a legitimate interest in such «private» choices, given the potential social, and in that sense *public*, consequences of private arrangements? More specifically, are decisions about what sort of child to have, and what means to use in the creation of a child, merely the flip side of decisions as to *whether* to have a child at all – that is, decisions about contraception and abortion? As bioethicist Tom Murray points out «advocates of procreative liberty fix on the free choices of autonomous adults. But abortion and contraception are means *to not have* a child... The not-so-flip side is the decision *to have* a child, to create a new person who will have interests, hopes and concerns of her or his own.»

2. *Contemporary Debates in Social and Political Theories*

Though coming in different forms, comprising philosophical traditions as varied as utilitarianism, Kantianism, and rights theory, the tradition of *liberalism* stresses the individual and his or her rights, often downplaying notions of duty and obligation to a wider social whole. Liberal thinkers assume, optimistically, that each individual's pursuit of self-interest will result in «good» for society as a whole. Those whose analyses begin with the free-standing individual as their point of reference, and the «good» of that individual as their normative ideal, are often called «individualists.» In the 19th century, this standard of individualism was articulated by John Stuart Mill in his classic work, *On Liberty* (1859).

By contrast, *communitarians* do not begin with the autonomous individual, but with a social context out of which individuals emerge. They argue that the pursuit of individual self-interest is more likely to yield a fragmented society than a «good» and fair one. Communitarian thinkers like Robert Bellah, Charles Taylor, Michael Walzer and others insist that rights, though vital, are not the individual's alone. Instead, individual rights necessarily flow from rights recognized by others within a community of a particular sort in which responsibilities are also cherished, nourished, and required of individuals.

The range of social and political debate between liberalism and communitarianism is reflected, deepened, and extended by contemporary *feminist theorists*. There is no single ethics or moral theory of feminism, though liberalism, with its vibrant individualistic strand, has been, in general, attractive to feminist thinkers. Indeed, the language of rights is a potent weapon against traditional obligations, particularly those of family duty or any social status declared «natural» on the basis of ascriptive characteristics. To be free and equal to men became a central aim of feminist reform. For an example, leading proponents of women's suffrage, in Britain and the United States, undermined arguments that justified legal inequalities on the basis of sex differences.

At times, feminist theory turns liberal equalitarianism on its head by arguing in favor of women's civic equality on grounds of difference. In the version of feminist ethical theory known as «the ethics of care,» appeals to women's liberation arises from their social location as mothers, using motherhood as a claim to citizenship, public identity, and civic virtue. To individualist, rights-based feminists, however, the emphasis on maternal virtue as a form of civic virtues is deemed a trap. They are convinced that only liberalism, with its more individualist construal of the human subject, permits women's equality and standing.

The division between feminists becomes evident over issues like in vitro fertilization (IVF), surrogate motherhood, sex selection, that is, the entire menu of real or potential techniques for manipulating, controlling, and altering human reproduction. One broad general tendency in feminist theory might be called pro-interventionist. *Pro-interventionists* welcome any development in reproductive technologies as positive because, as a promise to control nature, it potentially eliminates the distinction between male and female. Pro-interventionists, who applaud any technique that severs biological reproduction from the social identity of maternity, are heavily indebted to a liberal vision in which the self exists apart from any social order. This view of the self, in turn, is tied to one version of rights theory that considers human beings as self-sufficient, promoting a view of society that sees itself as organized around contractual agreements between individuals. In this view, the standard of evaluation concerning artificial reproductive technologies is self-sufficiency and control, paving the way for invasive procedures that break women's links to biology, birth, and maternal nurturing.

The pro-interventionist position, however, does not exhaust the richness of positions within feminist theories. On the other side of the spectrum, *non-interventionists* see reproductive technologies as a strengthening of arrogant human control over nature and thus over women as part of the «nature» that is to be controlled. This latter voice associated with feminism is rooted in a less individualistic, more communitarian frame. The non-interventionists ponder the nature of the many choices offered by new reproductive technologies. They wonder whether amniocentesis is really a free choice or merely a coercive procedure with only one «correct» outcome: to abort if the fetus is defective. (An outcome that, according to statistics, marks 9 out of 10 diagnoses of chromosomal abnormality in California, for an example, Down syndrome). They speculate whether new reproductive technologies are an imposition upon women who see themselves as failure if they cannot become pregnant. They insist that technological progress is never neutral. Because it requires the invasion and manipulation of women's bodies, progress must always be critically scrutinized and may need to be rejected.

I think these reflections, triggered by a «hermeneutics of suspicion» toward scientific development in general, are extremely important. In particular, they put us on guard against the possible exploitation of medical progress by powerful market forces. We do know that the practice of medicine does not take place in an *ideal* community of scientific discourse. Like every other social practice, it is culturally and historically determined, and as such, it can become subjected to mechanisms of reduction and alienation. In a situation where technology and the market forces have so great a role in molding and transforming our scientific and medical ends, even the reasons and arguments driving those ends can become merely technical, reflecting strategies for the achievement of goals whose value is measured by an instrumental rather than a specifically moral criterion.

Indeed, it is hard to miss the marketing and advertisement strategies associated with fertility clinics and service providers that, understandably, are eager to do what any business does best: sell to prospective customers. But what they are selling is packaged in the language of products and commodity. Take the example of *California Cryobank*, which pays donors $75 per specimen – with occasional gift vouchers and movie tickets thrown in. According to the gamete bank web page, basic information about donors – height, weight, IQ, occupation – comes free, but further information must be paid for. One can even get a facial-features report, listing such attributes as «nostril flare» (narrow, average or large), for $12, etc. As the *New York Times* recently suggests, in a fine article titled «Selling the Fantasy of Fertility,» artificial reproduction is a growing economic enterprise, with gross revenues of $4 billion per year, serving one in six infertile couples in the U.S. But what this enterprise is selling «is packaged in hope and sold to customers who are at their wits' end, desperate and vulnerable. Once inside the surreal world of reproductive medicine, you keep as long as your bank account, health insurance or sanity holds out» (*New York Times*, 11 September 2013).

3. *A False Alternative?*

Debates concerning reproductive technology and surrogacy inexorably lead us back into discussions of men, women, children, families, and the wider community. They also push us to question the exclusive reliance upon the notion, popularized by the 16th and 17th century contract theories of Thomas Hobbes and John Locke, that society is simply constituted by individuals for the fulfillment of individual ends, with social goods as aggregates of private goods.

This latter view must be integrated by one that offers a more rooted and historical picture of human beings. Perhaps, the primacy of rights and individual choice *is not* the self-evident starting point, «the original position,» as John Rawls calls it. Perhaps, a more nuanced understanding of the individual self is needed, one that, to quote philosopher Alasdair MacIntyre recognizes a «variety of debts, inheritance, rightful expectations, and obligations» which have priority because they «constitute the given of [our] life, the moral starting point.»[4]

Thus, our modern lack of roots and larger community ties should be viewed as a problem, in fact, an imperfect, rather than «progressive,» account of the human condition. To be cut off from a wider network of human relations, as well as from the past, is to deform present relationships and the identity of the self. Without a beginning that recognizes our essential sociality, there is no beginning at all. For sure, the world endorsed in this communitarian social model appears to be in tension with the individualist mindset of the one previously portrayed. Individualists sometimes claim that communitarians express little more than nostalgia for a simpler, pre-modern past. Communitarians, for their part, argue that the past presents itself as the living embodiment of vital traditions, in whose «effectual history» we stand as individuals concretely rooted in space and time.

I believe the «communitarian» model does not necessarily stand in opposition to the «liberal» model, in that it does not question the function of individual conscience and critical capacity in forging one's identity. The recognition of individual freedoms, such as freedom of speech, religion, association, and privacy are unquestionable values for any contemporary rendition of the relation between self and society. A society is a good society when it sustains freedom through the mutual respect its members show in their interaction with one another.

---

4   Alasdair MacIntyre, *After Virtue: A Study in Moral Theory* (Note Dame, IN: Notre Dame University Press, 1981).

This goes, first of all, to the realization that aiming at the *common good of a society* entails protecting, rather than eroding, a space for moral pluralism, hospitable to an interaction across differences, on the presupposition that the *public realm* is not just the neutral space to be conquered or won over, and that the members of an «open society» are not to be faced as enemies but as partners. To acknowledge such an a priori of communication of both epistemic and moral relevance has nothing to do with relativism: in fact, dialogue between moral agents, whether «strangers» or «friends,» to use the distinction *in vogue*, can only function on the presumption that any claim to meaning and truth be, at the same time, an attestation of freedom and respect for the other.

To that extent, liberalism and communitarianism not only stress two different dimensions of the same reality, but grow one on top of the cultural achievements of the other. As Charles Taylor recognizes «...the free individual with his own goals and aspirations... is himself only possible within a certain kind of civilization... it took a long development of certain institutions and practices, of the rule of law, of rules of equal respect, of habits of common deliberations, ... of cultural self-development, and so on, to produce the modern individual; and.. without these, the very sense of oneself as an individual would atrophy.»[5]

4. *Retrieving Civic Agencies: The Individual, the Family, and the State*

The above mentioned discussion has hopefully shown that the tension between liberalism and communitarianism is ultimately a discussion about the texture of our ethical polity. I would like to re-orient the tension in the discussion, by stressing the fact that what is at stake in it is not only the recognition of the individual's debt to the whole of society, but also the vindication of the function of those agencies within society, like family, schools, and religious communities, that mediate the moral identity of the self. The latter cannot be construed simply as the product of a self-defining indi-

---

5   Charles Taylor, «Interpretation and the Sciences of Man,» *Review of Metaphysics* 25 (1971) 1: 3-51.

viduality, facing the complexity of a pluralistic world from a position of neutrality. Yet a pluralistic civic order would seem to require diversity on the level of families as well as other institutions, which, in turn, promote and give rise to different moral identities and visions of the good. A host of questions opens up here concerning the public function of moral agencies such as families, schools and religious communities. What is the relationship between democratic theory and practice and the ethos that inspires those agencies?

Let me touch briefly upon one issue, namely, the issue of the family's relation to the larger society.[6] I believe family relations and responsibilities remain the best way to create human beings with a developed capacity to give ethical allegiance to the principles of a democratic society.[7] Because democratic citizenship relies on self-limiting freedom of responsible adults, a mode of child rearing that builds on basic trust and a sense of commitment is necessary. We do not choose our relatives. They are given to us and, as a result, we learn what it means to have a history. As philosopher William Desmond puts it, «... we are already given to be ourselves, before we give ourselves to ourselves; grown by relations to the others, before we can grow up to be ourselves; already reared and grown in relativity, before being grown up and giving ourselves in relation.»[8] This is why we need a moral language that helps us articulate the experience of the family and the loyalty it represents. Again, in the words of William Desmond, «...The family is not just a unit of biological procreation but the elemental metaphor for being together, in which the bonds of love exceed what people as their own can comprehend, in which the love can serve as a metaphor for the ultimate love of the absolute origin.»[9]

The intense obligations and moral imperatives nurtured in families may clash with the requirements of public authority, for exam-

---

6   A philosophical exploration of family can be found in William Desmond, *Ethics and the Between* (New York: SUNY Press, 2001), 385-413.
7   «The intermediation of the family is elemental, and if the seeds of ethical intermediation are not well sown at this intimate level, it is difficult to see our participation in a more public political space as being marked by genuine ethical openness to others. A public philosophy has to foster mindfulness of what is proper to the family,» William Desmond, *The Intimate Universal*, 174.
8   William Desmond, *Being and the Between* (New York: SUNY Press, 1995), 387.
9   *Ibidem*, 388.

ple, when young men and women refuse to serve in a war they claim is unjust because it runs counter to the beliefs, civic or religious, of their families. This, too, is vital for democracy. Keeping alive a potential locus for revolt, for particularity, for difference, sustains democracy in the long run. It is no coincidence that all twentieth-century totalitarian orders aimed to destroy the family as a locus of identity and meaning apart from the state. Totalitarian politics strives to require that individuals identify with the state rather than with specific others, including families and friends. Family authority within a democratic, pluralistic order does not exist in a direct homologous relation to the principles of civil society. To establish an identity between public and private lives and purposes would weaken, not strengthen, democratic life overall. Children need particular, intense relations with specific adult others in order to learn to make choices as adults. The child confronted prematurely with a «right to choose» is likely less capable of choosing later on.

To become a being capable of choosing alternatives, one requires a sure and certain place from which to start. For this reason, theorists representing the communitarian perspective are often among the most severe critics of contemporary consumerism, violence in streets and media, decline of public education, the rise in numbers of children being raised without fathers, or mothers, and so on. They insist that a defense of the family, that is, a defense of a normative ideal of mothers and fathers in relation to children and to a wider community, can help to sustain a variety of ethical and social commitments. Because democracy itself turns on a generalized notion of the fraternal bond between citizens, it is vital for children to have early experiences of trust and mutuality. The child who emerges from such a family is more likely to be capable of acting in the world as a complex moral being, one who is part of, and yet detached from, the immediacy of his or her own concerns and desires.

In Western bioethics, the notion of *solidarity* has recently emerged as the category able to strike a balance between the alternatives of collectivism and individualism. Such a notion plays an important function in a variety of issues, from reproductive rights to fair distribution of health care resources, to medical research and experimentation. A bioethics inspired by the notion of solidarity calls for a genuinely pluralist normative system that recognizes and sustains a mode of thinking equally distant from excessive privatization, on the one hand, and arrogant state control, on the other. Solidarity thinking pleads for a notion of democracy that entails a vision of tolerance and understanding of the importance of cultural traditions, the realization that the essence of democracy is the freedom which belongs to citizens endowed with a conscience. If such a vision were to prevail, the way in which our ethical dilemmas are adjudicated, including those emerging from bioethics, would be rich and complex enough to enable us to see the public and civic consequences of our private choices, even as it would guard against severe intrusion into intimate life from the outside.

Ethical dilemmas are inescapably political and political questions are unavoidably ethical. Bioethical dilemmas can never be insulated from politics, nor should they be. But the way in which such matters are addressed will very much turn on the social and political framework to which the ethicist, the medical practitioner, the patient, and the wider interested community are committed.

CHAPTER 2
METHOD AND MEANING
In Search of a Richer Bioethics

In this chapter, I approach bioethics in terms of a secular field of inquiry – secular in the sense that it is embedded in the concerns and dialogues of the world of our experience. All of the humanities and the life sciences contribute to this inquiry in an interdisciplinary way. I want, moreover, to turn attention from the past[1] to the present, to the crossroads that bioethics faces now, and to the future, or at least to the direction in which, in my view, bioethics should be headed. I want to argue for the importance of a sustained search for meaning in the context of ethics – a search for meaning that has been hindered in the medical context by the limited vision of positivist natural science, and in the framework of ethics by a preoccupation with the articulation and application of rules.

Such approach requires much more than simply «adding a little bit of meaning» to bioethical analysis and rule using. It entails a paradigm that functions, epistemologically, by turning first and primarily to experience and to the interpretation of experience – that is, to the discovery of the meaning of moral experience and the values and disvalues embedded in it – and then only secondarily to the study of the virtues, principles, rules, and practical judgments that embody this interpretation. I will examine several elements of an analysis of moral meaning, including the epistemological roles

---

1   For details on the origins of bioethics, see Warren Thomas Reich, «La Bioetica negli Stati Uniti,» in *Vent'anni di Bioetica: Idee: Protagonist, Istituzioni*, ed. Corrado Viafora (Padova: Gregoriana, 1990), 143-175. [There is an English translation of the book. See Roberto Dell'Oro and Corrado Viafora, ed., *History of Bioethics: International Perspectives* (San Francisco: International Scholars Publications, 1996)] See also «The Word "Bioethics": Its Birth and the Legacies of Those who Shaped It,» *Kennedy Institute of Ethics Journal 4* (1994):319-35; and «The Word "Bioethics": The Struggle over Its Earliest Meanings,» *Kennedy Institute of Ethics Journal 5* (1995): 19-34.

played by the ideas of care and attention, and I will illustrate the need for meaning in contemporary bioethics, suggesting an agenda for which the participation of theologians and religious scholars is crucially important.

1. *The Divorce of Norm and Meaning in Bioethics*

The problem of the search for meaning in bioethics is illustrated by the metaphor of the stethoscope. Richard Baron tells the story: «It happened the other morning on rounds, as it often does, that while I was carefully auscultating a patient's chest, he began to ask me a question. "Quiet," I said. "I can't hear you while I'm listening."»[2]

The «stethoscope metaphor» is emblematic of the inattention to meaning («not hearing») in bioethics brought about by the reductionistic focus (the mode of restricted «listening») in the methodologies of both modern scientific medicine and contemporary ethical theory. The mind-set created by modern scientific medicine, whereby medicine is strongly habituated to focus on («listen to») diseases in an «objective» way, has required it to be inattentive to («not hear») the sick person's experience of illness.[3] One can trace this mind-set historically to the influence of nineteenth-century medical scientists, particularly Wunderlich, Virchow, Helmholtz, and others, who proposed that practical clinical medicine be viewed as applied theoretical medicine. In this vision, clinical medicine came to be regarded as a source of questions to be addressed to a higher, theoretical discipline, pathologic physiology, which increasingly shaped clinical medicine. This led medicine to regard disease as an objective entity located anatomically or in a physiologic process. Careful scientific attention to the lesion or tumor has led to marvelous advances in modern medicine, but positivist natural science has also created a mind-set that has, in principle, excluded questions

---

2   Richard Baron, «An Introduction to Medical Phenomenology: I Can't Hear You While I'm listening,» *Annals of Internal Medicine* 103 (1985): 606-11, at 606.
3   Baron, «An Introduction to Medical Phenomenology,» 607-8.

of meaning that are highly significant to human well-being and to the ethical aspects of medicine. Husserl pointedly analyzed the implications of this worldview:

> The exclusiveness with which the total world view of modern man, in the second half of the nineteenth century, let itself be determined by the positive sciences and be blinded by the «prosperity» they produced, meant an indifferent turning away from the questions which are decisive for a genuine humanity. ...Fact-minded...science...excludes in principle precisely the questions which man, given over in our unhappy times to the most portentous upheavals, finds the most burning: questions of the meaning or meaninglessness of the whole of this human existence.[4]

What is needed is a medical mind-set that fosters the search for meaning – that is, the significance of illness, wellness, birth, dying, and so on for the sick or well person, for caregivers, for institutions, and for human society as a whole – in the context of an inquiry into moral values and disvalues, moral virtues and the like. A number of approaches have attempted to correct the medical mind-set and to move in this direction, including psychological, behavioral, and philosophical approaches. One such corrective approach is medical phenomenology, which shifts attention from the disease «out there» to the inner experience of illness «in here,» where the real human experiences of suffering and dis-ease occur. This approach takes seriously perceptions of the world as experienced rather than accepting scientific descriptions of the world as exhaustive of human knowledge on the topic.[5] The goal is to reunite the «objective» and abstract world of science with the subjective world of human experience and its (objective) interpretation as meaningful or meaningless.

The «stethoscope metaphor» also symbolizes the mind-set of the moral philosophy that has dominated and shaped much of the first generation of bioethical inquiry. Bioethics has used its «stethoscope» – its instrument for auscultating the moral problems of the

---

4 Edmund Husserl, *The Crisis of European Sciences*, trans. David Carr (Evanston, Ill.: Northwestern University Press, 1970 [1954]), 5-6; cited by Baron in «An Introduction to Medical Phenomenology,» 608.
5 Baron, «Am Introduction to Medical Phenomenology,» 608.

life sciences – to «listen» to a restricted language: the language of biomedical quandaries as well as that of the principles and rules ingredient in rational argumentation as applied to a determined case. The use of such a «stethoscope» strongly inclines one to be inattentive to significant moral voices that do not communicate in the language of quandary, do not create a challenge for ethical argument, or do not speak with the precision and articulateness that may be required in our intellectual culture for attracting the attention of serious ethical argumentation. Thus, both biomedical and ethical traditions have accentuated the problem of inattentiveness to the range of moral meaning that is significant and salient for moral reflection in bioethics.

2. *The Methodological Shift Regarding Norm and Moral Meaning in Bioethics*

A glance at the relatively brief history of epistemological developments in bioethics shows a methodological shift in the fundamental preoccupations of bioethicists. The scholars most commonly identified with being the first ones to elaborate a specifically bioethical body of literature – such as Paul Ramsey, Joseph Fletcher, and Van Rensselaer Potter – sought a horizon of meaning capable of sustaining ethical discourse that would respond to the value implications of technological developments in medicine and the life sciences. Those horizons of meaning – which inspired a moral-anthropological interpretive framework, whether theological or nonreligious humanistic – focused variously on the global moral meaning of nature, love, freedom, evolutionary progress, and so forth. A major shift occurred at the end of the 1970s and the beginning of the 1980s. Under the increasing influence of contemporary Anglo-American moral philosophy, bioethics developed a preoccupation with the elaboration of normative criteria («principles»)[6] which drew their justification from the perspective of a restricted cluster of concepts in political philosophy. This mor-

---

6   Confusion was caused by the use of the word *principles* (whose basic meaning is «sources») rather than *rules* to denote these moral criteria. For a comment on this terminology, see Warren Thomas Reich, «Introduction,»

al philosophical approach sought to create a consensus based on shared arguments that were divorced from the horizon of meaning and the meaningful narratives that initially inspired them.[7]

This methodological transition can be understood as a step toward the development of a rationally consistent discipline. It constituted a sort of «secularization» of bioethics not unlike the trend of secularization and «demythologization» of theology that had occurred in previous decades of this century. It was religion that had constituted the horizon of ultimate reference, and provided the necessary conceptual ethical tools, for the interpretation of the delicate moral problems (such as human experimentation and allocation of health resources) that emerged in the first decade of the field of bioethics.[8] However the religious foundations of bioethics were, to a considerable extent, couched in theoretical religious language, a major function of which was to elaborate a structure for religious (mostly Protestant and Catholic) identity, while also offering a moral vision intended to be useful for engaging in public moral discourse. Thus, by its very character, religious bioethics spoke in a language partly extraneous to the secular mind that was seeking normative guidance.[9]

Under the strong influence of the need for a consistent ethical basis for public policy formation, moral philosophy created for bioethics an arena of autonomous reflection by centering it on the

---

in *Encyclopedia of bioethics,* ed. Warren Thomas Reich, rev. ed. (New York: Simon & Schuster-Macmillan, 1995), xxviii-xxix.

7   Warren Reich points to the reliance of ethics and specifically bioethics on originary myths that create a horizon of meaning in «Alle origini dell'etica medica: Mito del contratto o mito di Cura?» in *Modelli di Medicina: Crisi e Attalità dell'Idea di Professione,* ed. Paolo Cattorini and Roberto Mordacci (Milan: Europa Scienze Umane Editrice, 1993), 35-59. Judith N. Shklar offered a thorough explanation of the reliance of ancient and modern political philosophy and ethics on creation myths in her «Subversive Genealogies,» *Daedalus* 101 (1972): 129-54.

8   See LeRoy Walters, «Religion and the Renaissance of Medical Ethics in the United States: 1965-1975,» in *Theology and Bioethics,* ed. Earl E. Shelp (Dordrecht, Netherlands: D. Reidel Publishing Company, 1985), 3-16.

9   An exception is found in the writings of Richard A. McCormick, who utilized language that reached well beyond his Catholic community to address in a useful way a number of public policy issues in bioethics. See, for example, his *How Brave a New World? Dilemmas in Bioethics* (Washington, D.C.: Georgetown University Press, 1981).

use of principles (rules) and the ethical theory that unites them, especially deontological and utilitarian theories.[10] The clumsy term *principlism* is now used to designate this approach.

It is interesting that this philosophical framework has precipitated far more explicit and systematic criticism than the religious orientation ever elicited. I believe this anomaly can be explained only by viewing jointly the inadequacies of both approaches, for the inadequacy of the first approach presaged and merged with the inadequacy of the second. In significant ways, both approaches shared similar assumptions; in particular, they were both principally onto-theological in their approaches. Both were based on the assumption that the core truths could be expressed in a few basic concepts (or beliefs or principles), from which the remainder of ethics and ethical judgments would follow, by application and/or specification. Thus, while one (the theological) approach embraced a set of meanings, both approaches tended to exclude, by their a priori assumptions, the pursuit of meaning on the part of the moral agent.

Consequently, the parallel shortcomings of both approaches are illumined by the stethoscope metaphor. While «listening for» the place for the insertion of concepts of sin and redemption, nature and grace, moral principles and principle-like virtues, theology was disinclined to «hear» the voice of the molested and to «read» the complex implications of new reproductive and transplant technologies. Freed of religious frameworks perceived as excessively sectarian in a pluralistic world, bioethics undoubtedly expected from the principles of moral philosophy an enduringly useful approach, only to discover – as noted above – that the stethoscope syndrome was found here, too. While «listening for» the signs and symptoms of the existence of quandaries to which its relatively few familiar concepts/principles could be applied, philosophy did not «hear» the voice and moral language of women, or the secular person's plaintive plea for the meaning of suffering, dying, and illness – a meaning that is crucial to moral decisions.

---

10   Tom L. Beauchamp and James Childress, *Principles of Biomedical Ethics* (New York: Oxford University Press, 1979).

There are now, of course, a number of «alternative» approaches[11] to bioethics. As I see it, all these approaches narrow the gap between normative ethics and the search for moral meaning. But it is not my purpose to argue for an «alternative» ethical theory.[12] In fact, the conviction which is the point of departure for my reflections prescinds somewhat from the normative preoccupation that characterizes the development of an ethical theory. My conviction is that ethics finds its starting point and its ultimate formative element in an experiential paradigm characterized by the search for the meaning of moral experience.[13] Normative ethics – which proceeds, for example, through «application» of rules and the instantiation of virtues – occurs within this larger interpretive framework.[14]

It must be acknowledged that this claim about meaning raises a number of questions: Was there a «golden age» in which meaning and life were one – and if so, has that been lost in our culture? Was there a connection between the meaning of life and ethics in the classical worldview – and if so, what led to the loss of that connection? Is it desirable to (re)discover within contemporary ethical reflection the component of meaning? Isn't attention to meaning a luxury that in fact distracts from a pursuit of practical ethics? Even

---

11   The term «alternative approaches is misleading, for it implies that the rule-based approach is fundamental and stable, while other approaches – which actually account for the majority of realms and levels of moral discourse – are reduced to being identified by that to which they are currently contrasted.

12   The various approaches that have not been sufficiently integrated into ethical and bioethical inquiry – and which I seek to unify around the central task of interpreting the meaning of moral experience – are nicely surveyed in *A Matter of Principles? Ferment in U.S. Bioethics*, ed. Edwin R. DuBose, Ron Hamel, and Lauence J. O'Connell (Valley Forge, Pa.: Trinity Press International, 1994).

13   For the articulation of this paradigm, see Warren Reich, «Bioethics in the 1980s: Challenges and Paradigms,» in *Biomedical Ethics: A Community Forum*, ed. H.M. Sondheimer (Syracuse, N.Y.: SUNY Upstate Medical Center, 1985), 1-35. And, more extensively, in Warren Reich, «Ein neues Paradigma: Erfahrung als Quelle der Bioethik,» in *Ethik in den Wissenschaften: Anriadnefaden im techniscen Labyrinth?* Ed. Klauss Steigleder and Dietmar Mieth (Tübingen: Attempto, 1991), 270-92.

14   European philosophical and theological language often uses the term *anthropology* (philosophical or moral anthropology) to incorporate the realm of meaning as a «necessary presupposition for the elaboration of normative criteria.» See Roberto Dell'Oro, «Antropologia ed etica: Oltre la bioetica nordamericana,» *Rivista di Morale* 106 (1995):203-20.

if it is desirable to retrieve the pursuit of meaning, why should we and how can we integrate the broader horizon of meaning into moral discourse, into practical moral judgments, and into our analysis of normative criteria.

A fleeting overview of responses to these questions, while not at all satisfying in their brevity, might at least help persuade the reader that meaning is, and always has been regarded as, an important part of human living; that reflection on meaning is an essential part of ethics; that ethics not based on the reflection on meaning is greatly impoverished and ultimately distorts the moral life; and that the ways of understanding moral meaning, as well as their certitude, change with various eras of human thought.

3. Recovering Meaning in Moral Discourse

In the West, it was the Greek mediation of meaning that resulted in classical culture – a culture that «breathed life and form into the civilization of Greece and Rome,...was born again in a European renaissance, [and] provided the chrysalis whence issued modern languages and literatures, modern mathematics and science, modern philosophy and history.»[15] The Greek's world could be understood; comprehension consisted in contemplating that reality and explaining it. For example, Socrates carried out an experiment that bears on meaning: In the early Platonic dialogues he puts questions to the Athenians that had as their purpose moving them from the primary, spontaneous level of meaning – where humans employ everyday language, invoking, for example, notions like courage, justice, or self-control – to the secondary level of meaning, where they could say what they mean by everyday language: how to define courage, justice, or self-control.

---

15 For these comments on the history and significance of meaning in classical and modern culture, I rely principally on Bernard Lonergan, «Dimensions of Meaning,» in *Collection: Papers by Bernard Longergan, S.J.*, ed. Frederick E. Crowe (New York: Herder & Herder, 1967), 252-67. Lonergan also comments on the theme of meaning in his «Theology in Its New Context,» in *Conversion: Perspectives on Personal and Societal Transformation*, ed. Walter E. Conn (New York: Alba House, 1978), 3-21. See also Charles Taylor, «Theories of Meaning,» in *Human Aging and Language: Philosophical Papers*, vol. 1 (Cambridge: Cambridge University Press, 1985), 248-92.

Classicist thought interpreted and then standardized the meaning of the world in major concepts such as nature (*physis*, regarded as an original principle) and happiness (*eudaimonia* in the sense of development of the essence of the individual). However, the same classicist tradition increasingly overlooked the lived and experienced dimension of human reality which is the constitutive function of meaning in human living.

A number of twentieth-century commentators have noted that, by and large, classical culture has passed away: «its norms of interpretation, and its ways of thought, its manner in philosophy, its notion of science, its concept of law, its moral standards...are no longer accepted.»[16]

This breakdown is true of the meaning of both science and philosophy. By their very nature, classically oriented science and philosophy concentrated on the essential while ignoring the significance of the accidental, on the universal while neglecting the particular, on the necessary while belittling the contingent, on certitude while failing to see how tentative their theories were. By contrast, modern science aims to understand not just the essential but all phenomena; and it claims probability, not certitude, for its positive affirmations. Philosophy and other human sciences increasingly seek to understand not simply the essence of humanness (e.g., body and soul, matter and form) and the unchanging meaning of human nature with its correlatively enduring natural moral law, but what the classical worldview regarded as the accidental (such as family and other human relationships), the particular (the history of peoples, cultures, religion), and the contingent (e.g., trust as an indispensable yet contingent aspect of human moral growth).

Furthermore, classical culture conveyed the conviction that the literal meaning of words and phrases somehow has priority, while figurative meaning is an ornament that makes the literal meaning more vivid and effective. Giambattista Vico is renowned for having put forward, in his *Scienza nuova,* the contrary view by proclaiming the priority of poetry. To proclaim the priority of poetry – and/or the priority of imagination – is a way of confirming that humans come to know and to express themselves through symbol and imagination before they know and can express conceptually (if ever

---

16   Bernard Lonergan, «Dimensions of Meaning,» 258-259.

they come to know and express conceptually) what these symbols literally mean. Much of the contemporary turn to inquiry into the meaning of experience is dependent on this question of the priority and role of preconceptual, imaginative language. Heeding the role of meaning has long been made difficult by the widespread philosophical assumption that since the Enlightenment one must divorce reason from both imagination and emotion. Yet even some of the leading Enlightenment philosophers quickly acknowledged that imagination is inherent in all knowledge and reasoning, and contemporary thought increasingly acknowledges the essential role of emotion and sentiment in human knowledge in general and the quest for meaning in particular.

The meaning and function of ethics have undergone the same stages in their development and ultimately the same recent shift in paradigm as have science, philosophy, and the other disciplines. So-called classical moral reflection – inspired by Aristotelianism and subsequently by medieval Thomism – was, by definition, a reflection on meaning: It proposed that reflection on the immediate criteria for acting was secondary in respect to the identification of a *telos* to be pursued.[17] In Aristotelian ethics, fundamental attention was not given primarily to concrete action and the norms that regulate it, but given more globally to the *praxis* that covers an entire existence and defines the destiny of life.[18] The good is only secondarily a law to be applied in a concrete case; *primum et per se*, it is the very horizon of the choices with which the agent designs his or her own moral space and the ultimate measure of his or her actions. The good bears the weight of individual moral decisions and unifies their meaning in the coherence of an overriding vision. This coherence can be seen only dimly and pursued with the liberty of a project, not with the cogent evidence of a mathematical formula. Yet, over time, classical ethics developed a deep cleavage. To make the notion of good applicable in both philosophical and theological contexts, it developed principles

---

17 The complex history of ethics' resistance to the question of meaning is traced by Alasdair MacIntyre in *After Virtue: A Study in Moral Theory*, 2nd ed. (Notre Dame: University of Notre Dame Press, 1984).
18 See Hans-Georg Gadamer, *The Idea of the Good in the Platonic-Aristotelian Philosophy*, trans. P. Christopher Smith (New Haven: Yale University Press, 1986).

and principle-like virtues whose formal philosophical function adopted more and more the characteristics of precision and certitude; yet their meaning was not precise, for ethics, regarded as practical science, is none other than the uncertain measure of freedom in the complexity of life.[19]

The classical era was an era in which meaning was commonly perceived and peacefully shared. Ethics was able to point the way to the realization of an end the truth of which was already established, for the *polis* or *societas christiana* did not yet know the split imposed by ideological pluralism. The modern epoch, which was born under the sign of religious division and anthropological pessimism, developed an ethic symbolized by a Hobbesian vision of all humans warring against all others. The purpose of ethics came to be the achievement of harmony and the promise of peace. Rational argumentation pertaining to a cluster of univocal ethical criteria that were to be universally applied came to represent the only significant purpose of moral reflection.[20] Indeed, following the Hobbesian vision of social chaos and the need for social consensus regarding coercive restraints, Engelhardt reduces the substantive content of ethics (and bioethics) to freedom/autonomy.[21]

In fact, however, modern ethics has been en-framed by meaning: Human experience has remained inscribed in the horizon of meaning and the individual conscience manifested the lived moral meaning of all individuals; but neither locus of moral meaning could find its way into philosophical ethics, for neither was judged amenable to public moral discourse. Increasingly distanced from the rnajor frameworks for meaning that were found in the classical era's natural teleology for the good and its religious eudaemonism,

---

19 Note Aristotle's explanation of ethics in the *Nichomachean Ethics*, trans. William David Ross (New York: Random House, 1941), and how it is viewed by Charles Taylor in his *Sources of the Self: The Making of Modern Identity* (Cambridge, Mass.: Harvard University Press, 1989).

20 See Hans-Georg Gadamer, Über die Möglichkeit einer philosophischen Ethik, in *Kleine Schriften* (Tübingen: 1967), vol. 1, 179-91. On the fundamentally different starting-points of «classical» and «modern» ethics, see also the instructive reflections of Ernst Tugendhat in *Probleme der Ethik* (Stuttgart: 1984).

21 H. Tristram Engelhardt Jr., *The Foundations of Bioethics* (New York: Oxford University Press, 1986).

modern ethics came to restrict itself minimally to the mere metaethical meaning of moral language itself,[22] and maximally to moral argumentation that was universally valid but significantly void of the content of moral meaning that might serve to articulate a web of socially shared convictions.[23]

The severely problematic question of the current direction to be taken by ethics and bioethics in particular arises from the situation in which ethics has increasingly become characterized by a rationality turned instrumental, due to the conviction that the pursuit of moral meaning is superfluous to the ethical enterprise. What is at stake in our choice of ethical roads to be taken in the next era of bioethics – and ethics more generally – is the very nature of moral reflection.

4. Retrieving the Pursuit of Meaning in the Practice of Contemporary Bioethics

The foregoing, brief historical survey highlights the importance of the recurring conviction that moral reflection does not begin with the analysis of arguments dealing with rules – that is, with normative ethical theories – but with a free and open look at the meaning of the experience of human life and destiny.[24]

What do we mean by the experience whose meaning is to be interpreted? It is not merely an objectively described empirical entity, for experience (etymologically, to survive or live through a crisis)

---

22 See, for example, Richard M. Hare, *Moral Thinking: Its Levels, Method, and Point* (Oxford: Clarendon Press, 1981)
23 I say «significantly» but not totally, because the ethical scrutiny of a small cluster of principles – while highly formal and abstracted from any attempt to root them in a horizon of meaning – constitutes a (restricted) form of interpretation of meaning, especially when this scrutiny is viewed against the background of the political experiences from which the principles indirectly arise. This would be true, for example, of *Principles of Biomedical Ethics* by Tom L. Beauchamp and James F. Childress, 3rd ed. (New York: Oxford University Press, 1989).
24 See Warren Thomas Reich, «Experiential Ethics as a Foundation for Dialogue Between Health Communication and Health-Care Ethics,» *Journal of Applied Communication Research* 16 (1988):16-28.

includes the notion of the agent's perception.[25] Experience is not fully separable from meaning. To use a phenomenological term, we are dealing with an ante-predicative experience, in which the original nucleus of what moral consciousness perceives is preserved in the unity it has prior to the distinctions brought about by faculties of reason, will, sentiment, and so on.

What do we mean by the meaning of such a reality? To apprehend the meaning of something is to understand and make a judgment about an experience (or about one sphere of the entire world of experience), so as to appreciate its significance and know how to live with and respond to its significance. The search for meaning is no novel recommendation in today's scholarly world. There is a blossoming of the search for meaning in human studies generally: for example, in women's studies (the meaning of women's lives, roles, perceptions, ways of reasoning); in ethnographic and anthropological studies (e.g., describing and translating cultural systems of meaning); in philosophical schools of thought (e.g., examining, through hermeneutical epistemology, the meaning of handicap, illness, wellness); and in theology (e.g., examining the significance of the experience of grace, of the lived ecclesial community. etc.).

The objection is sometimes raised, at least implicitly, that questions of meaning are only of secondary importance in contemporary ethics, particularly in clinical medicine, where the primary reality with which we must deal is the making of tough decisions in a concrete clinical, technical, and moral context. In this perspective, «gazing into the meaning of things» is an exercise not meriting much of our attention.

It does indeed make good sense to put meaning in a secondary place and give primacy, instead, to one's immediate reality, in situations in which infants and afflicted adults must be concerned with living or surviving, with learning (or relearning) how to speak, walk, hear, and eat – and how to differentiate and combine these

---

25 See, for instance, the important comments made by Gadamer on the concept of experience and the meaning of hermeneutic experience: Hans-Georg Gadamer, *Truth and Method*, trans. Joel Weinsheimer and Donald G. Marshall, rev. ed. (New York: Continuum Publishing, Co., 1994), 265-380. For an application of this concept in ethics, see Dietmar Mieth, *Moral and Erfahrung.Beiträge zur theologisch-ethischen Hermeneutik* (Freiburg i. Br.: Herder, 1977).

experiences in ever larger syntheses. But as mastery and use of language develop in the child or its broader use returns to the afflicted, we come to live, «not in the world of immediate experience, but in a far vaster world that is...mediated through meaning.»[26]

The task and challenge of incorporating the search for the meaning of moral experience in (bio)ethics is best understood by examining elements of moral epistemology that reach out for or actually incorporate the interpretation of meaning. I will briefly consider those elements individually and include examples of bioethical thought that have already taken a move in this direction, as well as examples that provide an agenda for the task that lies ahead.

5. Moral Decision Making

When the larger world of wellness, suffering, being struck with affliction, being sick, dying, parenting, nursing, and so on is not mediated in the decision-making process of clinical ethics – when, instead, clinical ethics mechanically relies on an algorithmic approach to solve its moral problems by utilizing a step-by-step procedure involving the standard concepts and the standard sources (advance directives, patient consent forms, values inventory, proxy consent forms, etc.) – clinical ethics ultimately creates an obstacle to good habits of moral reasoning and hinders the flourishing of bioethics, even if, ironically, such concepts and procedures produce the «right» answer for the question that was asked.

It must be acknowledged, of course, that an ethic which, in principle, aspires to the interpretation of all reality – or all relevant reality – is certainly never a fully closed ethic, nor will it ever offer the full certitude that some versions of classical ethics claimed. Nevertheless, an ethic based on the search for meaning is a manageable and useful approach to ethics, for it engages first one carrier or embodiment of meaning and then another, in dialogue – whether those carriers be found in language, symbols, history, art, anthropology, a person's incarnate meaning (her actual story, her actual way of life), story, theories, principles, and so on. Some promising inquiries into «carriers» that in fact shape contemporary clinical ethics include

---

26 Lonergan, «Dimensions of Meaning,» 25; see also 252.

works dealing with value-laden narratives of health-related experiences,[27] the uses of power by clinicians,[28] and, in general, the link between the theory and the practice of medical ethics in the clinical experience itself.[29]

Moral decisions are made in dialogue regarding meaning. The agent discovers metaphors to enlighten an entire range of behaviors, and images, models, and patterns to guide actions. Sometimes the actions are simply those of facing the ordeal of illness, of persisting faithfully to be present to the other, or of enacting a (secular) ritual – say, in transplant surgery or in companying the dying – as a means of preserving or creating moral meaning and «legitimizing» a continued reflection on moral, personal, professional, and spiritual meaning.[30]

Some authors argue that the dominant mode of judging and deciding is not formal logical argument but discernment, a sort of prudential or practical reasoning which deals with a complex set of epistemological factors through its elements of detecting, sensing, sifting, discriminating, comparing, connecting, and deciding.[31] This

---

27 See Arthur Kleinman, *The Illness Narrative: Suffering, Healing, and the Human Condition* (New York: Basic Books, 1988); Kathryn Montgomery Hunter, *Doctors' Stories: The Narrative Structure of Medical Knowledge* (Princeton, N.J.: Princeton University Press, 1991); and Rita Charon, «Narrative Contributions to Medical Ethics: Recognition, Formulation, Interpretation, and Validation in the Practice of the Ethicist,» in *A Matter of Principles? Ferment in U.S. Bioethics*, ed. Edwin R. Dubose, Ronald P. Hamel, and Laurence J. O'Connell (Valley Forge, Pa.: Trinity Press, 1992).

28 See Howard Brody, *The Healer's Power* (New Haven: Yale University Press, 1992).

29 See Richard M. Zaner, «Experience and Moral Life: A Phenomenological Approach to Bioethics,» in *A Matter of Principles? Ferment in U.S. Bioethics*, ed. Edwin R DuBose, Ronald P. Hamel, and Laurence J. O'Connell (Valley Forge, Pa: Trinity Press International, 1994), 211-39, and Glenn C. Graber and David C. Thomasma, *Theory and Practice in Medical Ethics* (New York: Continuum Publishing Co., 1989).

30 See William F. May, *The Patient's Ordeal* (Bloomington: Indiana University Press, 1991), 1-14; and May, *The Physician's Covenant: Images of the Healer in Medical Ethics* (Philadelphia, Pa.: Westminster Press, 1983), 13-25.

31 See James M. Gustafson, *Theology and Christian Ethics* (Philadelphia: United Church Press, 1974), and *Ethics from a Theocentric Perspective* (Chicago, University of Chicago Press, 1981). Ronald Carson has developed the notion of discernment in reference to bioethics in «Interpretive Bioethics: The Way of Discernment,» *Theoretical Medicine* 11 (1990):51-59.

mode of moral decision-making is relevant to all settings, but in particular to the sort of context in which, for example, parents must accept or reject their own initiation into the art of parenting their severely handicapped infant. When parents decide whether and how to bond with and then care for their child, they may not follow moral rules and principles at all; they are more likely to experience the possibility of living out an image of parenting they discover in the context of a major transition from loss of the envisioned child to attachment to the one they received. That image carries moral normative force, in the sense that it makes possible a spirited rejection or spirited emulation of values and virtues.[32]

6. Care: The Root of Striving for Meaning

The primary root of the pursuit of meaning in moral knowledge is the metaphysical notion of care. Although the idea has a long history in mythology, philosophy, literature, and the practice of the care of souls, little systematic attention has been paid to the important function played by care or concern in the history of (moral) knowledge.[33] Writing within this tradition, Kierkegaard explained that consciousness – as distinct from disinterested reflection – is inherently *concerned* with both the knowing subject and what is known.[34] For Heidegger, the self (*Dasein*) is care, in the sense that we understand and care about ourselves as beings-in-the-world because of our connectedness with being and action brought about by care. Thus, at the root of what it means to be a self, it *matters* that we can act: we care about it.[35]

---

[32] See Warren Thomas Reich, «Caring for Life in the First of It: Moral Paradigms for Perinatal and Neonatal Ethics,» *Seminars in Perinatology* 11, no. 3 (July 1987): 285.

[33] The idea to which I refer is part of what Warren Reich has called «the Cura tradition of care,» which is a major element in the history of the idea of care. See Warren Reich, «Care: I. History of the Notion of Care,» in *Encyclopedia of Bioethics*, ed. Warren Thomas Reich, rev. ed. (New York: Simon & Schuster-Macmillan, 1995), 319-31.

[34] Kierkegaard, *Johannes Climacus; or De Omnibus Dubitandum est; and A Sermon*, trans. Thomas Henry Croxall (Stanford, Calif.: Stanford University Press, 1958).

[35] See Martin Heidegger, *Being and Time*, new ed., trans. John Macquarrie and Edward Robinson (Oxford: Basil Blackwell, 1973).

Charles Taylor carries forward, more systematically, the foundational importance of care as the driving force toward meaning in his definition of the human agent as person: «Agents are beings *for whom things matter*, who are subjects of significance.» Because we are beings for whom things matter, we are, fundamentally, subjects who are open to matters of meaning or significance. Taylor continues: we evaluate self and life not essentially in light of fixed goals, but also and specifically in sensitivity to certain standards involved in peculiarly human goals.[36]

This theory of moral knowledge makes it not only relevant to, but even essential for, bioethics to consider such questions as these: How has a male-dominated culture shaped what it means to be a woman and the evaluation of her health needs? What does it mean for a woman to be a moral agent? What do women care about (say, regarding justice, abortion, caretaking roles, etc.)? Bioethics would be different and better if it were to acknowledge fully and integrate into its dialogue the voices of women in reply to these questions.[37] Similarly, the ethics of responding to aging and the elderly depends in great measure on discovering what it means to be an aged person.[38]

7. *The Nature of Moral Truth*

In contrast to the notion of truth as coherence, which has dominated our rationalist, argument-based approach to ethics, there is

---

36 Charles Taylor, «The Concept of a Person,» in *Human Agency and Language: Philosophical Papers*, vol. 1 (Cambridge: Cambridge University Press, 1985), 104-5. Alasdair MacIntyre, in his «Comments on Frankfurt» (*Synthese* 53 [1982]: 291-94), affirms the fundamental role of care for ethics: «The concept of *what we care about* assumes an importance in an area of our culture left vacant by the disappearance of a public theory of good or of the hierarchy of goods. "What is it important to care about?" replaces "What goods ought we to desire?"» (291).

37 See Susan Sherwin, *No Longer Patient: Feminist Ethics and Health Care* (Philadelphia: Temple University Press, 1992); and Karen Lebacqz, «Feminism,» in *Encyclopedia of Bioethics,* ed. Warren Thomas Reich, rev. ed. (New York: Simon & Schuster-Macmillan, 1995), 808-18.

38 An example of this sort of morally revealing inquiry is found in the work of Thomas Cole: *The Journey of Life: A Culture of Aging in America* (New York: Cambridge University Press, 1992); and *Voices and Visions: Toward a Critical Gerontology* (New York: Springer Publishing Co., 1993).

the often overlooked notion of truth as manifestation. The latter notion of truth involves the power of disclosure and concealment on the part of the known object and the related experience of recognition on the part of the knowing subject. In every true manifestation, there is a dialogical interaction between the object's disclosure and concealment and the subject's recognition through interpretation. That interaction is conversation.[39]

This notion of truth makes evident the relevance of a variety of texts that are correlative to a number of modes of moral knowledge in ethics. For example, narratives – which are the primary way of articulating, organizing, and communicating the sense human beings make of the world – are the prime locus of truth-as-manifestation;[40] and moral dialogue with a story is almost guaranteed, for the story itself is interactive before the listener interacts with the story. Whether the narrative is a primordial Greco-Roman myth of healing, an autobiography of a victim of sexual abuse, or an African-American woman's poem about aging, it can richly manifest (and instruct us regarding what it conceals about) the elements of a situation to be addressed by another's response.[41]

The truth manifested in a variety of «texts,» whether stories, shared experiences, or other forms, is usually initially perceived and interpreted, in conversation with the «text,» as antepredicative or preconceptual knowledge, perhaps later to be more extensively formulated and eventually employed in argument.[42] Even as antepredicative, however, moral knowledge is often operative and hence directly relevant to the moral discourse ingredient in ethics. Specifically, truths about the experiences of suffering, the harm done by neglect of suffering persons, and the good done by compassionate care of suffering persons – manifested through literature, psychol-

---

39   See David Tracy, *Plurality and Ambiguity: Hermeneutics, Religion, Hope* (San Francisco: Harper & Row, 1987), 28-29.
40   See Stephen Crites, «The Narrative Quality of Experience,» *Journal of the American Academy of Religion* 39 (1971): 291-311.
41   See Marian Gray Secundy, *Trials, Tribulations, and Celebrations: African-American Perspectives on Health, Illness, Aging, and Loss* (Yartmouth, Me.: Intercultural Press, 1992).
42   See David Burrell and Stanley Hauerwas, «From System to Story: An Alternative Pattern for Rationality in Ethics,» in *Knowledge, Value, and Belief*, ed. H. Tristram Engelhardt Jr. and Daniel Callahan (Hastings-on-Hudson, N.Y.: Hastings Center, 1977), 111-52.

ogy, autobiographical accounts, moral-anthropological models, and so on – can, through interpretive dialogue, give rise to a new (or renewed) ethic of responding to suffering.[43]

8. *Moral Agency and Intentionality*

The two previous sections, on care and the nature of moral truth, have both presumed the turn of ethics to the subject of morality, to the moral agent. An ethic that begins with meaning takes seriously the moral agent, who typically cares about the moral life, who stands, so to speak, behind moral language, but who has been lost behind the veil of anonymity created by a universal and impersonal ethic.[44]

The moral self or moral subject is recaptured, in part, by Husserl's notion of intentionality.[45] Since the ethics involves the study of all experiences and/or clusters of experiences that have a bearing on the moral life, ethics must (as Husserl claimed phenomenology must) involve the study of the objects of experiences, because experiences are *intentional*. Reference to an object is essential to experiences; and the reference of experiences to intentional objects consists of meaning. The content of this reference of moral experience to object is not accessible to the objectivity of an abstract rationality. It is «intuited» by thought that is synthetic, that takes feeling into account as part of rationality (as do Pascal's «reasons of the heart»), and that perceives the object in the unity of its meaning, as a world-for-me.

This is the world of moral orientation and of moral character, for without a consideration of the role of moral character as the subject of evaluation, it would not be possible to assess the meaning of the intentionality of our moral experiences. In this sense, then, character and virtue form an essential part of the arsenal of the pursuit of mor-

---

43 See, for example, Eric J. Cassell, *The Nature of Suffering and the Goals of Medicine* (New York: Oxford University Press, 1991); and Warren Thomas Reich, «Speaking of Suffering: A Moral Account of Compassion,» *Soundings* 72 (1989): 83-108.
44 See Derek Parfit, *Reasons and Persons* (Oxford: Clarendon Press, 1984).
45 Edmund Husserl, *Ideas: General Introduction to Pure Phenomenology*, trans. W.R. Boyce Gibson (New York: Collier Books, 1962), 235-60.

al meaning. Viewing the person as moral subject, Stanley Hauerwas argues for the primacy of character and narrative in philosophical and theological ethics. He regards character as the qualification of self-agency by the moral subject's beliefs and intentions.[46]

In light of these considerations, the role of moral character is an essential component in understanding, for example, the meaning of the experience of suffering due to blind overtreatment, the meaning of sexual molestation experienced as a shameful blow to one's self identity, and the meaning of suffering due to an untimely terminal illness.[47] Furthermore, however, to the extent that virtue ethics becomes a set of fixed ideals that function more or less as a hierarchy of ordered goods, virtue and character lose their capacity to serve the pursuit of moral meaning, for in such a view they have abstracted from the moral agent as intentional subject, no matter how appealing they may be to the motivations of some agents.

Similarly, casuistry can be viewed as embodying the intentionality of a community – its orientation to meanings, values, and goods. It is the art of making use of a moral community's memory by bringing to bear on decision-making the ways in which paradigmatic cases embody the wisdom of the community in rhetorical language that is only minimally discursive. Viewed as such, it can be a major instrument in the task of the interpretation of moral experience. However, to the extent that casuistry is viewed simply as a set of skills for the practical application of moral principles, divorced from the moral community in which it obtains its meaning, it loses its capacity to serve as a tool to be used in the search for moral meaning.[48]

9. An Ethic of Moral Response

An approach to ethics that provides the impetus to unify the search for moral meaning and normative ethics is an approach that

---

46 See Stanley Hauerwas, *A Community of Character: Toward a Constructive Christian Social Ethic* (Notre Dame: University of Notre Dame Press, 1981).
47 See William May, *The Patient's Ordeal*, 1-15.
48 See Albert R. Jonsen and Stephen Toulmin, *The Abuse of Casuistry* (Berkeley: University of California Press, 1988). My explanation of the interpretive function of casuistry will not be found in the Jonsen-Toulmin work.

Method and Meaning

is symbolized by what can be regarded as the first question to be addressed to the task of ethical reflection: What is going on around me, and what would be a fitting response to what I (we) find addressed to me (us) in this moral world of mine (ours)? The question necessitates a repositioning of the stethoscope of ethics, from its position over a narrowly concentrated problem to a position where it can hear the totality of the significance of the experience or sphere of experience that is addressing the moral agent.

This approach to ethics is sometimes called an ethic of response. Both H. Richard Niebuhr[49] and Charles Taylor offer a moral philosophy of response; for example, Taylor describes a person as «a being who can be addressed, and who can reply.» He calls the person a «respondent.»[50] The response approach to moral analysis is forcefully elicited by Levinas's image of the face of the stranger.[51] I may ignore or respond to the stranger; and if I respond it may be with words, silence, or actions that meet his or her needs. I may reject or welcome the stranger's world into my world, but I will ineluctably be morally affected by the encounter. The greatest danger is that we, both as moral agents and as ethicists commenting on morality, will succumb to the temptation to domesticate all reality[52] by subsuming the totality of what the Other represents into our familiar categories of moral thought. What is needed in bioethics is a firm focusing on the meaning of the stranger-subject before us – whether the different-feeling woman or man, the differently endowed fetus or handicapped individual, the different-skinned and different-speaking foreigner or native, or the hospital patient who has become a stranger by refusing to avoid the forbidden realm of death.

A response approach to moral interpretation and normative ethics is, I think, the most useful approach to developing a much needed

---

49 Richard Niebuhr, *The Responsible Self: An Essay in Christian Moral Philosophy* (New York: Harper & Row, 1963).
50 Taylor, «The concept of a Person,» in *Human Agency and Language: Philosophical Papers*, vol. 1 (Cambridge: Cambridge University Press, 1985), 97.
51 Emanuel Levinas, *Totality and Infinity: An Essay on Exteriority* (Pittsburgh: Duquesne University Press, 1969). Thomas Ogletree has situated the phenomenological thought of Levinas in the broader fields of philosophical and theological ethics in his work *Hospitality to the Stranger: Dimensions of Moral Understanding* (Philadelphia: Fortress Press, 1985).
52 David Tracy discusses this temptation in *Plurality and Ambiguity*, 51.

moral art of caring for the dying.[53] Principles of beneficence and the justification of withdrawal of treatment by appeal to «proportional» assessments in a calculus of benefits have only a limited reach in an ethics of dying; such moral language can conceal the larger question of moral response to the person who is attempting to have his or her own experience of dying. A response ethic in this area would focus on relational behaviors – on responses of companioning, communicating with, and showing compassion for the person who is in a unique situation of living through his or her dying days and hours.

10. *The Role of Attention*

Attention is the door through which we move to come into contact with moral experience and discover moral meaning; it is our window to the moral world. The word *attention* is taken from a Latin word whose radical meaning is «to stretch out to.» One stretches out to a work of art or another human being by attending to it, her, or him; in this way the agent perceives and interprets the other.

Simone Weil, a philosopher who regarded attention as a pivotal idea, explained that it entails, above all, a negative effort. It consists of suspending our thought, leaving it detached, empty, ready to be penetrated by the (manifesting) object, ready to receive the being one is looking at, «just as he is, in all his truth.»[54] It means waiting, seeking, being open to naked truth. For the realm of ethics, she rejects the adequacy of the Kantian notion of person. Attention to the person of the other is not at all helpful in discovering a useful moral norm; rather, one must exercise positive moral attention to the individual as individual. Her notion of justice is one bred of a compassion elicited in the exercise of attention.

Attention is the master key to an ethic that this essay has advocated – one chiefly characterized by the search for the meaning of mor-

---

[53] James Bresnahan makes an appeal for a contemporary art of dying (not an art of *caring for the dying*, such as I am advocating) in his «Death: Art of Dying: II. Contemporary Art of Dying,» in the *Encyclopedia of Bioethics*, ed. Warren Thomas Reich, rev. ed. (New York: Simon & Schuster-Macmillan, 1995), 551-53.

[54] *The Simone Weil Reader*, ed. George. A. Panichas (New York: David McKay, 1977), 51.

al experience. It is also key to the major methodological elements of the «experiential» ethic I have described: interpretation, radical care, truth-as-manifestation, moral agency, intentionality, character, and response. Consequently, I believe that bioethics, in an era that would devote itself to a study of the meaning of moral experience, must examine in a more fully developed way the role of attention in moral epistemology, and what difference attention would make in clarifying and developing bioethics in a mode of moral response.

Both the historical turn to the moral subject in ethics and the pressing need for a concerted turn to moral interpretation create a situation that breaks down the distinction between philosophical and theological ethics. While both of those disciplines make significant and separate contributions to special needs – for example, the needs to articulate the ethics of a particular religious community and to analyze particular moral concepts – the intellectual barrier dividing them breaks down in the arena of the public search for moral meaning.

In the contemporary cultural and intellectual setting that I have described, caring attention can be regarded as the starting point of, and the unifying factor for, all of ethics; and it is especially crucial for the development of a bioethics that would be responsive to our most pressing moral needs. Attention, however, is a contemplative sort of action; the contemplation it entails is contemplation of the moral significance of subjectivity. Both as regards its nature and its object, then, the attention required for interpretation is a spiritual act, in the sense that it entails regard for the very stuff of existence and its meaning, as well as regard for choices about one's existence – choices that imbue our world and our lives with the meaning that is symbolized by our basic moral language and responsiveness. Both philosophers and theologians have an important role to play in an ethic that is radically centered on a (secular) spiritual act.

The central role that attention to meaning should play in the future of bioethics convinces me that it is crucial that religious and theological scholars – for whom matters of meaning and interpretation are central to moral inquiry – participate more directly and more vigorously than they have in recent decades. Cooperation of this sort is needed for the development of a field of bioethics that is concerned with the meanings by which humans think out the possibilities of their own living and make choices among them.

CHAPTER 3
ON THE ULTIMATE, WHICH IS THE FIRST
Thinking Beyond (Bio) Ethics

1. *Introduction*

What does it mean to think metaphysically in bioethics, and is that possible at all? I mean, can we think beyond the terms of a reductive applied ethics that aims at a purely ontic rendition of questions of right and wrong? Could it be that the ontological ground upon which such considerations rest remains fully operative, even if elusive, an after-thought perhaps, yet determining what foregrounds our ethical preoccupations? With metaphysical thinking, then, I do not mean abstract thinking, the way a rationalistic philosophy might conceive. We live always on the basis of an implicit understanding of the ground of things, a ground that is either promising or threatening, gifting our daily endeavors with the joy of being, or leaving us in thrall of a different *mal de vivre*. Such understanding is most concrete: it colors our entire way of being in-the world, our perception of what it means to be, summoning our trust in the goodness of things, or, conversely, our suspicion toward it.[1]

One wonders whether the epistemic conditions that sustain such an elemental mindfulness are not already compromised by the legacy of modernity, whose effectual history feeds the projects of science no less than the practice of medicine, holding us bewitched to

---

[1] For a systematic articulation of metaphysics in the wake of postmodern deconstructionism, I am indebted to Leuven philosopher William Desmond. In his work, one finds the imposing quality of speculative thinking together with a more personal call to attention for the concrete and simple quality of experience, a kind of phenomenological «return to the things themselves», elicited by a language full of evocation and beauty. In my reflections I rely especially on William Desmond, *Being and the Between* (Albany, NY: SUNY Press, 1995), idem, *Ethics and the Between* (Albany, NY: SUNY Press, 2001) and idem, *The Intimate Universal. The Hidden Porosity Among Religion, Art, Philosophy, and Politics*, New York: Colombia University Press, 2016).

impossible dreams of trans-human enhancement.[2] Would the last man of Nietzsche still intone his song, when its melody echoes a twisted reverence, not for a final good, but for a dark origin, an ethos of valuelessness that says «nothing is good»?[3]

I want to think *beyond* bioethics, in the sense of getting at those questions bioethics tends to push aside, relegating them, at best, to benign superannuation, the useless pastime, supposedly, of meta-ethical devotees untouched by more pressing normative perplexities.[4]

Of course, if such supposition is false, ultimate questions will have to be treated as first, and this for the sake of the more ordinary concerns we implicitly acknowledge as deserving of our intellectual priority: the *ordo rerum* subverts any pre conceived notion of what we see as the *ordo disciplinae*, in which case a different reflection will be called for, and with it the retrieval of a deeper ground. What I call, with Desmond, «the primal ethos of life» entails an elemental attunement to the deeper sources of life, over, or against,

---

2   For a useful discussion of the issue from a theological perspective, see Ronald Cole-Turner, ed., *Transhumanism and Transcendence. Christian Hope in an Age of Technological Enhancement*, Washington (DC) 2011, and A. Verhey, *Nature and Altering It*, Grand Rapids 2010. More robust philosophical considerations can be found in N. Agar, *Truly Human Enhancement: A Philosophical Defense of Limits*, Cambridge (MA) 2014.

3   The nihilism in question is ultimately theological, rather than moral, though the death of God signals a metaphysical end, the end of the God of onto-theology, as Heidegger saw, which is not without ethical implications. Thus the last man, who is also the madman, famously adds to the declaration of God's death («Where has God gone?...I shall tell you. We have killed him») the following: «Whither are we moving now? Away from all suns? Are we not perpetually falling? Backward, sideward, forward, in all directions? Is there any up and down left? Are we not straying as through an infinite nothing?» F. Nietzsche, *The Gay Science. With a Prelude in Rhymes and an Appendix of Songs*, trans. by Walter Kaufmann, New York 1974, Book III, n. 125 «The madman», 181.

4   Analytic philosophy defends the relative separation of metaethics from normative ethics. Non-cognitivist versions of the former may, thus, coexist with strongly rationalistic renditions of the latter. R.M. Hare's universal prescriptivism is a point in case, most famously articulated in his *The Language of Morals*, New York 1952. The birth of bioethics is historically tied to a different appreciation for normative questions, and the assumption of so called principle of moral neutrality. The issue for me is not just metaethical, but more deeply metaphysical. At stake is not only the meaning of moral language, but of moral experience as such.

the ethos we super-impose upon them, thus clogging what is primal with constructions of our own making. The ethical discourse such constructionism engenders tends to focus more on the conditions for establishing moral consensus on what is right, than to articulate, in phenomenological faithfulness to the nature of things, the trust that nourishes our love of the good. Indeed, the good remains *incognito*, and this in spite of the seriousness of our ethical engagement with the main challenges of the day.[5]

I believe much more than issues of public consensus, reducible to the mini mal denominator of empty generalities, is at stake. Deeper questions of meaning, concerning the place of science and technology in a democratic society, the dignity of the human person, and the well-being of our ecosystem, call for a different mindfulness, beyond the pragmatic complacency of a content-thin ethical strategy.[6] It is not enough to keep the system open to the latest normative integration, in an endless exercise of reflective equilibrium, if such a system fails to address the deepest matters of our humanity. Brilliant moral theories might come too late, when ethics has already lost its soul.[7]

---

5   Thus Rawls, in a somewhat programmatic vein in an early work: «To establish the objectivity of moral rules, and the decisions based upon them, we must exhibit the decision procedure, which can be shown to be both reasonable and reliable». J. Rawls, «Outline of Decision Procedure for Ethics», *The Philosophical Review* 60 (1951) 177-198, at 177. For the distinction between right and good, see J. Rawls, *A Theory of Justice*, Cambridge (MA) 1971, 446-452.

6   As an example of the approach in question, see the classic text of T.L. Beauchamp – J.L. Childress, *Principles of Biomedical Ethics*, New York 1979, 2013[7]. The authors understand their project as a «work in theory», not a comprehensive moral theory, articulated on the basis of so called common morality (see 1-29, 351-429). Although there is only one universal common morality, there is more than one theory of it. On the topic, see the special issue of the *Kennedy Institute of Ethics Journal* 13 (2003).

7   I dispute the notion that coherence, as a necessary dimension of ethical theory, can be understood as a purely formal requirement. Reflective equilibrium is the attempt to create a coherence between moral principles and considered judgments, and this in light of changing circumstances: «What is required is a formulation of a set of principles which, when conjoined to our beliefs and knowledge of the circumstances, would lead us to make these judgments with their supporting reasons were we to apply these principles conscientiously and intelligently», J. Rawls, *A Theory of Justice* (cf. nt. 5), 20. See also Rawls' reference to reflective equilibrium in his later book, *Political*

A renewed attention to the primal ethos of life might provide us with a much needed passageway toward a richer bioethics, an opening, not just epistemological, but more deeply metaphysical, beyond the extremes of what seems to stall ethical discourse today: a univocal foundationalism, on the one hand, and an equivocal relativism, on the other. The former is the attempt to reduce the practical, and therefore dynamic, quality of the *ethos* to an abstract notion of the good, grounded in an essentialist understanding of human nature. A certain neo-scholastic tradition of natural law, as well as more recent versions of ethical essentialism, especially in the analytical mode, might be taken as examples of such an approach.[8] The other extreme is offered by a relativism that subverts the historicity of the ethos, and deconstruct anthropological constancies able to provide a basis for ethical judgment. [9]We must search for the promise of a different «middle» in the grounding of moral norms, and for a metaphysics of the good that enables the retrieval of a different ethical between.

I will begin with somewhat general claims about the predicament of con temporary medicine with respect to the oblivion of the

---

    Liberalism, New York 1996, 8, 381, 384, and 399. On the condition of contemporary bioethics, relative to a lack of questioning about moral meaning, see L. Kass, *Life, Liberty and the Defense of Dignity. The Challenge for Bioethics*, San Francisco 2002, 55-76, and G.C. Meilaender, *Body, Soul, and Bioethics*, Notre Dame 1995.

8   For an example, R.P. George, ed., *Natural Law Theory: Contemporary Essays*, Oxford-New York 1992. The discussion on natural law has been revamped, over the past thirty years, by Germain Grisez, John Finnis, Joseph Boyle, William May and Patrick Lee, among others, thought he relation between practical reason and the normativity of nature remains problematic in their account. From an analytic perspective, Q. Smith, *Ethical and Religious Thought in Analytic Philosophy of Language*, New Haven-London 1998, and my review in *Theological Studies* (1999) 379-380.

9   According to Stephen Toulmin, so called anti-foundationalism «shares in the conviction that all earlier questions for a comprehensive system of knowledge, based on permanent, universal systems of overarching principles, were misguided from the start, and are by now discredited. Claims to philosophical universality and permanence can be ignored: their only interest lays in the ways that they could serve as a «cover» for the collective interests of nations, social groups, or genders», S. Toulmin, «The Primacy of Practice: Medicine and Postmodernism», in R.A. Carson – C.R. Burns, ed., *Philosophy of Medicine and Bioethics. A Twenty Years Retrospective and Critical Appraisal*, Dordrecht 1977, 41-42.

search for meaning. In a second moment, I reflect, more specifically, on the metaphysical premises, or lack thereof, implicated in our public discussions on end of life and beginning of life issues. Finally, I offer a reconstructive attempt of an ethics defined by love of being, and the porosity bound up with a different opening to the generosity of the good.

2. *Modern Medicine and the Oblivion of Meaning*

The search for meaning in medicine might be illustrated by the metaphor of the stethoscope. Richard Baron, in a famous article for the *Annals of Internal Medicine*, tells the story: «It happened the other morning on rounds, as it often does, that while I was carefully auscultating a patient's chest, he began to ask me a question. «Quiet» I said. «I can't hear you while I'm listening»«.[10]

The stethoscope metaphor is emblematic of the inattention to meaning («not hearing») brought about by the reductionist focus (the mode of restricted «listening») in the methodologies of both modem scientific medicine and contemporary ethical theory. To start with, the mind-set created by modem scientific medicine has required for medicine *to be inattentive*, that is, not to hear, the sick person's experience of illness. Influenced by a positivist framework, 19th century medical scientists popularized the notion that practical clinical medicine should be viewed as a form of applied theoretical medicine. In the United States, the reformation of medical studies introduced by the medical educator Abraham Flexner, in the first part of the 20th century, completed the picture. Moreover, this happened as a result of modernity's understanding of scientific knowledge, which Gadamer poignantly describes as a capacity to produce effects. In the modem version of scientific knowledge, the mathematical-quantitative isolation of laws of the natural order provides human action with the iden-

---

10 R. Baron, «An Introduction to Medical Phenomenology: I Can't Hear You While I'm Listening», *Annals of Internal Medicine* 103 (1985) 606-611, at 606. See also W.T. Reich – R Dell'Oro, «A New Era for Bioethics: The Search for Meaning in Moral Experience», in A. Verhey, ed., *Religion and Medical Ethics. Looking Back, Looking Forward*, Grand Rapids 1996, 96-119. In an analogous phenomenological vein, see R. Zaner, *Ethics and the Clinical Encounter*, Englewood Cliffs 1988.

tification of specific contexts of cause and effects, together with new possibilities for intervention.[11] In relation to clinical medicine, such an idealization entails a tendency to reduce the *praxis* of medicine, with its matrix of subjective components and contextual features, to the detached objectivity of theoretical knowledge, and to interpret the healing process itself as a production of effects.[12]

Of course, one cannot question, in principle, the application of scientific reasoning to medicine. In trying to identify and explain the cause of symptoms, medicine employs probabilistic laws and rules, theories and principles, of the biomedical sciences. Concepts of normal and abnormal, for an example, are statistically derived concepts, based on scientifically validated norms of human biological functioning. In the attempt to classify symptoms as the manifestation of particular disease entities, medicine relies upon hypothetic-deductive and inductive reasoning. Moreover, in order to determine what can be done to remove or alleviate the cause of particular diseases, medicine appeals to prognostic knowledge about the course of the diagnosed disease, as well as efficacy and toxicity of relevant therapeutic possibilities.

And yet, in spite of its undisputable scientific basis, medicine resists final and complete reduction to science. Far from simply bringing different segments of scientific explanations into a unified theory, the specific goal of medicine consists in yielding a general understanding of illness with a specific medical decision on behalf of a concrete patient.[13] As a synthetic action, medicine is both theoretical *and* practical at the same time. Unlike the patho-physiology of disease, the medical act can be fully understood only hermeneu-

---

11   H.G. Gadamer, *The Enigma of Health: The Art of Healing in a Scientific Age*, Stanford 1996, 35; orig. German, Über die Verborgenheit der Gesundheit. Aufsätze und Vorträge, Berlin 1993.
12   M.W. Wartofsky, «What Can the Epistemologists Learn from the Endocrinologists? Or Is the Philosophy of Medicine Based on a Mistake?», in R.A. Carson – C.R. Burns, ed., *Philosophy of Medicine and Bioethics* (cf. nt. 9), 55-68.
13   I owe such a perspective to the philosophy of medicine of Edmund Pellegrino. See E.D. Pellegrino, «The Anatomy of Clinical Judgment: Some Notes on Right reason and Right Action», in H.T. Engelhardt – *al.*, ed., *Clinical Judgment. A Critical Appraisal*, Dordrecht 1979, 169-194; E.D. Pellegrino, «The Healing Relationship: The Architectonics of Clinical Medicine», in E. Shelp, ed., *The Clinical Encounter. The Moral Fabric of the Patient-Physician Relationship*, Dordrecht 1983, 153-172.

tically, as Gadamer suggests, through an act of interpretation that takes place within the sociological, cultural, and ideological matrix of a defined life-world. For this reason, medicine represents a peculiar unity of theoretical and practical knowledge within the domain of the modem sciences, «a peculiar kind of practical science for which modem thought no longer possesses an adequate concept».[14]

My point here should not be misconstrued. Careful scientific attention to the patho-physiology of disease, together with ever more extensive bio-techno logical applications, has certainly yielded marvelous advances in modem medicine.[15] Yet, its positivist reduction has also created a mind-set that brackets questions of meaning, themselves highly significant to human well-being and to the ethical aspects of medicine. The judgment of Edmund Husserl, while summarizing the development of modem sciences, offers at the same time a prophetic anticipation of the predicament of contemporary medicine:

> The exclusiveness with which the total world-view of modem man lets itself be determined by the positive sciences and be blinded by the «prosperity» they produced, meant an indifferent turning away from the questions which are decisive for genuine humanity. Fact-minded science excludes in principle precisely the questions which man finds the most burning: questions of the meaning or meaninglessness of the whole of human existence.[16]

The central task of ethics in medicine is to foster an *anamnesis* of the very questions medicine seems to bracket: the significance of illness and disease, of our human condition as embodied, of birth, suffering and death, and of the service to the ethos of generosity that sustains the healing professions. Ethics searches for a matrix of meaning supportive of human endeavor. Such matrix is not entirely constructed, even when it is the product of active investment on our part. Meaning can only be envisioned and recognized, and this in order to guide our projects towards their final goal. Would there be a hope of

---

14 H.G. Gadamer, *The Enigma of Health* (cf. nt. 11), 39.
15 L. Kass, *Life, Liberty, and the Defense of Dignity* (cf. nt. 7), 29-53.
16 E. Husserl, *The Crisis of European Sciences*, trans. David Carr, Evanston 1970, 5-6; orig. German, *Die Krisis der europäischen Wissenschaften und die transzendentale Phänomenologie Eine Einleitung in die phänomenologische Philosophie*, Haag 1962².

fulfilment at the end of our *praxis*, without a promise of meaning at the roots of our original constitution? The point is more Platonic than Aristotelean, for it calls for an archeology of the good as grounding any hope of teleological completion.[17] Therefore, the question of our attentiveness to what is *given* to us becomes paramount, in fact, more wondrous than any of our constructed achievements. We build for ourselves a «brave new world», whether techno logically or scientifically defined, but how grateful is our new *Gestell* to the home that birthed us, and within which we ultimately dwell?[18] Do we not run the risk of fashioning for ourselves a second ethos, without any connection to the primal ethos of life? As Desmond suggests:

> We are rooted in nature, but we risk denaturing ourselves in claiming to make ourselves according to a second nature. The second nature is not a second «yes,» a redoubled «yes» to the first «yes» at work in the *poiesis* of naturing and our *passio*. More often, it is a «yes» to a *conatus* that has deviated from the subtle insinuations of the now sunken matrix of fecundity.[19]

I want to show how such a «transvaluation»[20] emerges with respect to two issues, which are on the forefront of our public discussions today: the ethics of dying well, and the morality of artificial reproductive technologies.

---

17 On the archaeological nature of the good implied by a metaxological metaphysics, see William Desmond: «The metaxological turns to the otherness of the origin, for the beginning intimates the overdetermination of the good [...] this overdeterminate good is an agapeic good: out of its surplus it communicates», W. Desmond, *Ethics and the Between* (cf. nt. 1), 1-4, at 9. For a comment on Desmond's overall position, see C. O'Regan, «The Poetics of Eros: William Desmond's Poetic Reconfiguration of Plato», in *Ethical Perspectives: Journal of the European Ethics Network* 8 (2001) 272-302. For Levinas too, the Platonic vision of the good makes possible the exteriority of the Other as an irreducible moment of de-totalization. See E. Levinas, *Totality and Infinity. An Essay in Exteriority*, trans. Alphonso Lingis, Pittsburg 1969; orig. French, *Totalité et infini. Essai sur l'extériorité*, La Haye 1961.
18 The reference is to Heidegger's notion of «framework» in «The Question Concerning Technology». See M. Heidegger, *The Question Concerning Technology and Other Essays*, trans. William Lovitt, New York 1977, esp. 14-17; orig. German, *Die Frage nach der Technik*, 1954
19 W. Desmond, *The Intimate Universal* (cf. nt. 1), 327.
20 The language is appropriately Nietzschean (*Unwertung der Werte*), if the construction in question is, ultimately, an expression of will to power.

3. The «Good Death:» Endeavoring to Be and Letting Go

It is not easy to add entirely new perspectives to the vast bioethics literature that has emerged, over the years, on the ethics of dying well.[21] The passing of «aid-in-dying» laws in several states, whether defined by statutes, or as a result of popular referenda, simply stokes a fire that was never really extinguished, only kept alive under the embers of previously defined conceptual systematizations by new publicized cases of requests for «assistance» in dying.[22] California is only the latest story, for sure, not the last one. What is at stake in the conversation is the problem of articulating the conditions for a *good death* – and more specifically, for a good death when faced with the vulnerability of old age, terminal disease, and unbearable suffering. Such a task remains quite formidable both in relation to its philosophical foundations, as well as with reference to the analysis of specific ethical quandaries.[23]

The discussion about the ethics of dying centers on the resources of a restricted language game, defined by the conditions for a control of death.[24] At the heart of the reflection is the use of medical technology, of the medical power to prolong life. Such power has

---

[21] For a philosophical articulation of the main bioethical issues, e.g., the distinction between assisted suicide and euthanasia, the problem of causation and the moral relevance of the distinction between killing and letting die, intention and foresight, and the entire question of the principle of double effect, see T.L. Beauchamp, ed., *Intending Death: The Ethics of Assisted Suicide and Euthanasia*, Upper Saddle River 1995. Recently, with particular reference to the European experience, D.A. Jones – al., *Euthanasia and Assisted Suicide. Lessons from Belgium*, Cambridge 2017.

[22] For a recent overview the legal landscape, J. Keown, «Legal Issues at the End of Life», in I. Carrasco de Paula – P. Pegoraro, ed., *Ageing and Disability*, Rome 2014, 203-213. More broadly on theoretical questions concerning public policy, especially with respect to the legal principle of the inviolability (sanctity) of human life, J, Keown, *The Law and Ethics of Medicine*, Oxford 2012, and J. Keown, *Euthanasia, Ethics and Public Policy*, Cambridge 2016².

[23] For an analysis of dying well in relation to the larger spectrum of anthropological and ethical issues concerning ageing and disability, see M.-J. Thiels, ed., *Ethical Challenges of Ageing*, London 2012. Also I. Carrasco de Paula – P. Pegoraro, ed., *Assisting the Elderly and Palliative Care*, Rome 2015.

[24] For a historical analysis, which highlights the particularity of our contemporary attitude toward death, see the classic works of P. Ariés, *Images of Man and Death*, Cambridge (Ma.) 1985; orig. French, *Images de l'homme devant la mort*, Paris 1982.

pushed the limits of our technical possibilities, creating an imbalance between what *can* and what *should* be done, thus leading to the question of the quality of life endorsed or maintained by a particular treatment. To use the distinction made famous by philosopher James Rachels, there is a difference between «being alive» and «having a life».[25] But how to decide on the boundary between the two? If «having a life» depends on a perspective of value regarding what renders life livable or worth living, do we end up in the trap of a subjectivism that dispenses with all criteria? More over, how to reconstruct an objectivity that obtains for the medical act itself, beyond two equally false alternatives: either turning medicine into a function of patient's individual preferences, or reducing it to maintenance of purely biological mechanisms. In the former case, the *lex suprema* of the medical act is conflated with what the patient wants (*voluntas aegroti*); in the latter, the well-being of the patient (*salus aegroti*) is confused with stubborn insistence on the biological functioning of individual organs. In both cases medicine loses sight of its ultimate end, namely, the integral good of the patient.[26]

The debate on the good death – and thus on the normativity of dying – seems paradoxical, in that it unfolds on the premise of a suspension, a bracketing placed on the meaning of death. One often speaks of the ethics of dying, of «dying well», but without always knowing in relation to what. Bioethics claims to provide normative criteria. It does so, however, on the presupposition of suspending any symbolic horizon capable of saying what death *is*, what it represents for the person. Of course, one cannot but be pleasantly impressed by the formal elegance and analytical consistency with which the moral principles of a presumed common morality, logically shared by all rational agents, are put into play, when faced with the most complicated ethical conflicts.[27] So called

---

25 J. Rachel, *The End of Life: Euthanasia and Morality*, New York 1986.
26 At stake in the discussion is the viability of the foundational principle of Hippocratic medical ethics, i.e., beneficence, in its relation to patient's autonomy. For a classical, and still very valuable, articulation of the questions, see E.D. Pellegrino – D.C. Thomasma, *For the Patient's Good: The Restoration of Beneficence in Health Care*, New York 1988.
27 T.L. Beauchamp – J.F. Childress, *Principles of Biomedical Ethics* (cf. nt. 6). The approach in question, known as «principlism», has been the subject of sharp criticism in the debate over the method of bioethics, both in the

«principlism», with its mantra of beneficence, non-maleficence, autonomy, and justice, in turn specified by the rules of proportionality, informed consent, etc., might constitute a helpful point of reference when tackling the many conflicts faced by health professionals, patients and their families, in the different clinical settings. Yet, when ethical «principles» and «rules» are employed mechanically, as if in a kind of a priori framework, such elegant bioethical theory looks more like a game without any grasp on reality. In this version of ethics, «substantial» rationality gives way to «formal» rationality, reflection on ethical content dissolves in sheer proceduralism.[28]

How could such an approach provide recommendations toward a truly good and dignified death? What if ethical formalism were to betray, in the end, a lack of any points of reference, if not resentment, as Nietzsche might suggest, when faced with the void of sense? Morality, in this case the morality of a good death, would only be smoke and mirrors, a nihilistic enchantment. How can we leave unaddressed the existential aspects of death and dying considered in their experiential value, that is, as dimensions of our journeying (*experior*) toward the end? Likewise, how to pass over in silence those dimensions that speak to the trial (*peiros*) entailed by the agony of passing, the physical pain, the loneliness of suffering?

A good death cannot be envisaged as anything but the fulfillment of a good life, and this with reference to a life that will inevitably age. Death will be considered good when it succeeds in expressing the meaning of living, understood as living well. That living well

---

United States as well as in Europe. See, E.R. Dubose – *al.*, ed., *A Matter of Principles? Ferments in U.S. Bioethics*, Valley Forge 1994; H. ten Have, «Approcci europei all'etica della medicina clinica», in C. Viafora, ed., *Comitati etici. Una proposta bioetica per il mondo sanitario*, Padova 1995, 91-118; G. Khushf, ed., *Handbook of Bioethics. Taking Stock of the Field from a Philosophical Perspective*, Dordrecht 2004.

28 For a criticism of bioethics as a purely secular field of investigation, autonomous with respect to any moral substantial premise of content, therefore, as a purely formal endeavor, see J.H. Evans, *Playing God? Human Genetic Engineering and the Rationalization of Public Bioethical Debate*, Chicago 2002. The analysis of Evans, which turns on Max Weber's distinction mentioned above between formal and substantial rationality, refers to the discussion on genetics and genetic research. Nevertheless, in its basic meaning, it could apply to the entire field of bioethics, as it has developed in the United States. For a more generous account of the history of bioethics as public discourse, see A.R. Jonsen, *The Birth of Bioethics* (cf. nt. 4), 352-376.

can and must end, at times even tragically, and why – these are questions that do not belong to the discipline of ethics per se. Yet ethics cannot even begin to reflect on its proper criteria if not because it lets itself be challenged by the existential perplexity such questions entail. I believe such questions have a metaphysical quality to them: they interrogate our attitude toward being as such, and to the meaning of things.[29]

I find Desmond's distinction between *conatus* and *passio* helpful in this context.[30] *Conatus essendi* is a way of standing before things defined by the endeavor, the effort to be. We do not choose such a posture. We are already endowed with it by virtue of our relation to the world, though such posture may take up a certain primacy on account of our emphasis on doing or acting. The *conatus* is the defining posture of modernity, in whose larger narrative the «effort to be» makes the appearing of things conditional upon a subjectivity that posits and determines. Being *is* insofar as it responds to the (transcendental) forms of its apperception by a subject that measures and rules any phenomenic presence.[31] Such a posture has obvious epistemological importance, which Descartes and Kant will fully unpack. Given the ambiguity that marks the appearing of things – thus the doubt about them, the *cogito* tries to recover an irrefutable certainty, starting no longer from the promise of meaning that inhabits reality, but from the subjective certainty that defines the very act of thinking. Such act must necessarily presuppose – and beyond all doubts no less – the existence of the cogito who thinks. In this way, however, one sees a shift, a «Copernican revolution» in the relation of subject and object, and, moreover, in the priority of the former over the latter.

---

29   Thus, von Hildebrand, commenting on Pascal (*Pensées*, VI, Frag. 347) writes: «And so (Pascal) alludes to in a singular manner the contradictory nature of the metaphysical situation of mankind, in part due to the fallenness of life, yet also to the ineffable survival of his personal condition after death», D. von Hildebrand, Über *den tod*, St. Ottilien 1980, 33.

30   The distinction is central throughout *The Intimate Universal*, but see also W. Desmond, «Pluralism, Truthfulness, and the Patience of Being», in C. Taylor – R. Dell'Oro, ed., *Health and Human Flourishing. Religion, Medicine, and Moral Anthropology*, Washington (DC) 2006, 53-68. The distinction can be properly understood only in light of the complete work of Desmond. For a study of Desmond's thought, see T. Kelly, ed., *Between System and Poetics: William Desmond and Philosophy after Dialectic*, Burlington 2007.

31   W. Desmond, *Ethics and the Between* (cf. nt. 1), 17-47.

This «anthropological turning point» entails important cultural consequences, and conditions our way of thinking about nature. We no longer «undergo» nature, so to speak, but actively shape it according to heuristic models, which, reducing nature's complexity to mathematical univocity, enables us to describe and empirically verify it. For sure, the book of nature has much to say still, but will do so because the «spectator scientist» sets the conditions to prevent its hiding, thus forcing nature to yield its secrets, as Galileo suggests in *Il Saggiatore*.[32] All this presupposes a neutralization of being to *mathesis universalis*. One thinks of the meaning of such neutralization with respect to the distinction between primary and secondary qualities, or the Cartesian reduction of the human body to *res extensa*, now become a mechanism separate from the mind. If true reality can only be rendered in mathematical terms, then it is imperative to bring the unverifiable pathos of things back to the dianoetic precision of scientific formulae. This holds true also for the subject, whose emotional complexity will have to be reduced, now, to the act of «thinking clearly and distinctly». In the words of Spinoza: «*non ridere, non lugere, neque detestari, sed intelligere*».[33]

So much for the epistemological significance of the *conatus*, whose implications, however, extend to ethics, and, more specifically, to the ethics of dying. I would say the following: the subject who fashions reality also grants value to it. This is so because the neutralization of being, with respect to the object, entails something like a rebound effect, a kind of «contraction of value» – especially with Kant – in favor of the subject. Only the person possesses an intrinsic value: as a good in-itself, it is never to be treated as a means,

---

32 «Nature loves to hide» had said Heraclitus, but the modern gaze is more akin to an act of unveiling. The forcing of nature also signals the end of teleology: *Naturam finem nullum sibi praefixum habere, et omnes causas finales nihil nisi humana esse figment* («Nature has no fixed goal and all final causes are but figments of the human imagination»), B. Spinoza, *Ethics*, trans. Samuel Shirley, Indianapolis 1992, 59.

33 B. Spinoza, *Theological-Political Treatise*, trans. Samuel Shirley, Indianapolis 2001, 1, 4. Nietzsche correctly interprets the spirit of Spinoza's quotation when he says: «What does knowing mean? *Non ridere. non lugere, neque detestari, sed intelligere!* Says Spinoza, so simply and sublimely, as his wont. Nevertheless, what else is this *intelligere* ultimately, but just the form in which the other three things become perceptible to us all at once?», *The Gay Science* (cf. nt. 3), Book IV, n. 333 «What Does Knowing Mean?», 261.

only as an end. Unlike nature, understood now as phenomenal field open to endless manipulation, the person is not neutral; rather, being the source of absolute meaning, it becomes the condition for the very possibility of meaning's attribution. But, as William Desmond points out:

> Here is the sting. The subject cannot live with this devaluation of otherness, and even less with the devaluing of its own valuing. It will not be passive to this. It will be active. The subjectification of value inevitably leads to the primacy of self-activity that impresses itself on the other...We witness the recoil of the subject on itself out of the hiding of neutrality it had schemed for itself. There is no escape from itself, but now when it awakens again to itself, it has been transformed into a more *radically self-assertive subjectivity*.[34]

In this paradigm, the good death is the *humanized* death, death lived as chosen, not as undergone or endured. «Choosing death» is to determine it, the way our choice determines the theoretical models that grant access to reality as such. To be «the measure of all things» is to be greater than death. Thus, the latter will be neutralized, if not in its inevitability of fact that inexorably happens, at least, in its dramatic quality of experience to be resisted.[35] The effort to be, the *conatus essendi*, is a struggle against death, the attempt to indefinitely postpone it, or else, to anticipate it «rationally», as in the case of euthanasia or assisted suicide. Absolute passivity is not worthy of man.

This paradigm is not without important emphases. The efforts to humanize death, above all through the contributions of medical and scientific research that aim at the treatment and management of pain, are essential part of our modem way of relating to death.

---

[34] W. Desmond, *Ethics and the Between* (cf. nt. 1), 29 (emphasis in the original).
[35] According to Leo Scheffcyzk, this is the final outcome of Heidegger's reflection, which, if on the one hand, recognizes death in its inevitability of fact that occurs, on the other hand, tends to overcome it «trans-subjectively». See L. Scheffcyzk, «Die Phänomenologie des Todes bei Dietrich von Hildebrand und die neuere Eschatologie», in J. Seifert, ed., *Truth and Value: The Philosophy of Dietrich von Hildebrand*, Bern 1992, 271. Heidegger's effectual history, however, has been important with respect to theological reinterpretations. For an example, K. Rahner, *On the Theology of Death*, New York 1961; and L. Boros, *The Moment of Truth. Mysterium Mortis*, London 1969.

An inhumane and dehumanizing death can not be good. Yet, this paradigm contains also the seeds of a possible degeneration. It risks thinking of a «good death» according to the logic of scientific-technological control and neutralizing planning, which, in the paradigm of modernity, renders a life worth living. The truth is another: we can pro long life, eliminating from it all pain and suffering, but will never succeed in «managing death». Death will always come, an unexpected surprise and an expression of the heteronomy of nature, even more striking now, because it seems to contradict the autonomy by which we attempt to completely define ourselves. The separation, the dualism of person and nature, constitutes the condition of possibility for controlling death, but it can also lead to conflicting results: a technological effort that de-personalizes nature; or a will to power that de-naturalizes the person, reducing it to self-determining rationality. The *epoche'* on any search for the meaning of death is the inevitable result of both these developments, stemming from the same root.

I ask: what if something else companioned, more originally (*co-natus*), the vector of intentionality that drives our own effort and strive for control? What if a more radical openness, perhaps even intimacy, to reality sustained our θαυμάζειν, the astonishment at the fact that being *is*, when it could also *not* be. In wonder, we take up our residence in the between, attuned to the saturation of meaning that dwells in things, in *their* value, and in whose hospitality we build a world for ourselves.[36] The *conatus essendi* can only be a derivation, of course possible and legitimate, of a more original *passio essendi*, of an undergoing (*passio*) that also becomes a «passion» for being. With respect to the previous par-

---

36 The turn to «givenness» is, of course, central for phenomenologically inspired thinking. This goes beyond differences among phenomenological schools, whether «realistic» or «transcendentally» defined. For an example of the former, see the beautiful book of E. Kohak, *The Embers and the Stars: A Philosophical Inquiry into the Moral Sense of Nature*, Chicago – London 1984. Marion's position is an attempt to overcome the dichotomy in question on the basis of a reversal of the transcendental position itself, in which intuition exceeds intentionality. Before coming to itself as self-determining, the subject is already called into reciprocity, already «appealed to». See J.L. Marion's essay on «The Saturated Phenomenon», in J.L. Marion, *The Visible and the Revealed*, trans. Christina M. Gschwandtner and others, New York 2008, 18-48; orig. French, *Le visible et le révélé*, Paris 2005.

adigm, the *passio essendi* bears with it the recognition that we are not the origin of meaning. Only because originated, can we attribute meaning to things, and do so on the condition of a previous attunement (Heidegger's *Stimmung*) to the promise of meaning that already inhabits things. In this paradigm, there is no separation between being and value, fact and meaning, for being is, intrinsically, promising and valid, good and beautiful. Of course, we produce and make, search and fashion, yet do all this without bracketing the charged sensuousness of the world in which we dwell. Our activity perfects nature, it acknowledges in being a reserve of meaning to make our own and bring to fulfillment. The receptivity at work in this is clearly not a form of passivity either, for it is to a consciousness and to its active intentionality that the meaning of things discloses itself. And yet, the activity of consciousness rests on the inexhaustible mystery of things (with a bow to Gabriel Marcel), on their endless and never to be reduced profundity, which makes itself known, because it opens itself up, because it reveals itself. In this perspective, our relation to being is a relation of trust, rather than doubt, of promising proximity, rather than distancing suspicion. With respect to freedom, we come to recognize that its task is indeed to do and build, to fashion the world, but only because, prior to this, the world was «let be». Thus, the task of freedom is essentially «responsorial», in fact, a responsibility, beyond the autonomy that finds fulfilment in will to power, seized at another's expense. It is, rather, a freedom that lets things be, in the generosity of love and giving.[37]

The relation to death unfolds within this context, and according to the same logic. The humanization of death will be possible on the condition that death be accepted, not suppressed or censured. Death is, after all, part of the human experience, an event whose

---

[37] This inevitability entails a demystification of the modern ideal of autonomy, a recognition that, in the long run, Kantian autonomy degenerates into will to power, as in Nietzsche. On this reading, see already R. Guardini, *Das Ende der Neuzeit. Ein Versuch zur Orientierung*, Würzburg 1951, and H. de Lubac, *Le drame de l'humanisme athée*, Paris 1945. For a different, more positive interpretation of modernity, with respect to the ideal of autonomy and its possible Christian reinterpretation, see J. Schwartlaender, ed., *Modernes Freiheitsethos und christlicher Glaube*, München 1981, especially the contributions of Schwartländer, Honecker, Kasper, and Böckle.

significance cannot be anticipated; a disclosure, in fact, a total revelation of meaning, both promising and significant.[38] The «passivity» implied by death is, therefore, an expression of the more general receptivity of life: «There is a passivity without which man could not be man. Part of the reason for this is the fact we were born, that we were given-birth-to. Here there follows the fact that we are loved. So, too, is the fact that we die».[39]

The acceptance of death is still bound to an act of preparation on our part, one that opens up for us a space of creativity. We all die, yet, we face death *differently*. In the same way in which life requires its own special art, accomplished daily in the cultivation of virtues, so, too, does death require a kind of art, the *ars moriendi*. Death is a threshold toward which we journey together, as if in pilgrimage, comforted by prayer.[40] In a Christian framework, death is, at bottom, an eschatological event, one which belongs to the personal narrative of each and every human being, yet also points to a trans-historical fulfillment, to definitive communion with God beyond the limits of

---

38   From a Christian point of view, this acceptance concerns both suffering and death, and yet not in the sense of a masochistic passivity. Klaus Demmer writes: «In the end, the Christian faith is anything but an ideology of suffering. Even for the Christian, suffering does not possess value in and of itself, and therefore it is never sought for its own sake. Rather, one accepts it, almost as an anticipation of death, which, too, must be accepted», K. Demmer, *Leben in Menschenhand. Grudlagen des bioethischen Gespräch*, Freiburg 1987, 146. On the topic of suffering, see the reflections of M. Scheler, «The Meaning of Suffering», in M.S. Frings, ed., *Max Scheler (1874-1928): Centennial Essays*, The Hague 1974, 121-163. From a theological perspective, see D. Soelle, *Leiden. Annehmen und widerstehen*, Freiburg 1973.

39   E. Jüngel, *Death: The Riddle and the Mystery*, London 1975, 85. The point is also made by Levinas, deserving of broader exegetical attention and interpretive articulation, well beyond the limits of a single quotation. Still, here is one, as a *donne à penser*: «[...] the subjectivity of the subject, its very psyche, (is) a possibility of inspiration. It is the possibility of being the author of what has been breathed in unbeknownst to me, of having received, one knows not from where, that of which I am the author. In the responsibility for the other we are at the heart of the ambiguity of inspiration», E. Levinas, *Otherwise Than Being or Beyond Essence*, trans. Alphonso Lingis, Pittsburgh 1998, 148-149; orig. French, *Autrement qu'être ou au-delà de l'essence*, La Haye 1974. «Inspiration» is existing «through the other and for the other, but without this being alienation», E. Levinas, *Otherwise Than Being* (cf. nt. 39), 114-115.

40   On prayer as a dimension of the *ars moriendi*, see W. Reich, «L'arte del prendersi cura del morente», *Itinerarium* 4 (1996) 31-43.

history: «Birth and death are thresholds and transitions, and as the radical transition of birth is creation, the radical transition of death may not be nothing, but resurrection».[41]

My reflections, at this point, would have to become more attentive to the particularity of cases and situations, thus letting anthropological considerations merge more smoothly into the ethical. The passage is not without difficulties, the logical pitfall implicated by the so called naturalistic fallacy being only one of the potential missteps. A more daunting task consists, in my opinion, in the difficulty to articulate a nimble casuistry, which takes into account nuances of contexts and diversity of clinical situations. Though a universal experience, death reserves for each of us a more intimate invitation. To prepare for the ultimate journey, we must face *the fact* that we die alone,[42] even when surrounded by others, given over, in the most radical way, to the mystery of our own singularity.[43] For example, the condition of patients who lost their autonomy to cognitive disability will impose ethical challenges that are different from those of patients who never possessed such discretionary autonomy. Likewise, we will have to distinguish criteria defined by substituted judgment standards from those based on rules of beneficence or non-maleficence, as in the case of best interest assessments by surrogate decision makers. Such a detailed casuistry will have to find more adequate treatment elsewhere. What is relevant, in this context, is the realization that the ethical quality of one's death depends on embracing, rather than rejecting, the inevitable passivity entailed by life's ontological condition. Such condition is not chosen, but given, all the accomplishments of our own making notwithstanding.

Consider the anticipation of treatment decisions in advance directives. In the materiality of the «letter», one such document may betray a different attitude of «spirit»: the acceptance of death, in the logic of the *passio essendi*, or the other, more desperate effort, in-

---

41 W. Desmond, *The Intimate Universal* (cf. nt. 1), 363.
42 Is this something analogous to a Kantian *Faktum der Vernuff*? In which case, the facticity of death would be the ultimate test confronting our freedom, now, no longer a freedom of choice, only a freedom to let go.
43 There is also an ethical singularity that is thrown into relief by the ontological singularity in question. On this, already K. Rahner, «On the Question of a Formal Existential Ethics», *Theological Investigations*, II, Baltimore 1961, 217-234.

deed the *conatus*, to deny it. In articulating our personal preferences for this or that treatment, we might only exorcise our fears, reassuring ourselves that in managing its terminal phases, we'll grant the mute alterity of death a face we can at least recognize. Let there be silence after that: if something more is to be had, it will be according to our own measure, beyond all feelings of dependence, and without surrendering to any kind of heteronomy, such as the one imposed by a treatment we do not want.[44]

I am not rejecting the preparation of advance directives, only the potential abstractness that might accompany their drafting, when such *preparatio mortis* bespeaks an exercise of freedom that stubbornly decides and plans.[45] Still, death will come to us in a future that is *ad-ventus*, both indeterminate and indeterminable. In trusting abandonment, death must be let happen, for we cannot escape it. In this light, the attempt to control death can become a paradox, especially when we fail to see how the scrupulous articulation of our directives, for an example, in relation to treatment decisions, points more to the radical alterity of death, than to our presumed capacity to domesticate it. Of course, we can clearly state what we want, even with a document that possesses the power of binding others to our wishes; yet, it will always be *others* who are charged with the task of respecting our desires and carrying them out: *their* decision will be, eventually, beyond *our* control.

I think of this paradox with regard to «aid in dying». The request for assisted suicide, now legal in eight jurisdictions in the U.S., will be heralded as an instance of self-determination, and in the name of a «death with dignity». And yet, while asserting their own autonomy for the last time, patients who choose to die must still abandon themselves to someone else, who, providing the lethal

---

[44] On the ethical challenges of surrogate decision making, see A.E. Buchanan – D.W. Brock, *Deciding for Others. The Ethics of Surrogate Decision Making*, Cambridge 1990.

[45] Relying upon Dietrich Bonhöffer's distinction between «resistance» and «surrender», see P. Cattorini, «Tra resistenza e accettazione: indicazioni etiche per superare accanimento vitalistico ed eutanasia», in P. Benciolini – C. Viafora, ed., *Etica e cure palliative. La fase terminale*, Roma 1998, 77-87. For a physician's narrative of care of patients at the end of life, specifically from the perspective of palliative care, see I. Byock, *The Best Care Possible: A Physician's Quest to Transform Care Through the End of Life*, New York 2012.

cocktail, does for them what they can no longer do on their own.[46] It is clear that, if the language of dying, even in the ethically exemplary case of filling out advance directives, is articulated by the patient against the back drop of an *epoche'* on the true meaning of death, the doctor and the healthcare personnel, in turn, cannot but share this «conspiracy of silence», in which that which cannot be spoken about, will inevitably be passed over in silence.[47] Of course, doctors will have their own reasons for resisting talking about death with their patients, and will express their will to power in their very unique way. For an example, by hiding themselves behind the technological imperative to fight death until the end, by opting for an aggressive treatment that has become futile, or by stirring up for patients and their families an impossible hope of recovery, which is but a mask of fear, the doctors', no less.[48] The bracketing of the reality of death, as well as the privatization of the criteria for dying, renders any solidarity with the dying impossible: the last word in this predicament can only be the loneliness of the dying.[49]

4. *Assisted Reproduction and the Exploitation of the Body*

The discussion on questions of assisted reproductive technologies provides a second example.[50] A look at the cultural context

---

46   Michael Maret defines euthanasia as «the paradoxical figure of autonomy». See M. Maret, *L'euthanasie. Alternative sociale et enjeux pour l'éthique chrétienne*, Paris 2000, 71-100. For an articulation, both anthropological and theological, of the dialectic of activity and passivity in dying see K. Demmer, «Handeln als Einüben des Sterbens. Ein Kapitel theologischer Anthropologie», in A. Holderegger, ed., *Das Medizinisch assistierte Sterben. Zur Sterbehilfe aus medizinischer, ethischer, juristischer und theologischer Sicht*, Freiburg 1999, 175-191.
47   J. Katz, *The Silent World of Doctor and Patient*, Baltimore, 2002.
48   For an empirical analysis of the relation between doctor and patient with respect to death, see the instructive study of K.K. Curtis – M.G. McGee, «An Overview of Physician Attitudes toward Death and Dying: History, Factors, and Implications for Medical Education», *Illness, Crisis and Loss* 8 (2000) 341-349.
49   N. Elias, *The Loneliness of the Dying*, New York 2001.
50   See the useful articulation of the various ethical issues, together with a very substantial bibliography, in the 2004 Report of the President's Council of Bioethics, *Reproduction and Responsibility: The Regulation of New Bio-*

in such technologies have developed shows quite clearly that their evolution has led to nothing less than a deconstruction of procreation. From an integral experience of human relationality, endowed with specific phenomenological characteristics, procreation has now been reduced to reproduction, a process of technical making, guided by the logic of calculative rationality. Such a deconstruction, subtle as it may be, entails also a redefinition: not only of parenthood, but of human identity *tout court*. Though originally born of a commitment to the alleviation of infertility, artificial reproductive technologies have progressively surpassed, if not abandoned, their original therapeutic intent, taking on, rather, unquestionable eugenic features. The search for a remedy to pathological conditions in both women and men has morphed into the search for the perfect progeny, a development in line with the logic of neutralizing commodification pursued by the market: like things, gametes, embryos, the wombs of women, and so children also, have now a price, rather than a dignity. In the end, artificial reproductive technologies have revolutionized the dynamics driving the appropriation of personal identity, the bond between generations, and the meaning of the historical links that tie them together.

The discussion about artificial reproductive technologies does not take place in a vacuum. It is nourished by recessive premises. Consider, among others, the question of embodiment, a topic that hardly surfaces as relevant in contemporary bioethics. The focus on normative dimensions, already found as dominant in the conversation about end of life issues highlighted above, tends to push to the side premises of a deeper philosophical nature, unquestionably central to any ethical reflection.[51] Thus, what one encounters as serious suggestions for policy proposals on artificial reproduc-

---

technologies (available at https://bioethicsarchive.georgetown.edu/pcbe/reports/reproductionandresponsibility/exec_summary.pdf). Also M. Warnock, *Making Babies: Is there a Right to Have Children?*, New York 2002. More recently, S. Wilkinson, *Choosing Tomorrow's Children: The Ethics of Selective Reproduction*, Oxford 2010. For an analysis attentive to anthropological and theological dimensions, see P. Lauritzen, *Pursuing Parenthood: Ethical Issues in Assisted Reproduction*, Bloomington 1993.

51 On the predicament of contemporary bioethics, and the need for anthropological integration see C. Taylor – R. Dell'Oro, ed., *Health and Human Flourishing* (cf. nt. 30).

tive technologies do fly in the face of elementary considerations about our embodied condition; as such, they hardly withstand even the lowest bar of philosophical justification. Take as an example the proposal submitted by the international committee on artificial reproductive technologies of the World Health Organization, in October of 2016. The proposal in question would change the WHO's previous definitions of infertility and disability, to now include single women and men who are unable to have children due not only to a medical condition, but also to either the inability to find a suitable sexual partner, or the lack of a sexual relationship that might bring about conception. What this means is that, under the new definition, the WHO would deem single men and women as equally infertile and disabled as heterosexual couples unable to conceive a child on account of a recognized medical pathology. Subsequently, their condition of infertility will count as «disability». As a disability, it should deserve, so goes the argument, publicly-funded provisions of reproductive technologies, including gamete donation and surrogacy.

A more serious example is the case recently publicized in the American news, concerning the Food and Drug Administration (FDA) approval of an *in vitro fertilization* technique, which uses DNA from three people in an at tempt to prevent certain illnesses, like muscular dystrophy and respiratory problems.[52] The United Kingdom's fertility regulator, the *Human Fertilization and Embryo Authority*, already changed its own laws, in December 2016, to permit the procedure. And last January 2017, the announcement came from Ukraine, that a child had been successfully produced with such mitochondrial transfer technique. Most commentators, especially scientists and doctors, welcome the advent of yet another technological fix to a congenital predisposition with an attitude of unquestionable awe. On the other hand, the more critically minded, among them ethicists, are willing to grant that some moral problems for this «three parent baby» solution do exist after all: doubts about safety are raised, together with the fear of unforeseen eugenic slippery slopes. Strangely passed over in silence, though, remains the most obvious question, «whose child will this baby be»?

---

52 See K. Tingley, «The Brave New World of Three-Parent I.V.F.», *The New York Times*, June 27, 2014.

Of course, experts are quick to rebut this preoccupation as scientifically naive, if not totally unfounded: they reassure the concerned public that because the female donor of healthy mitochondrial DNA to the defective biological mother provides, in the end, a very negligible genetic contribution, she should not be described appropriately as a «parent». However, when considered from another angle, namely, that of the personal identity of a child thus produced, the question «whose child will this baby be»? comes to the fore as actually very serious.

This is so because personal identity is now imperiled by what I would call «an ambiguity of belonging»: for the child so produced, the embodied matrix of traceable biological debts represents more an opportunity for doubt, than a condition for self-identification. Lack of evidence about one's distinct genetic lineage may turn the trust in the source that gives to be, under normal circumstances the syngamy of two genomes, into puzzlement about one's own origin and identity.[53]

Examples could be multiplied *ad infinitum*. Like others, the two I mentioned cannot fail to raise concerns. At stake are recessive premises about the body, embodiment, and the «embodied self» that drive these technologies in the first place, and, more in general, our understanding of medicine's goals. Furthermore, the development of artificial reproductive technologies, especially in their most extreme expressions, stands squarely within the legacy of a dualistic anthropology, itself resting upon the broader attitude toward being previously portrayed. Anthropology always reflects a specific view of metaphysics, of what does it mean «to be», and the mechanization of the body brought about by modernity will be better understood, when seen within the horizon of the more general neutralization of reality modernity inaugurates.[54] As neutral, the natural order has no language of its own, no deeper message to convey to an observer willing to see, or to listen. This is so because a deep perplexity has

---

53 For a stimulating analysis of the way in which biotechnology redefines embodiment, see M.J. Thiel, «La corporéité face à la maladie et la mort», in S. Müller – al., *Exploring the Boundaries of Bodyliness. Theological and Interdisciplinary Approaches to the Human Condition*, Göttingen 2013, 1-13.
54 On this R. Dell'Oro, «Embodiment as Saturated Phenomenon: Medicine, Theology, and Some Metaphysical Premises of Modernity», *International Journal of Philosophy and Theology* 2 (2014) 69-84.

now replaced the ancient wonder about the inherent value of being, more, about the inherent goodness of being. In this view, the subject has become the *only* source of value in an ethical sense: the good is not, as in the classical definition, «what everyone wants» (*bonum est quod omnes appetunt*); rather, what we want, we call the good (so Hobbes and our contemporary versions of moral contractualism inspired by him). Whether responding to the necessity of a rational ordering of duty, as in the Kantian version of autonomy, or the maximization of value in a network of effective powers, as in the calculative prudence of utilitarian rationality, the moral self of modernity emerges in its absolute centrality. Moreover, the moral self stands before the good as a dis-embodied self, auto-nomous because separated not only from what it sees as the heteronomy of nature, including that of the body, but also from the heteronomy of larger claims to social solidarity, as in the various versions of individual liberalism. The modem self, as Alasdair McIntyre and Charles Taylor have so eloquently highlighted, is, in the end, the «unencumbered self», an atomistic individuality that fails to recognize the embodied nature of communal and historically defined ties.[55]

In his latest encyclical letter, *Laudato Sii*, Pope Francis reminds us that the problems inherent in the modem paradigm cannot be denied any longer, for an example, in the face of the current ecological crisis.[56] And yet, it is not only an explicitly Christian-inspired anthropology what raises doubts about our current predicament. The debate within feminist theories on the ethics of artificial reproductive technologies, among others, suggests something of the tensions intrinsic to the modem understanding of body, procreation, and parenthood as neutral practices, as such entirely open to endless manipulation. For sure, *pro-interventionist* feminist thinkers tend to welcome developments in reproductive technologies as positive. They promise to

---

55   The term is Sandel's, and it refers to Rawl's notion of agency implied by the original position. See M. Sandel, «The Procedural Republic and the Unencumbered Self», *Political Theory* 12 (1984) 81-96. The genealogical dimensions of such notion have been explored by A. McIntyre, *After Virtue. A Study in Moral Theory*, Notre Dame 1981, 2007³. Charles Taylor speaks of the «buffered self». See C. Taylor, *A Secular Age*, Cambridge (MA) 2007, 27.
56   *Encyclical Letter Laudato Si of the Holy Father Francis on Care of Our Common Home*, at https://www.vatican.va/content/francesco/en/encyclicals/documents/papa-francesco_20150524_enciclica-laudato-si.html

control nature, and to re-define the meaning of gender constructions, relative especially to the distinction between male and female. In this view, invasive procedures that break women's links to biology, birth, and maternal nurturing can only further a feminist agenda of self-sufficiency and control.[57] On the other hand, *non-interventionist* feminist thinkers see reproductive technologies differently: a strengthening of arrogant human control over nature, and thus over women as part of the «nature» that is to be controlled. They see new reproductive technologies as an imposition upon women who look at themselves as failure, if they cannot become pregnant. They insist that technological progress, requiring the invasion and manipulation of women's bodies, must always be critically scrutinized with a kind of hermeneutics of suspicion, especially when the market becomes the ultimate mechanism for the exploitation of the body.[58] Indeed, it is hard to miss the marketing and advertisement strategies associated with fertility clinics and service providers, which, understandably, are eager to do what any business does best: sell to prospective customers, and this in the language of products and commodities:

> The danger is that the bodies of the couple and the child – now conceived as a product – are seen too much under the light of serviceable disposability. Indeed, embryos are disposable if they are not serviceable. It is manipulation, not participation, but also manipulation through a kind of participation: the exploitation of life is beneficiary to the gift of life that forgets the giftedness it exploits. These manipulations are ominous with respect to the deeper participation of the human being in energies of fecundity that come to it from beyond itself and that take it, help it partake of what is, beyond itself.[59]

There is more to body, procreation, and parenthood than our technical rationality assumes. There is an irreducible otherness to them that

---

57 For a synthesis, A.M. Jaggar, «Feminist Ethics», in L.C. Becker – C.B. Becker, ed., *Encyclopedia of Ethics*, New York – London 2001[2], 528-539. Classical works of feminist ethics on the issue includes S. Sherwin, *No Longer Patients. Feminist Ethics and Healthcare*, Philadelphia 1992, and R. Tong, *Feminine and Feminist Ethics*, Belmont (CA) 1993.
58 On this, B. Duden, *Disembodying Women. Perspectives on Pregnancy and the Unborn* Cambridge (MA) 1993. Also H. Haker, *Haupsache gesund? Ethische Fragen der Pränatalund Präimplantationsdiagnostik*, München 2011.
59 W. Desmond, *The Intimate Universal* (cf. nt. 1), 327.

reflects the personal presence of the embodied person, an ontological incommunicability that resists any constructive pretense. Moreover, to recognize the embodied condition of our being-in-the-world, to grant its radical otherness, is to abide by the symbolic reminder of our being-given-to-be. In the flesh that nourishes our joy and suffering, pain and pleasure, lies the trace of the source that releases us into being, the subtle allusion, most often forgotten, at times denied, of the gift that we are, not from ourselves, but from «an-other».

5. *Thinking Beyond: (Bio) ethics in the Love of Being*

How to address the predicament of contemporary bioethical thinking? Whether in the areas of reproductive technologies or at the end of life, I find normative questions wanting: not so much for lack of proper ethical discernment, or failure to address questions with a sense of nuances. It is more the overall agnosticism about the ground of our ethical perplexities what leaves one with a feeling of inescapable dearth. As if there were more: beyond the «overlapping consensus» of regulatory frameworks that are supposed to put our moral differences to rest, leaving us content with peaceful agreements negotiated by political cunning. Mindfulness about things that matter to us the most – the wondrous mystery of life giving, the frightening inevitability of our demise, our coming into being, our passing into nothing, keep us perplexed still, puzzled about the meaning of it all.

The metaphysical retrieval I am calling for is a return to a different awareness of the *intrinsic* value of being, the source that gives in the dawn of life, but also the night into which everything will eventually disappear. Is such a source love able and worthy of trust, or a dark origin engendering horror? If the latter, it will have to be reduced to the forms of our manipulative domestication. To think so is to be in thrall to a metaphysical, albeit nihilistic, premise, for a dark origin still pervades our sense of things, even when it precipitates them into the ground-less abyss.[60] As it happens, not few in bioethics are less than shy, when it comes to judging life as

---

60 The reader will forgive my linguistic exhibitionism, as I play with the semantic proximity of the two German words in question, i.e., *Grund*, ground, and *Ab-grund*, abyss

worth or not worth living, whether in the conditions of near-death neurological impairments, or in those of prenatal deformity.[61] But what does it mean to prefer the logic of me-ontic annihilation to the logic of life-affirming openness? Who can say that *not* to be born is better than being given a chance at living? Isn't the wisdom of the Greek Silenus, which both Schopenhauer and Nietzsche are fond of quoting, echoing somewhere in the distance: first, if it is possible, best not to be at all, and, second, if in being, best not to be, as soon as possible. One will then argue that every person is severely harmed by the very fact of being «thrown» into existence, that in bringing any person into existence one impermissibly harms that person.[62]

My reflections are grounded in a different confidence[63]: that the primal ethos of life is loveable and worthy of trust, an *agapeic* origin that is also the issue of the good. And so our ethical thinking will be true when it rests on the premise (or promise) of a just rapport with the good. To be on a par with the claim the good makes is to be more profoundly attuned to the generosity of its self-giving (*bonum diffusivum sui*), attentive to an offering of grace that is born (or re-born) of a porous opening to what (or who?) gives us into being. To be is to be gifted: our disposition to receive already subtends any endeavor to be, mindfulness of the *cum* already companions the *co-natus*, and this in terms of relativity to both the deeper metaphysical sources of being, as well as the demands of daily otherness, from those more intimate, implicated in the proximity of family and friends, to those awaken by a more universal generosity: to the unknown stranger, the handicapped child, the immigrant foreigner.

The «reconstructed» ethos of contemporary bioethics is blind to the sources of value that nourish the primal ethos of life. Phenom-

---

61 A point in case: the recent developments in common law concerning the notion of «wrongful birth».
62 I hear an implicit reference to Heidegger's *Geworfenheit*. The more obvious conclusion of the «anti-natal» view in question is that it is always wrong to have children. A less obvious, yet still plausible conclusion is that it would be better if humanity became extinct. Along those lines, D. Benatar, *Better Never to Have Been: The Harm of Coming Into Existence*, New York 2006. For a critical assessment, see the review of E. Harman, *NOUS* (2009) 776-785.
63 Yes, there is a faith involved in this, *cum-fides*.

enologists speak of *Wertblindheit*,[64] and perhaps this is an appropriate, if somewhat technical, way to put it: blindness to the sheer givenness of being as good, now reduced to neutral thereness available for endless manipulation. Being springs from an origin that gives without boundaries, out of a love that is unconditional, a love that lets be in pluralized creation, saturated with aesthetic worth.

I speak of creation here, not to immediately qualify the issue as theological. One should resist the attempt to re-colonize public discourse in the name of a political use of theology, born of resentment toward a secular bioethics that has marginalized religious voices. The issue is more deeply philosophical in nature: the task of a theologically mindful bioethics may not be achieved without unclogging the resistances to «think beyond», recognizing the hyperbolic signs at the heart of being itself.[65] This requires philosophical *finesse* more than proselytizing ardor. As von Balthasar suggests, «in order to be a serious theologian, one must also, indeed, first, be a philosopher; one must – precisely also in light of revelation – have immersed oneself in the mysterious structures of creaturely being».[66] Desmond puts the matter in terms of the porosity that we *are*, both in relation to what we have received, and in terms of our *own* openness beyond ourselves: «We are porosity because we are first received in being: given to be, before we are self-surpassing, or porous in a derived sense to what is beyond ourselves. We are in being as idiotic singulars, but at the heart of the idiotic selving is this intimate porosity that is the mark of our being creatures: emergent as what we are from no-thing – created from nothing».[67]

---

[64] Especially D. von Hildebrand, *Ethik* (GW II), Stuttgart 1973. On «value blindness» and Modernity, see J. Schmucker-von Koch, «Wertblindheit als Signatur der Moderne: Zum Verhältnis von Recht und Sittlichkeit bei Dietrich von Hildebrand», in J. Seifert, ed., *Truth and Value* (cf. Nt. 35), 141-152.

[65] The sheer fact of our very being, with the contingency it entails, is already hyperbolic: it «throws us above», in an exercise of transcendence that is a reversal of our existential fallenness.

[66] H.U. von Balthasar, *Theo-Logic: Theological Logical Theory, Volume I/ Truth of the World*, trans. by Adrian J. Walker, San Francisco 2000, 8; orig. German, *Theologik. Erster Band: Wahrheit der Welt*, Einsiedeln 1985. With specific reference to the interplay of theology and philosophy in ethics, see K. Demmer, *Moraltheologische Methodenlehre*, Freiburg 1989, 119-178.

[67] W. Desmond, *The Intimate Universal* (cf. nt. 1), 211-212.

Recovering a sense of the worth of beings entails an articulation of respect on our part: respect for other human beings, respect for the givenness of creation. This too is necessary, if not that, *qua* human expression, respect remains ambiguous, even contradictory: in the language of the 1999 National Bioethics Advisory Commission, which drafted the first document on embryo experimentation for the purposes of stem cell extraction, «respect for the embryo» can be reconciled with the intention to destroy and use it.[68] The matter then, is deeper. At stake is not only an *ethical* attitude, but an *ontological* love, love of being as worthy to be and to be affirmed. That we exist and live in the opening of such love, in the *passio essendi* that generates our ontological *complacentia* toward being, orients all our endeavors, the striving of our *conatus*, in the direction of an affirmation of otherness. Porosity beyond the atomistic individuality of «unencumbered selfhood» is more than an exercise in autonomous self-determination. The question of what limits the latter is very much at stake in the tension between liberalism and communitarianism, which defines much of contemporary ethical discourse. For sure, the world endorsed in the communitarian social model appears to be in tension with the individualist mind set of liberal thinkers. Individualists even claim that communitarians express little more than nostalgia for a simpler, pre-modern past. But does the «communitarian» model necessarily stand in opposition to the «liberal» model? The recognition of individual freedoms, such as the freedom of scientific research and clinical experimentation, are unquestionable values for any contemporary rendition of the relation between self and society. A society is a good society when it sustains freedom

---

68 National Bioethics Advisory Commission, «Ethical Issues in Human Stem Cell Research. Volume 1» (https://repository.library.georgetown.edu/bitstream/handle/10822/559364/nbac_stemcell1.pdf?sequence=1&isAllowed=y.) The logic behind the reasoning must be unmasked as fallacious, when appealing to a Kantian justification: persons, so goes the argument, presuppose moral agency. Since embryos are incapable of moral agency, they should not be recognized as persons, i.e., object of respect. Such an argument fails to see that for Kant, respect for a person is rationally grounded in the intersubjective character of the categorical imperative. Thus, it presupposes moral agency as a matter of *necessity*, rather than discretionary attribution. The humanity *in* the person is the transcendental condition of possibility of her moral agency, not the other way around.

through the mutual respect its members show in their interaction with one another. This goes, first of all, to the realization that aiming at the good of society entails protecting, rather than eroding, a space for moral pluralism, hospitable to an interaction across differences, on the presupposition that the *public realm* is not just the neutral space to be conquered or won over, and that the members of an «open society» are not to be faced as enemies but as partners: dialogue among moral agents, whether «strangers» or «friends», to use the distinction *in vogue*, can only function on the presumption that any claim to meaning and truth is, at the same time, an attestation of freedom and respect for the other. To that extent, liberalism and communitarianism not only stress two different dimensions of the same reality, but grow one on top of the cultural achievements of the other. As Charles Taylor recognizes, the free individual with his own goals and aspirations is himself only possible within a certain kind of civilization. It took a long development of certain institutions and practices, of the rule of law, of rules of equal respect, of habits of common deliberations, of cultural self-development, and so on, to produce the modem individual. Without these, the very sense of oneself as an individual would atrophy.[69]

This is true. However, one must go beyond the potential accommodation of two reciprocally implicated social models. If what is at stake in the debates of bioethics is ultimately the full extent of our porosity to the good, then the question is not only retrieving the relativity of autonomy to otherness, but «to open up» autonomy, even the autonomy of social intermediation, beyond itself, toward a more generous freedom, an agapeic freedom that responds to the value of being in its unconditional worth. Such freedom is irreducible to serviceable disposability, whether predicated on contractarian interest, or utilitarian maximization of social value: «The agapeics transforms the social space of our between-being, consecrates it into a neighborhood of love wherein neighboring, as a «being beside,» is neither simply passive nor simply active [...] We receive and do ourselves in the agapeic neighborhood».[70]

---

69 See the impressive reconstruction in C. Taylor, *Sources of the Self. The Making of Modern Identity*, Cambridge (MA) 1989.
70 W. Desmond, *The Intimate Universal* (cf. nt. 1), 411.

Can such a freedom beyond autonomy be recognized without reference to an agapeic God, a source of endowing freedom that is also an enabling of social intermediation? Desmond alludes to an «antinomy of autonomy and transcendence», thus seeming to offer a negative response, for the God of autonomy is only a practical postulate, not an endowing source. But to be bound to an agapeic God is not to be in bondage: «The enabling of social power is given but now understood as gifted by a surplus generosity, ultimate in itself and calling human beings to imitate and to enact this generosity in finite life. This is not a matter of our erotic self-transcendence, it is a communication of transcendence itself into the midst of our transcending, which now no longer can just circle around itself».[71]

Will contemporary ethical discourse heed the call to such a freedom, breaking the spell that has bewitched its reasoning into the vicious circularity of will to power, affirming only itself, only to destroy itself? I suggest that the opening can be occasioned by a porosity to a theological contribution, itself sustained by a robust metaphysics, which calls public discussions on moral questions to the suspension of preconceived judgments and dogmatisms of any kind, opening our eyes to a deeper vision of what is good for us, because worthy to be affirmed in itself.

---

71   W. Desmond, *The Intimate Universal* (cf. nt. 1), 417.

CHAPTER 4
REPRODUCTIVE CLONING, MARRIAGE AND FAMILY
Ideological Premises and Forgotten Issues

1. *Introduction*

Without any doubt, the topic of cloning has most recently received the most attention within the field of bioethics.[1] It is my purpose, in this chapter, to explore some aspects of the ethical judgment on the practice of cloning, and to do so with particular attention to its repercussions upon the fundamental human experiences that define marriage and family.

Because of this particular focus, my analysis will concentrate especially on the ethics of cloning to produce children rather than on cloning for bio-medical research. By the latter, I understand the production of a cloned human embryo formed for the purpose of research or for extracting stem cells. The ultimate goal of such activities is to gain scientific knowledge in the hope of developing cures for human diseases. Since 1998, when researchers were able, for the first time, to isolate human embryonic stem cells, many scientists have come to believe that these versatile cells, capable of becoming any type of cells in the body, hold great promise for understanding and treating many chronic diseases and conditions. Some investigators also believe that stem cells derived from cloned human embryos, produced explicitly for such research, might prove uniquely useful for studying many genetic diseases and devising new therapies.[2]

---

[1] The relevant literature will be quoted in the course of the article. A well-reasoned and extremely thorough bibliographic analysis of the literature on cloning up to 2002 can be found in «Bioethics and Cloning», *Scope Note 41* and *42*, in: *Kennedy Institute of Ethics Journal* 12 (2002), 305-324, and *Kennedy Institute of Ethics Journal* 12 (2002), 391-407. An unquestionable turning point in the reflection is represented by the exhaustive report of the President's Council on Bioethics: *Human Cloning and Human Dignity*, New York: Public Affairs, 2002.

[2] From a strictly scientific point of view, the debates as to whether embryonic stem cells hold greater promises of clinical application than «adult» stem

Cloning to produce children, on the other hand, is a form of asexual reproduction, that is, the production of individuals who are genetically identical to an already existing individual. In the process of reproductive cloning, the nucleus of a mature, unfertilized egg is removed or de-activated, in order to be replaced with the nucleus obtained from a somatic cell of an adult organism. If and when the egg begins to divide, the little embryo is transferred to a woman's uterus, where it initiates a regular pregnancy. Almost all the hereditary material of a cell is contained within its nucleus; thus, the re-nucleated egg and the individual into which it develops are genetically identical to the organism that was the source of the transferred nucleus.

One can see that, although the term cloning might be applied indifferently to either reproductive cloning or to cloning for research purposes, there is a clear distinction between the two. Such distinction must be maintained for the sake of clarity on the nature of the practices involved, as well as in view of an ultimate ethical judgment. However, moral reflection cannot fail to recognize that a distinction, though valid in theory, may not be viable in practice. Suffice it to say that the possibility to link, or, conversely, separate our judgment on the two types of cloning stands or falls with the practical viability of the distinction itself.

2. *The General Condition of the Field of Bioethics*

The present contribution to the ethics of cloning will be framed within an anthropological perspective that, at least implicitly, borrows from the Christian theological tradition. I intend to speak from

---

cells is still open. Hematopoietic stem cells from bone marrow, the placenta, or umbilical cord blood in live births have been found promising. Similarly, new advances have been announced in isolating mesenchymal cells from bone marrow and directing them to tissue formation. For an overview of the scientific *status questionis*, C.M. Verfaille: «Multipotent Adult Progenitor Cells: An Update»: in: The President's Council on Bioethics: *Monitoring Stem Cell Research*, Washington, D.C.: The President's Council on Bioethics, 2004, 295-307. On the possible ethical advantages of adult versus embryonic stem cell research, see P. Lauritzen: «Report on the Ethics of Stem Cell Research», in: The President's Council on Bioethics: *Monitoring Stem Cell Research*, 237-272.

a particular point of view, yet with the ultimate goal of making sense beyond the limits of specific ideological boundaries. This is not an easy task. At present, bioethics seems to exhibit a certain resistance to the integration of anthropological perspectives larger than those defining the limits of the pragmatic consensualism currently dominating the field.[3]

Indeed, a look at the relatively brief history of epistemological developments in bioethics clearly shows a methodological shift in the fundamental preoccupation of bioethicists. Scholars most commonly identified with the pioneers in the field of bioethics, contributing to what Albert Jonsen has termed «the birth of bioethics», sought a horizon of meaning that would respond to the value implications of technological developments in medicine and the life sciences.[4] Such a horizon had a pluralistic character: it inspired moral anthropological interpretations in a theological fashion, as well as generally humanistic, when not explicitly non-religious, hermeneutics.[5]

As a relatively new field of study characterized by an interdisciplinary approach to ethical questions raised by advances in biotech-

---

3   The best example of such an approach is T. Beauchamp/J. Childress: *Principles of Biomedical Ethics*, New York: Oxford University Press, 2001. The book, first published in 1979, is now in its 8th edition. Through the various editions of the book, the authors display an apparent concern for the unfolding of the methodological debate in bioethics. On the other hand, their commitment to a principle-based approach remains unshakeable, in spite of mounting external criticism. Moreover, the commitment to a principle-based approach in bioethics extends beyond the work of the authors mentioned above as it represents a larger theoretical gesture of mainstream Anglo-American bioethics. See, for an example, R. Veatch: *A Theory of Medical Ethics*, New York: Basic Books, 1981, and H.T. Engelhardt: *The Foundations of Bioethics*, New York: Oxford University Press, 21996.
4   See A. Jonsen: *The Birth of Bioethics*, New York: Oxford University Press, 1998. One might think of scholars like Paul Ramsey, Josef Fletcher, Hans Jonas, Daniel Callahan, and Warren Reich. On the question of the origins of bioethics, particularly with reference to the importance of theoreticians at the Kennedy Institute of Ethics at Georgetown University, see J.K. Walter/E. P. Klein (eds.): *The Story of Bioethics: From Seminal Works to Contemporary Explorations*, Washington, D.C.: Georgetown University Press, 2003.
5   Most recently, on the methodological debatre G. Khushf (ed.): *Handbook of Bioethics: Taking Stock of the Field from a Philosophical Perspective*, Dordrecht: Kluwer, 2004. Also J. Sugarman/D. P. Sulmasy (ed.): *Methods in Medical Ethics*, Washington, D.C.: Georgetown University Press, 2001.

nologies and the life sciences, and because of its inherent complexity, bioethics relies upon the contribution of different perspectives on the issues it faces. Thus the interaction of theology, philosophy, sociology, and public policy has contributed to a richer understanding of the implications of medical research for society at large.

However, in the struggle to find a specific methodology for the dialogue across different disciplines, bioethics tends to reduce the contribution of philosophical ethics to the reconstruction of the conditions for moral consensus in society at large. Although bioethics encounters questions of ultimate meaning as it investigates issues like cloning, genetic engineering, assisted reproductive technologies, euthanasia, and assisted suicide – to name but a few, it fails to address such questions in their broader ethical significance. Taking moral pluralism as an obstacle toward the sharing of any «thick» notion of the good within society, bioethics expunges as philosophically uninteresting any attempts to ground such notions upon a coherent anthropological basis. In so doing, it replaces questions of moral meaning with questions of procedure; secondly, it reduces the task of ethics to a logical analysis of moral argumentation, one that functions meaningfully as long as it remains within the framework of a «thin» common morality.[6] The solution adopted by Thomas Beauchamp and James Childress in their widely used textbook, Principles of Biomedical Ethics, might be taken as exemplary. The textbook reflects a concern with the elaboration of normative criteria (so called principles of respect of person, beneficence, non-maleficence, and justice) that draw their justification from the perspective of a restrictive cluster of concepts in political philosophy. This approach seeks to create a consensus based on shared arguments that are, in principle, divorced from the horizon of meaning and the meaningful narratives that initially inspired them. Under the strong influence of the need to provide a consistent ethical basis for public policy formation, moral philosophy builds for bioethics an area of autonomous reflection centered on the use of principles and rules, and the ethical theories that unite them, especially deontological and utilitarian theories.[7]

---

6   See K. Wildes: *Moral Acquaintances: Methodology in Bioethics*, Notre Dame: University of Notre Dame Press, 2000.
7   The term «principlism» was eventually used, in the wake of critical remarks by philosophers Clouser and Gert, to designate this approach. See K.D.

In the end, the field of bioethics mimics in its fundamental approach the wider theoretical strategy pursued by political liberalism. For the latter, ethics cannot rationally justify «ideologies of human fulfillment» with their relative notions of the «good life». At best, it can provide criteria for the positive interaction of individual autonomous agents within a community of moral discourse. Insofar as those criteria are the result of consensual agreement, they can exert normative force only in relation to the rightness of the actions they prescribe. However, the goodness of the ends implicitly pursued by these actions, the kind of moral personalities and moral societies they yield, remain beyond the scope and limits of ethics.[8]

3. Bioethics and the Cloning Debate

Leon Kass comments critically on the inherent value, or lack thereof, of such a prevalent approach in today's bioethics, specifically on the value of previously mentioned principles, when applied to particular cases: they translate mainly into concerns to avoid bodily harm and do bodily good, to respect patient autonomy and secure informed consent, to promote equal access to health care and provide equal protection against biohazards. So long as nobody is hurt, no one's will is violated, and no one is excluded or discriminated against, there is little to worry about. The possibility of willing dehumanization is out of sight and out of mind.[9]

Clouser/B. Gert: «A Critique of Principlism», in: *The Journal of Medicine and Philosophy* 15 (1990), 219-236. Most recently B. Gert/K.D. Clouser: *Bioethics: A Return to Fundamentals*, New York: Oxford University Press, 1997. For a recent general assessment of principlism see T. Beauchamp/D. De Grazia: «Principles and Principlism», in G. Khushf (ed.): *Handbook of Bioethics*, 55-74.

8 See J. Rawls: *A Theory of Justice*, Cambridge, MA: Harvard University Press, 1971; also J. Habermas: *Moral Consciousness and Communicative Action*, Cambridge, MA: MIT Press, 1991, and J. Habermas: *Justification and Application: Remarks on Discourse Ethics*, Cambridge, MA: MIT Press, 1993.

9 L. Kass: *Life, Liberty, and the Defense of Dignity: The Challenge for Bioethics*, San Francisco: Encounter Books, 2002, especially 56-76. As an example, consider the 1999 document on stem cell research by the National Bioethics Advisory Commission. See *Ethical Issues in Human Stem Cell Research*, vol. 1: *Report and Recommendations of the National Bioethics Advisory Commission*, Rockville, MD: National Bioethics Advisory Commission,

Implications for the current debate on cloning follow from such premises. For sure, over the past six years, the prospect of human cloning has been the subject of considerable public attention and intense moral debate, both in the United States and around the world. Since the announcement in February 1997 of the first successful cloning of a mammal (Dolly the sheep), several other species of mammals have been cloned.[10] Although a cloned human child has yet to be born, and although animal experiments have had low rates of success, the production of functioning mammalian cloned offspring suggests that the eventual cloning of humans must be considered a serious possibility. Indeed, in an article published for the journal *Science* in February 2004, South Korean scientists reported to have created human embryos through cloning, proceeding afterward to the extraction of embryonic stem cells.[11]

The announcement of the South Korean scientists echoed that of American researchers who, in November 2001, claimed to have produced the first cloned human embryos. These embryos reportedly reached only a six-cell stage before they stopped dividing and died. In addition, several fertility specialists have announced their intention to clone human beings.[12]

In the United States, the cloning debate has produced, in addition to legislative interventions by Congress and Senate banning cloning to produce children, three national-level reports by, respectively, the National Bioethics Advisory Commission (1997), the National Academy of Sciences (2002), and the President's Council on Bio-

---

1999. The authors of the document can, ultimately, agree on *safety* as the only moral constraint against the practice of reproductive cloning.

10 See I. Wilmut/A. E. Schnieke/J. McWhir *et al.*: «Viable Offspring Derived from Fetal and Adult Mammalian Cells», in: *Nature* 385 (1997), 810-813. For the impact of the animal discovery on the potential application to humans see I. Wilmut: «Application of Animal Cloning Data to Human Cloning», paper presented at *Workshop: Scientific and Medical Aspects of Human Cloning*, National Academy of Sciences, Washington, D.C., 7 August 2001.

11 See W.S. Hwang/Y. J. Ryu/J. H. Park *et al.*: «Evidence of a Pluripotent Human Embryonic Stem Cell Line Derived from a Cloned Blastocyst», in: *Science* 303 (2004), 1669-1674.

12 See J. Cibelli/A. Kiessling/ K. Cunniff *et al.*: «Somatic Cell Nuclear Transfer in Humans: Pronuclear and Early Embryonic Development», in: *The Journal of Regenerative Medicine* 2 (2001), 25-31. For a complete history of the scientific attempts in cloning up to the year 2002, see chapter 4 of President's Council on Bioethics: *Human Cloning and Human Dignity*.

ethics (2002). The gravity and urgency of the questions surrounding human cloning, however, have not abated, in the United States as in other parts of the world.[13] In fact, such questions continue to dominate the scholarly discussion in bioethics and that of the public at large. This is for good reason: the prospect of human cloning lies not only at the center of our fictional imagination. It has also become the latest fruit for our biomedical science and technology to reap. Indeed, creating and manipulating life in the laboratory is the gateway to the *Brave New World* in fiction and in fact. Yet, whereas some scientists tend to downplay the momentous significance of the decision to clone a human being, seeing it as just another step in the triumphal procession of modern medicine, to the majority of lay people the real possibility of cloning represents more like a wake up call. As if from a dogmatic slumber bestowed upon humankind by the Baconian dream to «conquer nature and relieve man's estate», we are finally awakened to the gravity of basic moral and political questions: What does it mean to treat nascent human life as raw material to be exploited or as a mere natural resource? What does it mean to blur the line between procreation and manufacture? What are the goals of the project for the mastery of human nature? What are its limits?

The reason these questions have been and continue to be central – which is also the reason bioethics has been an object of such lively public interest and concern – is quite obvious: these questions involve some of the most important aspects of our humanity and raise the deepest challenges about what it means to be human and to act accordingly.

---

13 For a synthesis of the different aspects of the cloning debate see G. McGee (ed.): *The Human Cloning Debate*, Berkeley, CA: Berkeley Hills Books, 1998, and G. Pence (ed.): *Flesh of My Flesh: The Ethics of Cloning Humans, A Reader*, Lanham, MD: Rowman & Littlefield, 1998. In reference to the question of cloning for biomedical research rather than reproductive cloning see P. Lauritzen (ed.): *Cloning and the Future of Human Embryo Research*, Oxford: Oxford University Press, 2001. For religious perspectives on cloning see R. Cole-Turner (ed.): *Human Cloning: Religious Responses*, Louisville, KY: Westminster John Knox Press, 1997. Most recently, the ethical and theological issues of both types of cloning are viewed within the larger context of «the new genetic medicine» by T. Shannon/J. Walter: *The New Genetic Medicine: Theological and Ethical Reflections*, Lanham, MD: Rowman & Littlefield, 2003.

Yet, our public debate tends to neglect these larger questions.[14] We find ourselves reacting with a kind of *ad hoc* strategy to the latest biotechnological possibilities, but we fail to see their meaning within a larger horizon. Surrendering to the incommensurability of moral ideas in a context of social and cultural pluralism, we lack faith in the possibility of reconstructing such a horizon as the common framework of human goods we wish to preserve and defend.[15]

4. On the Way to Cloning: Ideological Premises and Unchallenged Practices

There is no denying that, over a period of thirty years, we have become accustomed to new practices in human reproduction: not just *in vitro* fertilization but also embryo manipulation, embryo donation, surrogate pregnancy, and pre-implantation genetic diagnosis, to name but a few.[16] New technologies of human reproduction have clearly undermined the justification and support that biological parenthood gives to monogamous marriage. The case of surrogate motherhood brings home the point, for here «normal» kin

---

14   The complaint can be found especially in L. Kass: *Life, Liberty and the Defense of Dignity*. Also G. Meilaender: *Body, Soul, and Bioethics*, Notre Dame: University of Notre Dame Press, 1995. Historically, the concern for the larger anthropological impact of the perfecting cloning techniques was raised by theologians and philosophers. Yet, most recently, philosophers seem to have bought into a bioethical analysis that reduces itself to questions of benefit and harm. Compare, as an example, the differences in philosophical sensibility between H. Jonas: «Lasst uns einen Menschen klonen: Von der Eugenik zur Gentechnologie», in H. Jonas: *Technik, Medizin und Ethik*, Frankfurt: Insel Verlag, 1987, 162-203, and D. Brock: «Cloning Human Beings: An Assessment of the Ethical Issues Pro and Con», in: M. Nussbaum/C. Sunstein: *Clones and Clones: Facts and Fantasies about Human Cloning*, New York: Norton & Company, 1998, 141-164.
15   This goes to the question of how to find a common ground of ethical principles in the notion of «common morality» and, theoretically, to the debate on so called «principlism» as a viable normative framework for bioethics. See A. Jonsen: *The Birth of Bioethics*, Oxford: Oxford University Press, 1998, 325-376.
16   On the entire question of artificial reproduction see the recent report President's Council on Bioethics: *Reproduction and Responsibility: The Regulation of New Biotechnologies*, Washington, D.C.: The President's Council on Bioethics, 2004.

relations are confounded. Who is the mother, one could ask: the egg donor, the surrogate who carries and delivers the baby, or the one who rears the child?[17]

Moreover, changes in the broader culture make it very difficult to express a common and respectful understanding of sexuality, procreation, nascent life, family as well as to give a convincing meaning to terms like motherhood, fatherhood, and the links between the generations.[18] The judgment on these cultural changes is, of course, very complicated and one should refrain from cheap generalizations. For sure, to understand the meaning of phenomena like the sexual revolution, feminism, or the gay rights movement, would entail engaging in a cultural hermeneutic that goes beyond moralistic stereotyping. At the same time, these phenomena are nonetheless univocal in their own fundamental ideological thrust and point in a very specific direction.

The sexual revolution has tended to expunge procreation from the fundamental meaning of the sexual experience, questioning the inherent connection between sex and transmission of life in procreation. But the disconnect between sex and procreation can work in the direction of a radically new interpretation of the meaning of procreation as well, resulting in the latter having no necessary connection to sex either.[19]

Natural heterosexual differences have also come under intense scrutiny in their functioning as objective conditions of our un-

---

17 For a thorough ethical and theological assessment see K. Demmer: «Das ethische Umfeld der assistierten Zeugung: Ein interpretierender Bericht», in: *Gregorianum* 82 (2001), 87-128. Also P. Lauritzen: *Pursuing Parenthood: Ethical Issues in Assisted Reproduction*, Bloomington: Indiana University Press, 1993.
18 On the challenges of a public moral argument for the Christian tradition see L. Sowle Cahill: *Sex, Gender, and Christian Ethics*, Cambridge: Cambridge University Press, 1996, especially 217-254. Also L. Sowle Cahill/ D. Mieth: «The Family», vol 4 of *Concilium* (1995); K. Demmer: *Angewandte Theologie des Ethischen*, Freiburg: Herder, 2003, 149-194; and E. Schockenhoff: «Krisenerscheinungen der Familie: Zur Basis des sozialen Zusammenlebens», in: *Die politische Meinung* 376 (2001), 11-16.
19 On such a disconnect, see already the instruction Congregation for the Doctrine of the Faith: *Donum Vitae*, 1987. For a nuanced commentary of this instruction, see E.D. Pellegrino/J. P. Langan/J. C. Harvey: *Gift of Life: Catholic Scholars Respond to the Vatican Instruction*, Washington, D.C.: Georgetown University Press, 1990.

derstanding of sexuality, now being reduced to mere «cultural construction». One can see, then, that if male and female are no longer normatively complementary and significant in the process of reproduction, children do not even need to come from the union of sperm and egg. Indeed feminist theorists espousing a pro-interventionist attitude toward techniques of artificial reproduction see them as a promise to control nature and to potentially eliminate the distinction between male and female.[20] There is no denying that the existence of human life in the laboratory, outside the confines of the generating bodies from which it sprang, challenges the meaning of our embodiment. When in the summer of 1978, Louise Brown, the first human being to be conceived with in vitro fertilization, was born, many people rejoiced at the wonder of a couple who, thanks to science and technology, were blessed with a child of their own. In the succeeding twenty-five years, thousands of people have rejoiced as the Browns did at the same experience. The desire to have a child of one's own is indeed a deep-seated human desire, and the satisfaction of this desire, by the relief of infertility, is one major goal achieved by in vitro fertilization and embryo transfer.

Of course, one still wonders at the exact meaning of the dream that such a technology makes possible. How do we understand the meaning of «having a child»? Does «having a child» refer primarily to the fact of gestating or bearing? Or does it concern the act of rearing a child without possibly any of the former experiences, as in the case of adoption? And moreover, under what conditions can the experience of having a child of our own be protected from an

---

20 On the support for radical women's reproductive freedom and as an example of an early «feminist» call for new reproductive technologies see especially S. Firestone: *The Dialectic of Sex: The Case for Feminist Revolution*, London: Women's Press, 1983. With time, however, the position of feminist thinking on reproductive technologies became more nuanced. The concerns raised by feminism address the question of human life and bodies being reduced to «commodities», the fact that in vitro fertilization and other reproductive technologies might exploit women only to compensate for men's infertility, and further that because women's desire for children is itself socially conditioned, artificial reproductive practices end up reinforcing women's oppression. For an example of feminist literature going in this direction, M.A. Warren: *Gendercide: The Implications of Sex Selection*, New York: Rowman & Allanheld, 1985.

attitude of parental projection, itself only human, that sees the child only as a mere possession, as the extension of the parents' desire for something to be had?[21]

At the same time, there is no denying that to have a child of «one's own» is what leads most people to choose in vitro fertilization over adoption: a couple's desire to embody, out of their conjugal union, a child who is «flesh of their flesh». In spite of the intricacies affecting the moral judgment on such a clinical use of in vitro fertilization, it is nevertheless difficult to deny that the goal such a technique pursues on a general level is indeed a worthy one. This is so because a couple seeking a child derived from their flesh celebrate in so doing their self-identification with their own bodies and acknowledge their self-transcendence in the direction of the child.

In vitro fertilization techniques introduce an obvious separation between sexuality and procreation, at least at the level of the individual act of love making. However, when it is carried out within the context of a married couple, it is ultimately an affirmation of the transmission and importance of lineage and connectedness as central dimensions of human embodiment.[22]

But what about all other uses, involving third parties, to satisfy the desire to have a child in which the meaning of «having a child of one's own» is not so unambiguous any longer? Statistics prove that the need for extramarital embryo transfer is real and large, probably even greater than that for intra-marital ones.[23] In these cases, quite widespread in fertility clinics around the world, the principle truly at work in bringing life into the laboratory is not to provide married couples with a child of their own, but to provide a child to anyone who wants one, by whatever means possible.[24] Life in the labora-

---

21 On the danger of a manipulative conception of the child see already K. Demmer: «Ein Kind um jeden Preis? Anmerkungen zur laufenden Diskussion um die extrakorporale Befruchtung», in: *Trierer Theologische Zeitschrift* 94 (1985), 223-243.
22 This in spite of the recognition that a general moral judgment on in vitro fertilization, even when homologous, might remain problematic.
23 See the data provided by the Report of President's Council on Bioethics: *Reproduction and Responsibility*.
24 See J. Robertson: *Children of Choice: Freedom and the New Reproductive Technologies*, Princeton: Princeton University Press, 1994.

tory allows those who donate or sell sperm, eggs, or embryos, or those who would bear another's child in surrogate pregnancy, to declare themselves independent of their bodies. Here the reality of our human, embodied nature is grossly misinterpreted: the rational will that affirms its autonomy in the signing of contractual agreements does so while using its own body as a tool to be sold for a price. But human procreation is not simply an activity of our rational wills in which the body partakes by accident as *res extensa* to be handled at whim. Procreation is a complete activity precisely because it engages us bodily, erotically, and even spiritually. The severing of procreation from sex, love, and intimacy is thus inherently dehumanizing, no matter how good the product.[25]

A culture that trivializes the profound meaning of the body in the person by severing biological reproduction from the social identity of maternity and paternity and, ultimately, from the intimacy of marriage and family is willing to accept the morality of cloning. Indeed, for such a culture the clone is the ideal emblem: the ultimate single-parent child. Moreover, cloning fits nicely into the general ideological *Denkform* of a political liberalism that recognizes primacy of rights and individual choice as the self-evident starting point in the definition of the person. In such a perspective, we envisage ourselves, if only through a «veil of ignorance», as islands detached from one another, independent of a variety of debts, life-worlds, and traditions to which we belong, pure projects projected forward to the future, with no history and memory of the past.

The ethics of liberalism sees cloning in the context of rights, freedom, and personal empowerment, a new option for exercising an individual right to reproduce or to have the kind of child they want. For those who hold this outlook, the only moral restraints on cloning are adequate informed consent and avoidance of bodily harm.[26] If no one is cloned without their consent, and if the person

---

25 See L. Sowle Cahill: *Sex, Gender and Christian Ethics*, 73-107. Also L. Sowle Cahill/M. Farley (ed.): *Embodiment, Morality, and Medicine*, Dordrecht: Kluwer, 1995.
26 As an example, M. Roberts: «Human Cloning: A Case of No Harm Done?», in *Journal of Medicine and Philosophy* 21 (1996), 537-554.

eventually cloned is not physically damaged, the liberal conditions for licit, hence moral, conduct are definitely met.[27]

I submit that it is very problematic to subsume the human experience of reproduction and the intimate relations of family life under the language of rights and individual empowerment. Because such notions belong primarily to a political and legal genetic context, signaling an adversarial and individualistic understanding of self and society, they tend to undermine, rather than recognize, the deep connections between «private» decisions about reproductive choices and their «public» impact. Although they belong, by definition, to the «private» sphere, the experiences of bearing and rearing a child, together with family life and its bond to the covenant of marriage, do take place within a particular communal milieu and share in the cooperative and duty-laden character of social life.

5. Toward a Broader Anthropological Framework

The thrust of my argument, so far, is that human cloning is, at least, *partly* continuous with those reproductive technologies that have made it more difficult for both parents and children to honor and affirm the bond between the generations and accept as a given the lines of kinship that locate and identify them. Such an analogy between cloning and, say, artificial insemination by donor or surrogate pregnancy, though not completely accurate from a strictly *logical* point of view, has nevertheless a powerful *rhetorical* meaning: it forces us to rethink the general morality of practices now taken for granted in light of their very real developments. Indeed, those who favor cloning see it as an extension of existing techniques for assisted reproduction and for determining the genetic make-up of children. Like them, cloning will be regarded as a neutral technique, with no inherent meaning or goodness, but subject to multiple uses,

---

27 In this vein, see especially J. Robertson: *Children of Choice*, J. Robertson: «Embryos. Families and Procreative Liberty: The Legal Structure of the New Reproduction», in: *Southern California Law Review* 59 (1986), 939-1041, and J. Robertson: «The Question of Human Cloning», in: *Hastings Center Report* 24 (1994), 6-14.

some good, some bad.[28] As a neutral means, the morality of cloning will entirely depend upon the goodness or badness of the ends it pursues as well as on the intentions and motives of the cloners. In this perspective, the ethics (of cloning) must be judged only by the way the parents nurture and rear their resulting child and whether they bestow the same love and affection on a child brought into existence by a technique of assisted reproduction as they would on a child born «in the usual way».

But can such a perspective be truly adequate in judging the important moral questions at stake in the discussion on cloning? The following reflections by Leon Kass point in a different direction:

> The technological, liberal and meliorist approaches all ignore the deeper anthropological, social and, indeed, ontological aspects and meanings of bringing forth new life. To this more fitting and profound point of view, cloning shows itself to be a major alteration – not to say violation – of our given nature as embodied, gendered and engendering beings and of the social relations built on this natural ground. Once this perspective is recognized, the ethical judgment on cloning can no longer be reduced to a matter of motives and intentions, rights and freedoms, benefits and harms, or even means and ends. It must be regarded primarily as a matter of meaning: Is cloning a fulfillment of human begetting and belonging? Or is cloning rather, as I contend, their pollution and perversion?[29]

The discussion on the ethics of cloning should point in a direction altogether different from the one that is currently being pursued by contemporary bioethics. The profound mysteries of birth, individuality, and the human meaning of the parent-child relationship need to be unpacked beyond the reductive optic of science and the ethics that engages it without critical distance. An awareness of the fundamental asymmetry defining the outlook of science and ethics thus becomes the necessary antidote against methodological simplifications that seem to have no appreciation for the specificity of both. Whereas scientific rationality tends to be «instrumental» as it measures its criteria of progress against the standard of quantitative improvement, ethical

---

28 See L. Silver: *Remaking Eden: Cloning and Beyond in a Brave New World*, New York: Avon Books, 1997.
29 L. Kass: *Life, Liberty and the Defense of Dignity*, 152.

rationality looks at progress from a very different perspective, one in which the non-disposability of the person's dignity takes central stand against all forms of manipulation and reduction.[30]

The particular challenge of *theological* anthropology to the public articulation of moral discourse is to question any such attempts to de-humanize the reality of the person. Although one might encounter within moral theology different interpretive models on the relation between faith and moral reason, it remains undisputable that faith generates a particular anthropological pre-understanding.[31] Faith produces in the believer a *specific* horizon of meaning, something like a «system of anthropological coordinates» (Edward Schillebeeckx), which molds the believer's self-understanding and the world of human experience at large.[32]

One of the central elements in such a Christian horizon of meaning is the intrinsic *dignity* of the human person, which is at the root of the principle of autonomy and also of the notion of freedom. The dignity of the human person flows from the faith in a personal God as creator, who posits the human being as «the event of a free, unmerited and forgiving, and absolute self-communication of God» (Karl Rahner). The Christian tradition has further articulated such an understanding of the human person as the image of God (*imago Dei*), who participates in the new creation of Jesus Christ. Closely connected to this is the awareness of the singularity of one's existence as well as the unrepeatable nature of one's personal history.

---

30 On the difference between the *Denkform* of science and ethics see K. Demmer: «Stillstand in der Bioethik?» in: *Freiburger Zeitschrift für Philosophie und Theologie* 51 (2004), 191-219, at 201.
31 One might think here of the different assessment of the analogy of faith and reason between Catholic and Reformed theologies, but also of the polarity within the Catholic moral tradition itself between an «autonomous morality in a context of faith» and «an ethics of faith». For an overview of the debate E. Gillen/R. Simon: *Fonder la morale: Dialectique de la foi et de la raison pratique*, Paris: Seuil, 1974, and E. Gillen: *Wie Christen ethisch handeln und denken: Zur Debatte um die Autonomie der Sittlichkeit im Kontext katholischer Theologie*, Würzburg: Echter, 1989.
32 The importance of anthropological presuppositions to the moral judgment on genetic intervention is well illustrated by J.J. Walter: «Human Germline Therapy: Proper Human Responsibility or Playing God?», in R. Cole-Turner (ed.): *Design and Destiny: Religious Views on Human Germline Modification*, Cambridge, MA: MIT Press, 2005, forthcoming.

A theological anthropology defined by the notion of *imago Dei* opens up a special understanding of the uniqueness of each person, of her absolute singularity and incommunicable mode of existing. Though many human beings have existed in the course of history and are now existing, every person is, as it were, the only one. Every person is a *universale concretum*, a concrete whole, in which there is certainly included the nature of the species with its general characteristics, but it is also true that this nature is appropriated by the subject in an absolutely singular way, so as to transcend that nature. Thus Romano Guardini defines the singularity of the person as «the fact that she exists in the form of "belonging to herself" (*in der Form der Selbstgehörigkeit*).»[33]

The theological framework of imago Dei and individual uniqueness provides the context within which an ethics of cloning can be articulated in terms of an ontology of sexual reproduction.[34] Sexual reproduction, that is, the generation of new life from two complementary elements, one female, one male, is the *natural* way of all mammalian reproduction. By nature, each child has two complementary biological progenitors, thus stemming from and uniting exactly two lineages. Moreover, in natural generation, the precise genetic constitution of the resulting offspring is determined by a combination of nature and chance, not by human design: each human child shares the common, natural, human genotype; each child is genetically akin to both parents; each child is also genetically unique. Although these are only biological truths about our origins, they do have a powerful *symbolic* meaning, for they insinuate deep truths about our identity and our human condition. Every one of us is equally human, and that means, equally enmeshed in a particular familial nexus of origins, equally individuated in our trajectory from birth to death and, given the necessary biological premises, equally capable of participating with a complementary other in the renewal of such human possibility through procreation.[35]

---

33 R. Guardini: *Welt und Person: Versuche zur christlichen Lehre von Menschen*, Würzburg: Werkbund Verlag, 1955, 128.
34 On the relation between moral norms and sexual anthropology the best book remains, in my opinion, the classic of A. Guindon: *The Sexual Language: An Essay in Moral Theology*, Ottowa: University of Ottowa Press, 1977.
35 See L. Kass: *Life, Liberty and the Defence of Dignity*. See also the interesting observations of J. Ferrer: «Reflexiones éticas a propósito de la clonación», in:

Moreover – and this is particularly relevant in relation to the issue of cloning – our *genetic individuality* is also symbolically relevant. It shows forth in our distinctive appearance through which we are individually recognized and named; it foreshadows exactly the unique, never-to-be repeated character of each human life.

Human societies virtually everywhere have structured childrearing responsibilities and systems of identity and relationship on the basis of these deep *natural* facts of begetting. Within a wide spectrum of sociological variations, all *cultures* have historically exploited these facts to create for everyone clear ties of meaning, belonging, and obligation.

Asexual reproduction, on the other hand, represents a radical departure from the *natural* human way, confounding all normal understanding of father, mother, sibling, grandparent, and the like, together with the moral relationships tied to that understanding. It becomes even more of a radical departure when the cloning process presents the following characteristics: the resulting offspring is a clone derived from a mature adult to whom it would be an identical twin; second, the process occurs not by natural accident, as in natural twinning, but by deliberate human design and manipulation; third, when the child's genetic constitution is pre-selected by the parent or by the scientist.

In relation to these three dimensions, the process of cloning is vulnerable to three correspondent objections, all of which stand upon the following premise: given the current rate of the technology's success, any attempt to clone a human being would constitute an unethical experiment upon the resulting child to be, a most blatant violation of the first principle of medical research, *primum non nocere*.

However, even when successful, so to speak, cloning threatens confusion of *identity and individuality*. The clone may experience concerns about his distinctive identity not only because he will be, in genotype and in appearance, identical to another human being but because he may also be the twin to the person who is his «father». Unlike normal identical twins, a cloned individual will be burdened with a genotype that has already lived. How could such an individual be fully a surprise to the world? True, one must recognize that, because the genotype is not, by itself, the whole of a

---

Gregorianum 79 (1998), 129-148, at 143.

person, the cloned being is not necessarily destined to be the person from whom he came. Yet, how can one avoid seeing such a child in continual comparison with the original version? After all, to look like the original was the reason for cloning him in the first place.[36]

Cloning, and this is the second objection, represents a decisive step, though not the first one as I noted earlier, toward transforming procreation into manufacture and toward the increasing de-personalization of the process of generation.[37] I believe the emphasis here should not be on the fact that «natural is better» because «nature knows best». The indiscriminate appeal to nature may hide the fact that biological criteria function as true indicators of humanity only when interpreted and understood in their own transparency to a previous anthropological project. In this sense, an ideology that preaches «back to nature» *sine glossa* may be as de-humanizing as the ideology of technological production it intends to defy. The real problem is that any child whose being, character, and capacities exist owing to human design does not stand on the same plane as his makers. *Homo faber* will always transcend the product of his making as a superior, not as an equal. To use a Hegelian formula, the clone will stand before the cloned as a *being-in-itself*, yet never at the same time as a *being-for-itself*, that is, as a truly free human being.

In light of this last observation, cloning, not unlike other forms of eugenic engineering of the next generation, represents a form of despotism of the cloners over the clone, a form of violence between persons which is the more problematic when thought in relation to the proper meaning of the parent-child relations, of what it means to have a child. Think of what happens when a couple chooses to procreate: they are saying yes to *having* a child only because they allow that child to be, a new life in its novelty. Thus in saying yes to having a child, they are letting that child be whatever he or she will turn out to be.

The power of sexuality is, at bottom, rooted in a strange connection to mortality, which sexuality simultaneously accepts and

---

36 On the «right to a genetic identity» see K. Demmer: *Angewandte Theologie des Ethischen*, Freiburg: Herder, 2003, 218-220.
37 The danger was already foreseen by P. Ramsey: *Fabricated Man: The Ethics of Genetic Control*, New Haven: Yale University Press, 1970.

tries to overcome. In accepting our finitude, in opening ourselves to our replacement, we tacitly confess the limits of our control.[38] By procreating, we embrace the future, but we do so only because we dispose ourselves, in an attitude of gratefulness, to recognizing our children as *other* from ourselves: not our property, or our possession, but free creatures living their own lives, not ours. Our children's genetic distinctiveness is the natural symbol, once again, of the deep truth that they have their own life to live. Though coming from the past, they take an uncharted course into the future: truly, where they are going, we will be able to follow only for a while. This is why, in the end, cloning contradicts the entire meaning of the parent-child relation.

6. Conclusion

I believe the three objections I raised are only initial hints for the elaboration of a broader judgment on the immorality of cloning. As such, they do not tell the whole story, yet. But they were not meant to exhaust the ethics of cloning to start with. The 2002 report of the President's Council on Bioethics, entitled *Human Cloning and Human Dignity*, may be a good suggestion for those who want to reflect further.[39] The conclusions of that report, though ambivalent about the morality of cloning for biomedical research, clearly point in the direction of a global legal ban on cloning to produce children. As evidenced by my reflections, I wholeheartedly agree with something like an «international» public policy going in this direction. There remain questions, however, beyond the law, questions of meaning for which the strategies of procedural thinking must surrender to a deeper mindfulness. Is our current bioethical reflection *on a par* with such a challenge? This must necessarily remain an open question for, as one might imagine, it does suggest another topic altogether.

---

[38] On the relation between morality and sexuality, see already Sigmund Freud.
[39] President's Council on Bioethics, *Human Cloning and Human Dignity* (Washington, D.C., 2002).

CHAPTER 5
THE ETHICS OF HUMAN GENOME EDITING
Some Philosophical *Prolegomena*

Opening

In this chapter, I offer some philosophical *prolegomena*. My contribution is less about the particular issue of human genome editing, and more about the prevailing bioethical discourse that frames it.[1] It is about the limits of such an approach and the need for a broader integration of perspectives. The alternative in question will presuppose, as its starting point, the critique of a principle-based approach, as we all know, the dominant methodology framing not only the general features of bioethics, especially American bioethics, but also the many areas of *public* ethical engagement, including the area that is under our investigation.

The reference to the public is important, to begin. Here a reductive notion of public argumentation rests on specific limitations imposed on both complex conceptual premises and semantic sophistication. The result is a kind of imperial univocity dominating the field. I refer to it as the *orthodox* way of bioethics, in that it imposes a notion of ethical argumentation reduced to evidentiary plausibility, as such potentially intelligible without any deeper philosophical commitment, whether anthropological or metaphysical, let alone theological.

This seems too little, not only with respect to the theoretical seriousness of public argumentation, including its requirements for a different conceptual «thickness.» Puzzling is also the inability to see the existential weight of what is at stake in the debate: not just the rightness of individual practices, judged in a piecemeal fashion, and with reference to statistical acceptability, but the good of our own lives in an integral sense.

---

[1] For a good review of the literature of the topic, see John H. Evans, *The Human Gene Editing Debate* (New York: Oxford University Press, 2020).

In this light, the question is not only whether a society can feel comfortable with the introduction of practices hitherto considered unethical, or newly appearing on the moral radar thanks to new scientific and technological advances. The deeper question might be whether the practices at stake in the public debate do contribute to enhance the human quality of our deepest experiences, or compromise it forever. How do they affect our own understanding of the *humanum* broadly conceived, the *authentic* good of the human condition?

For sure, one ought to distinguish between statements of public policy and broader expressions of philosophical rumination. Public policy statements, whether in the form of legally binding documents, regulatory reports, or more general recommendations, cannot fully convey the conceptual «thickness» of ethical discourse, nor should they. Public policy aims to convey the consensually defined position of specific social bodies. Its function is to offer a *minimal* basis for a conversation to begin, not to exhaust *all* the conceptual resources that might inform and define the complex matrix of philosophical premises, analytical arguments, and prudential conclusions on a specific action or practice.

Still, the tendency I see in bioethical discourse today is to collapse the potential richness of the latter into the relative «thinness» of the former. Human genome editing is a case in point. Reference to the principles of beneficence, non-maleficence (safety), autonomy, and justice, though differently orchestrated in their application, are ubiquitous in the statements of official bodies. They seem to have a *prima facie* validity not only in terms of their normative weight, but also with respect to a taken for granted linguistic shallowness. Thus, non-maleficence will be essentially reduced to safety, beneficence to social benefit, autonomy to «expressive individualism» (Robert Bellah), justice to fair distribution of resources, etc.

The call for a moratorium on reproductive germline modification» by the 2017 *National Academies of Science, Engineering, and Medicine* report (NASEM) rests on such principle-based strategy.[2] As sociologist John Evans has illustrated, the public discus-

---

2   National Academies of Sciences, Engineering, and Medicine, *Human Genome Editing: Science, Ethics, and Governance* (Washington, D.C.: National Academies Press, 2017).

sion seems less driven by the quality of ethical arguments, Weber's «substantive rationality», and more by a pragmatic willingness to accept developments on the slippery slope of scientific development. The only proviso is that the sliding down the slope be justified in the name of beneficence, autonomy, or fairness in the distribution of opportunities.

Thus, what functioned as *barriers* in genetics yesterday[3] can quickly turn into less stringent *speed bumps* today, once the conditions for moving down the slope are finally met in the name of «relieving suffering» or «bodily autonomy.» In this logic, a deontologically grounded «ban» can turn into a temporary «moratorium,» waiting to be further removed by a consequentialist calculus on the potential benefits for the greatest number, or the unjustified encroachment on the freedom of the few.

Take the case of pre-implantation genetic diagnosis (PGD). If one accepts PGD for the *selection* of certain genetic traits (negative eugenics), one will very soon justify the *modification* of genetic traits as well (positive eugenics). In this case, «procreative liberty» will function as the «grease,» to keep the metaphor going, on the slope toward full and unencumbered reproductive autonomy (consult legal scholar John Robertson).

In the application of CRISPR-Cas9 technology, NASEM advocated that germline interventions be restricted to «preventing serious disease or condition, and editing genes that have convincingly demonstrated to cause or to strongly predispose to the disease or condition.» The report supported «converting such genes to versions that are prevalent in the population, and are known to be associated with ordinary health with little or no evidence of adverse effects.»

With these conditions, NASEM not only pulled down in one sweep the previously established somatic/germline barrier invoked for gene therapy applications; it also overturned the disease/enhancement barrier. Furthermore, it introduced a shift in the argument used to justify germline modification, since the intervention

---

3   Buchanan *et al.*, in their highly influential book, refer to them as «moral firebreaks.» See Allen Buchanan, Dan W. Brock, Norman Daniels, and Daniel Wikler, *From Chance to Choice: Genetics and Justice* (New York: Cambridge University Press, 2000)

to stop an embryo from having a «disease» is actually not about disease, health, or the reduction of suffering, that is, beneficence toward the children that are born, but about the parental desire for genetically related children.

One can see, in light of the last case mentioned, that, in addition to problems concerning the reductive notion of the values in question, there is also the issue of defining the normative strategy for privileging them in the process of application: which principle should prevail and why? In technical terms, what guides the «balancing» process, given the potentially equal weight of different principles at stake? Does beneficence overrides justice, or should autonomy prevail over safety?

What moved over the barrier in the germline gene editing endorsed by NASEM was not the initial commitment to beneficence, but rather the new dominant value of autonomy: «the desire to have genetically related children may arise from a variety of factors ... and precluding access to this technology could be regarded as limiting parental autonomy.» If people want to have children with a particular treat (genetic relatedness), then they have a right to do so.

So much for the general quality of ethical arguments and the logic with which they function! I imagine the pioneers of bioethics looking on to the field, today, many of them gone by now. I wonder what they see. I mean, how would they judge our methodological instabilities, the spectacle of widespread conceptual shallowness, too thin to provide meaningful signposts to our ethical quandaries?

Our scholarly fore fathers (and mothers) might ask us to step back, to retrieve a more *critical* stance: without fearing the disapproval of the scientific community; without subservience to the allures of public visibility, short lived in the show of media-fed self-importance. Chocked by the impelling demands of technological advancements, we have become unable to breathe properly. I mean, breathe *deeply* in a philosophical sense. What would taking a deep breath entail?

I submit that, at a minimum, it might entail retrieving a broader horizon of meaning for our own reasoning. It might also call for a different attunement to what, later on, I will refer to as the primal ethos of life. The first calls for an epistemological recharging. The second for renewed metaphysical sensibility.

1. The Search for Meaning

What of the horizon of meaning? A horizon conditions our seeing, frames our vision. It also opens our imagination: it is with us as we move within the world, follows us when we falter, recalling us back to a wider perspective, when our eyes are cast too low and our energy weakens. A horizon stands as the elusive, ultimate line in an open space; ever removed, always beyond, yet constantly present. It escapes rigid fixing, for that would prevent our journeying beyond the landscape presently offered, caging our vision within the boundaries of what we already know, what we have already experienced. A horizon, thus rigidly conceived, would only give us the measure of past accomplishments, all the while limiting our future aspirations.

Is the horizon of bioethics just a set of ethical principles in a «foundationalist» fashion? I mean, a set of norms – like the mantra of autonomy, benefice, non-maleficence, and justice, defined once and for all, beyond space and time? If so, why those points of reference and not others? Following the lead of European scholars of bioethics, why not embrace more expansive notions, like dignity, vulnerability, and solidarity?

We might take for granted the necessity of *all* the principles in question, their ability to map out a normative geography for our ethical preoccupations. If we accept them as *determinate* boundaries, what do we make of the process of *determining* that grounds their validity? *Qua* grounding source, the determining will already be beyond the determinate principles, as their condition of possibility. Furthermore, it will be a dynamic, rather than static, condition; historically situated and context-bound, not timeless and abstractly universal.

The horizon of meaning I seek is more than a normative theory, nor is it simply a «work in theory,» as Beauchamp and Childress have claimed in the latest iterations of their widely read *Principles of Biomedical Ethics*. The notion of a «work in theory» seems to convey the fact that at stake here is the activity of *practical* reason, and the less than precise contours of *praxis*, an old realization, already envisioned by Aristotle.

But the retrieval of the specific features of practical rationality, a task of enormous importance in itself, is only half the story. To abandon the ideal of a rationalist ethics, whether calculative, in a utilitarian fashion, or transcendental, in a deontological vein, is to be brought back to the reality of incarnate agents, of human beings inhabiting historically defined spaces, not just abstract rationalities functioning in an ideal «kingdom of ends.»

The interplay of ethics and philosophical anthropology is here at stake. I stand by a notion of ethics that emerges from broader anthropological commitments. To ask how we ought to behave is to do so, implicitly or explicitly, on the background of presumed notions of human flourishing and fulfilment.

One conclusion to draw from the previous insights is the sense of critical distance gained by envisaging our ethical predicament against such horizon of meaning. Whether we think of it as a «constitutive» condition, in a more metaphysical sense, or as a purely «regulative» one, a kind of practical postulate in a Kantian fashion, a horizon of meaning can provide our ethical compass with a steady sense of direction, a focal point unconditionally there.

The methodological broadening I am calling for is not new in contemporary moral philosophy. It finds expression in the attempt to recast the relation between rightness and goodness, and in the retrieval of eudaimonistic ethical theories. Whereas the first attempt offers a more nuanced understanding of the conceptual basis for the judgment on individual actions and practices, the second provides a content-full theory of the moral life, together with a teleological orientation toward an «ideology of human fulfilment.»

A moral philosophy defined by the separation of rightness and goodness assumes that judgments on the morality of actions and practices depend on the exclusive justification of the former. Moral discourse pertains to the sphere of what is «right.» The question of how such justification might contribute to the «good» or, conversely, detract from it, remains elusive, in fact, unimportant. When brought back within the limits of a finite, «secular» enterprise, supposedly the only legitimate in a democratic moral community, the task of ethical justification reduces to a strategy of intersubjective universalizability. As an ethical criterion for actions or for practices, rightness is the result of consensus in a community of moral discourse defined by rules of freedom, fairness, and complete transparency of presuppositions. What is right will be, by definition, procedurally justified against the backdrop of «common morality.»

The philosophical work of much contemporary bioethics is done mostly at this level. It is a secular exercise, not devoid of theoretical acuity, in the articulation of right actions and practices. The good remains *incognito*: impossible to define, on a purely empirical basis; too elusive to grasp in its metaphysical meaning. The four principles of bioethics will be played out, then, according to strategies of balancing or specification, without explicit reference to any underlying theory of the good.

In reality, what is only recessive remains no less effective. A theory of the good will be smuggled through, if only implicitly, either in its utilitarian version, with a tilt toward social beneficence, or in the deontological one, with a preferential option for autonomy and justice. A different strategy, however, one that is anthropologically charged with notions of human fulfilment and flourishing, remains out of this methodological picture.

A bioethics defined by the separation of rightness and goodness becomes increasingly problematic. The discomfort toward such strategy of argumentation is felt by a moral philosophy concerned not only with the justification of actions and practices, but with the *telos* of moral experience as defining the entirety of a life project, whether of individuals, of communities, and, ultimately, of society as a whole.

A moral life is more than the sum total of individual acts, seen in their relative isolation and requiring step-by-step justification. When understood in light of such holistic integrity, the moral rightness of individual actions and practices escapes isolated measure, for it is the authenticity of a life project (Charles Taylor), or the moral integrity of a community, what is being enhanced, its goodness relative to broad ideals of human fulfilment and happiness.

With this *teleological* shift in the direction of a eudemonistic ethics there comes a different understanding of the process of moral *justification* as well. To justify an action against criteria only procedurally measured is to rely upon an objectivity external to moral agency. On the other hand, the justification of a moral life requires personal approbation, for at stake is what Kierkegaard calls «subjective truth,» i.e., the personal investment in the call of values and the credibility of their appeal for a moral subjectivity. The truth of a life project points to internal commitment and free disposition. To justify an action is to explain its congruence with a universal criterion. To justify a life is to understand its ultimate meaning for the moral subject whose life it is.

2. Metaphysical Recharging

The probing toward ultimacy seems formally defined, at this point. Yet, the horizon is only an epistemological category, conditioning what we see and how. Is such category still caught in the dualism of subject and object? How does the horizon come into vision in the first place? How does it condition what it opens up as a field for our sight, not only its extension at the boundary, but what comes in the middle, between us and it?

Consider the issue with respect to the problem of normative application. Granted the eudemonistic thrust of our ethical thinking, relative to ideals of human flourishing and fulfilment, how do we pass from the general (the horizon) to the particular (the case in the middle)? What determines the movement in question, and who can reassure us that bridging the gap in the process of application will not end in downfall?

If the horizon is a *meaningful* horizon, determining the ethical quality of individual actions and practices, what gives meaning to the horizon at all? Why is such horizon a horizon of meaning, and not of absurdity, a nihilistic edge against which all our moral efforts come to nothing? Were the latter the inevitable destiny of our ethical laboring, our criteria for right and wrong would still remain insecure, the ground of our judgments shaky, in spite of methodological earnestness, in spite of the seriousness of our ethical projects.

To speak of ethical projects is to *pro-ject*, to endeavor in acts of construction entirely contingent upon our self-determination and will. The good envisioned in such case would still be our good, better a good *for* us. What of the good *in itself*? What of the good that is not just the horizon of our eudemonistic projects, but the source of their value, as such already beyond the dualism of subject and object entailed in the predicament of application? What is intrinsically valuable would ground the horizon, the vision, and the process of determining that brings the horizon to bear on the quality of what is in the middle.

We need something like hyperbolic categories to clarify the point, images beyond concepts. Think of the Platonic metaphor: the good is «like the sun» that throws light on everything, thus opening up a field of vision. The sun is not just a horizon, but a source that makes possible for us to envision what is there, against the edge of its ultimate extension. When night falls, darkness overtakes both, the horizon and the middle space of vision, precipitating the latter into a colorless neutrality, an obscure thereness without limits or measure.

I am calling for a metaphysical recharging in our ethical thinking, and this especially with respect to issues such as human genome editing. The development of the field of (American) bioethics is said to result from a shift in focus by philosophers in the analytic tradition, from metaethics to normative ethics. The former kept scholars busy for much of the first half of the twentieth century, only to engender a distance from real life issues, a distance that became unsustainable, finally, by the late 1960s. As Stephen Toulmin put it in a famous article, medicine «saved the life of ethics» by giving philosophers a different sense of the importance and urgency of their work. The new problems, among others, those brought about by medical and technological advances, stimulated a «thinking on the concrete» that consigned previous philosophical questions to the proverbial Humean flames. Though not metaphysical or theological in nature, such questions were still abstract enough, according to Toulmin, to provide little help to the mounting ethical perplexities of the day.

The doubt on the entire philosophical enterprise soon became patent: the focus on meta-ethical concerns about language and semantics, when moral problems called with pressing urgency for a different mindfulness, was less than adequate. Parsing the meaning of moral terms and the variety of language games warranting their functioning seemed playful enough for thinkers in the ivory tower. In the face of ethical issues bordering on the tragic, requiring immediate conceptual clarification and responsible thinking, a moral philosophy entirely detached from the real world looked increasingly like a reprehensible exercise in evasion. The move to practice and a different style of thinking made every other philosophical concern seem useless, if not irresponsible.

I do not deny the enormous work in «applied ethics» generated by such «turn to the concrete,» nor undermine its value. However, pitting the theoretical against the practical as alternative modes of thinking has not only thrown into the open the painful sting of a rationalistic wound at the heart of analytic philosophical discourse. It has *de facto* expanded the wound, leading to a further rationalist ripple effect on the retrieved passion for the practical.

Insofar as it conceives of moral language in its semantic abstraction, resting on the separation of meaning, relative to a subject, and truth, as an objective dimension of reality, meta-ethics operates with a concept of reason that has already detached itself from the ontological depth of being. Marked by the premises of such epistemological dualism, meta-ethics becomes, in its essence, an anti-metaphysical exercise. But so will also be the counter-movement of thought toward the practical, when it operates on the very same presuppositions. If the rebound effect in the direction of the concrete does not reach far enough, into the primordial ontological intimacy of mind and being, it cannot completely heal the rationalistic wound; at best, it provides a temporary soothing, lacking the curative quality of an effective therapy.

A normative ethics defined by suspension of meta-ethical premises, neutral with respect to the recessive sources of its engagement with reality, is bound to falter, sooner or later, on the very prescriptions it recommends for the moral life. For at stake is not only the nature of meta-ethical options, poised between cognitive or non-cognitive interpretations of the moral language. It is the metaphysical mystery grounding our opting in either direction, already resting on a more fundamental disclosure of being, offering itself in the call of goodness and the transcendental distinction between right and wrong, good and bad.

The retrieval of such ground will not be possible, if the ground is understood as a fixed, external principle. I mean a principle defined in its meaning as a pure propositional truth, demanding mechanical correspondence to reality, a naïve *adequatio ad rem*. We do not have «the truth» of the ethical, as if in a ready-made list of eudemonistic platitudes; even less does anyone owes a priori «the truth» of the *humanum*, in a packaged set of anthropological tenets.

The ground sought is more like an *event* of progressive disclosure offered to our own searching, not a thing-like presence determining a priori its findings. Such an a priori presence would paralyze the search, rather than energize our desire for the good.

On the other hand, the *grounding* energy of desire for the good, articulated in the quest for what is right in practice, sustains our confidence in the truth that, progressively, takes shape in its mysterious contours, giving itself to us as a promised gift. Such truth is not just the result of our own constructions, the endpoint of our ethical endeavors.

When it comes to the problems of bioethics, the happening of the moral truth is conditional upon the giving of the good of life (*bonum diffusivum sui*): life that opens itself to us, offers itself as a gift from the origin, grounding our efforts, mending our faults, aiding our perplexities in light of ever newly emerging quandaries. The metaphysical trust in being as full of promise possibilizes our endeavors and constructions. Our ethical thinking emerges as the fruit of a more primordial confidence (*cum-fides*) in the goodness of things, rescuing our thinking from conceptual superficiality and nihilistic incantation.

3. Theology in Disguise?

I do want to push philosophical ultimacy to the point where it merges with the preoccupations of theology. The work of earlier practitioners of bioethics, many of them theologians, was precisely of such nature.

I submit that what made their contributions so substantive might have something to do with their background premises, informed by metaphysical confidence and thick anthropological notions, rooted in theological categories, yet open to public argumentation and discussion. Theirs was not just a *confessional* exercise, driven by missionary intention. Nor was the quality of their presuppositions lost in the «rational» accessibility of the normative solutions they suggested. Their moving from a perspective of ultimacy was relevant to the way they understood what is at stake in the middle.

Of course, theological thinking does not operate in a vacuum. It stands within a moral tradition with all its twists and turns, arising in the life of faith communities, defined by a twofold commitment: to the truth of an original event of absolute disclosure, and to the progressive articulation of its ethical implications for the present. If the first commitment signals something like a retrospective move, the second points in the direction of a constructive endeavor, with progressive, in fact, subversive character.

The work of theology rests on the interpretive discernment of historical narratives, texts, and tradition, and the unrelenting work of cultural appropriation, both dialogical and dialectical, requiring philosophical intermediation. A matrix of ethical truths drives the hermeneutical process in question. I think of the confidence in the goodness of life; the glory of creation and the trust in the intrinsic value of being; the recognition of the inherent dignity of all human beings, rooted in their personal status; the equality of human beings, beyond race, gender, and social location, which requires attentiveness to the obligations of justice; the recognition of the vulnerable and marginalized as special in the eyes of God; the summoning to the solicitude of social solidarity; the call for love of neighbor and even enemy in the generosity of communities of agapeic communication.

A matrix is not just a collection of unrelated moments, operating independently of one another, available for a «pick and choose» exploitation according to circumstance and taste. It is a framework marked by logical coherence and ontological depth, kept open by faith in God's presence to the world, as such irreducible to the closed immanence of conceptual categories. A theological matrix is more than a system of thought. Born of a lived encounter, it is a saturated phenomenon of an interpersonal nature, shattering the totalitarian pretense of a priori intentionality and purely representational thinking.

The work of theology unfolds in the task of philosophical intermediation, which tries to unpack the meaning of the ethical matrix in question. This also with respect to the field of genetics. Here we confront the challenge of scientific and medical advances, calling into question our metaphysical trust in the ultimate nature of life and the anthropological meaning of human experience in all its relevant phases.

A bioethics theologically informed entails a kind of reversal of perspectives with respect to the prevailing methodological orthodoxy, which, on the other hand, tends to see any hospitality to religious content by philosophy as a dangerous promiscuity, unfaithful to the secular purity of its premises. Such «heterodox bioethics» stands between two opposite extremes: the claim to neutrality of moral discourse, articulating a purely secular account of ethics, and shunning hospitality to theological contributions, on the one hand; the colonizing pretense of theological discourse to entirely subsume secular reason under its religious premises, on the other.

It threads a middle course of a different nature, calling for a «companioning» of philosophy and theology (so William Desmond) in which neither pretend to englobe the other. An encounter based on dialogical, rather dialectical, intentions. If the latter, only in a metaxological sense (Desmond again), keeping open the space for generous intermediation, rather than close mediation.

Theology offers reasons to support a eudaimonistic and communitarian turn in ethics: it recognizes the primacy of the good, the relevance of virtuous moral agency, its rootedness in community, the need to articulate criteria for rightness of actions and practices in light of full narratives of human fulfilment and flourishing. At the same time, theology underlines the *teleological* orientation of ethics with a more original confidence in the *archeology* of the good. The good given in the gift of life grounds the search for a final good. As for the search of truth, resting on the original intimacy of knowledge and being, the desire for the good relies on the original communion of will and being (so Thomas): our freedom could not find the good, had it not already a sense of the meaningful character of its searching. It is worth searching for the good, because an original participation in it companions the seeking all along.

An ethics of life (*bio-ethics*), theologically grounded, reflects such ontological confidence (*cum-fides*) in all its articulations. The matrix of ethical insights flowing from the «primal ethos of life» unpacks all the implications of the original disclosure, redefining the meaning of all ethical principles and norms:

Consider the potential impact of a theologically defined bioethics with respect to *trans-humanistic projects*, not extraneous to some iterations of human genetic engineering. Here the pretensions to shape and redefine the natural conditions of human life are most clearly thrown into relief in their hubristic character:

> We are rooted in nature, but we risk denaturing ourselves in claiming to make ourselves according to a second nature. The second nature is not a second «yes,» a redoubled «yes» to the first «yes» at work in the *poiesis* of naturing and our *passio*. More often, it is a «yes» to a *conatus* that has deviated from the subtle insinuations of the new sunken matrix of fecundity.[4]

---

4   William Desmond, *The Intimate Universal: The Hidden Porosity Among Religion, Art, Philosophy, and Politics* (New York: Colombia University Press, 2016), 327.

Also the inadequacy of traditional notions of beneficence, non-maleficence, and autonomy becomes obvious: their operating in a metaphysical and anthropological vacuum makes the process of their application, whether in terms of balancing or specification, impossible. How could a re-shaping of human nature, aiming to overcome its current predicaments of finitude and mortality, find *any* ethical point of reference within a principle-based framework?

Transhumanism, as an exercise in genetic engineering, is literally «beyond good and evil.» For good and evil are still categoreal determinations of a more original disclosure of the *humanum* in its transcendental dimensions. Why should such disclosure be respected in its given, rather than constructed, character? Why should it be recognized in its intrinsic value? Why *reverence* toward life, rather than hatred of its current existential conditions, calling for its constant overcoming? The trans-humanist project can only make sense against the backdrop of a correlative «trans-valuation» of values (Nietzsche), already beyond the normative limitations imposed by *any* ethical principle.

A theologically informed ethics is defined by a *love for life*, a responsorial doing that finds ethical measure in the primal ethos of life. Such measuring will be at the same time a «being measured,» an operating in the fold of the *given* ethos of life, not an autonomous exercise, taking place in a vacuum. The directionality of true humanism, therefore, will require something like a return to interiority (Augustine), more than a wandering into impossible spaces of anthropological creativity; it will be an *intra* rather than a *trans*-humanism.

What of human genome editing then? What of the promise of improved health and potential eradication of disease, what of the dream in the progress of science and medicine toward an ever more sanitized, perhaps *better*, version of the world? Can we distinguish between hubristic pretense, mindless of limitations, and honest commitment to the alleviation of suffering, still calling for inevitable compromises between risks and benefits, the courage to chip away at the security of precautionary measures in the hope for what lies ahead, toward futuristic benefits awaiting us in return?

Do not read my reflections as an invitation to eschew ethical discernment in favor of reactionary posturing or, worse, paralyzing anguish. I framed my contribution as a philosophical prolegomenon.

But forward thinking can either mean proceeding impatiently toward a progress defined by lack of reverence for life and the dignity of human beings, or journeying, in fear and trembling (the lesson of Hans Jonas!), toward a future attentive to the fragile nature of life, and the love of being that surpasses any manipulative pretense and exploitation.

Can such love of being, the love of life, find universal articulation? How can it be framed within the limits of a so called *public* argument? At stake in this final question is the problem of moral pluralism, the dialogue among different moral traditions, the reasons that ground the promise of *better* ethical alternatives.

Can ethical excellence be fully domesticated within the limits of univocal reasoning? What if the «reasons of the good» are never good enough? What if they are impossible to reduce to univocal determination, not because of lack, but because of their excessive, over-determinate quality? If so, they would be like Pascal's «reasons of the heart,» reasons of which reason knows nothing.

The testimonial character of Christian ethics requires, in the end, witness to the good more than minding of ethical quandaries, the saint more than the speculative hero. But the poverty of theology, like that of philosophy, will not castigate the dynamism of the spirit into critical submission, in Kantian fashion; nor lead theology to reject its own limits in a dialectical movement of progressive self-surpassing, as in Hegel. It will rather open any rationalistic self-imposed limitation or, conversely, self-proclaimed pretention, to a different recognition: that the adventure of thought finds its completion beyond itself, in a transcending movement of self-giving, into «works of love» (Kierkegaard again?), into the vast, yet promising, ocean of life itself.

CHAPTER 6
ON THE ETHICS OF TWINS SEPARATION
The Forgotten Case of Babies Jodie and Mary

On August 8, 2000, conjoined twins Jodie and Mary were born at St. Mary's hospital in Manchester, England. Their separation, imposed by British legal system against the parents' wishes, raised, at the time of the decision, a host of ethical questions.[1] Indeed, rather than putting an end to the debate surrounding the twins, the legal verdict contributed to setting in motion a reflection which has not abated yet.[2] For good reasons. To start with, the case was emotionally charged and polarized the attention of many. The attention and concern surrounding the conjoined twins went well beyond the sensationalism of a biological oddity; it reminded us of medicine's unique power to function in our society as main catalyst of moral reflection and discussion.

It may be time to reap the fruits of the past discussion and to analyze the case from a situation of objective distance. This is what I intend to do in these pages. Before, however, I will review the main elements of the case. Because, I believe, the case can be considered paradigmatic, it may be worthy to linger on its details in order to be able to better assess the main components of the ethical judgment. In particular, given the reliance of the parents and of public opinion upon religious considerations against the separation, I will discuss some of the problems concerning *this* particular case from the perspective of Catholic moral theology.

---

1   The focus of this chapter will not be on the legal dimensions of the case. For an assessment of the legal perspective see, among others, G.J. Annas, «Conjoined Twins: The Limits of Law and the Limits of Life,» in *New England Journal of Medicine* 344 (2001), n. 14, April.
2   «The Hastings Center Report» hosted alternative perspectives in the January-February 2001 issue. See A.J. London–L.P. Knowles, «The Maltese Conjoined Twins: Two Views of Their Separation,» in *Hastings Center Report*, 48-52.

1. The Facts

Conjoined twins have caught the imagination of people for centuries.[3] Mary and Eliza Chulkhurst, also known as the Biddenden Maids, were born around 1100 and are the first documented case of conjoined twins. In the sixteen century, the French surgeon Ambroise Pare, believing that conjoined twins were «contrary to the common decree and order of nature» struggled to determine the cause of conjoined twins. Perhaps the most famous pair of conjoined twins were Eng and Chang Bunker. Born in Siam (now Thailand) on May 10, 1811, the Bunker Twins were attached by a five-inch connecting ligament near their breastbones. Billed as the «Siamese Twins,» they had a long career as a public exhibition; yet they managed to marry sisters, father 21 children and become successful businessmen and ranchers in Wilkes County, North Carolina. They lived for 64 years.[4]

Conjoined twins are a rare form of identical twins, occurring perhaps in one out of every 75,000 to 100,000 births. The twins originate from a single fertilized egg, so they are always identical and of the same sex. The developing embryo starts to split into identical twins within the first two weeks after conception but then stops before completion, leaving a partially separated egg which continues to mature into a conjoined fetus.

Conjoined twins are generally classified in three ways, the classification referring to the point at which they are joined: 73% are connected at the mid torso (at the chest wall or upper abdomen); 23% at the lower torso (sharing hips, legs or genitalia); 4% at upper torso (connected at the head). However, this basic classification can be combined to more closely define individual cases.

---

[3] In myths and legends twins have been worshiped as gods and feared as monsters. The Greek and Roman god Janus had two faces, one young, one old. Centaurs, a combination of horse and man, may have been inspired by *pygopagus* twins who often have four legs (in this condition twins are likely position back-to-back and usually have a posterior connection at the rump). A Common heraldic symbol, the Double-Headed Eagle, is common throughout Central Europe. For documentation see N. Segal, *Entwined Lives: Twins and what They Tell Us About Human Behavior* (New York: Dutton, 1999).

[4] The story is recounted in I. Wallace and A. Wallace, *The Two: A Biography* (New York: Simon and Schuster, 1978).

The birth of two connected babies can be extremely traumatic and approximately 40-60% of these births are delivered stillborn with 35% surviving just one day. The overall survival rate of conjoined twins is somewhere between 5-25% and historical records over the past 500 years detail about 600 surviving sets of conjoined twins with more than 70% of those surviving pairs resulting in female twins.

Conjoined twins with separate sets of organs have greater chance for surgery and survival than those sharing the same organs. The first successful operation to separate conjoined twins did not occur until 1912, although attempts had been made as early as 1689. The operation is risky, complicated and lengthy, involving a team of surgeons, anesthetists and physicians. Two of the most well publicized cases of twins separation in recent history took place at Children's Hospital of Philadelphia. In September, 1974, chief surgeon, chief surgeon Dr. C. Everett Koop led a team of twenty-three doctors in the dramatic separation of a 13-month-old ischiopagus girls from the Dominican Republic, Clara and Altagracia Rodriguez.[5] Though not life-threatening – they had crossed ureters and shared only a liver and part of the colon – their close connection meant that they could not walk or sit properly. Eventually, the five hour surgery was a complete success. On June 29, 1993, Angela and Amy Lakeberg were born in Indiana. The babies were joined at the chest and shared a heart and liver. They could not survive together so it was decided to separate them. Ultimately sacrificing one for the other. The surgery took place at Children's Hospital in Philadelphia when the girls were seven weeks old. Angela, the stronger of the two, was chosen to be the survivor. However, Angela never went home. After ten months in the hospital, she died of pneumonia.

The questions raised by the Lakeberg case are not dissimilar from those confronting us in the case under discussion. Insights gained from the experience of past cases represent a reservoir of accumulated wisdom and reflection that can assist in the articulation of ethical judgment on what confronts us in the present. However, the reference to a paradigmatic case must be driven by the specific rules of *analogical* reasoning. The similarity of cases cannot be

---

5   *Ischipagus* twins are joined by the coccyx (lowest part of the backbone) and the sacrum (backbone immediately above the coccyx).

claimed without recognizing the differences that distinguish them. Good casuistry prevents ethical thinking from falling into the trap of equivocal comparisons.[6] As the British Court of Appeals recognizes, this case «is truly a unique case.» The following is a general description of the case.[7]

On August 8, 2000, Jodie and Marie were born at St. Mary's Hospital in Manchester, England. Jodie and Mary are conjoined twins. They each have their own brain, heart, lungs, and other vital organs, and they each have arms and legs. They are joined at the lower abdomen but are capable of lying flat on their back. At first glance they appear as if they were one single trunk with a head and limbs at both ends. Their spines are fused but their legs are independently formed and criss-cross each other. At birth, Jodie was active and breathing voluntarily with a good heart and chest movement and moving all four limbs. In Mary's case, there was a minimal response from the cardiopulmonary system before it failed. The medical team soon realized that Mary's heart and lungs were so poorly developed that she was totally dependent on Jodie for oxygen and blood circulation. While Jodie's system did collapse from blood poisoning shortly after birth, her heart and blood are reported to be fully functioning. She is said to have the same mental awareness as other newborn children. On the other hand, Mary's mental state is unclear: she has a very primitive brain, though during court hearings, doctors said that she was moving her limbs and had opened one of her eyes. Although surgery would be complicated, they can be successfully separated. However, the operation will kill the weaker twin whose lungs and heart are too deficient to oxygenate and pump blood through her body. Had she been born a singleton, she would have died shortly after her birth. She is alive

---

[6] On the analogical character of casuisty see Albert Jonsen and Steven Toulmin, *The Abuse of Casuistry: A History of Moral Reasoning* (Berkeley: University of California Press, 1988).

[7] Since the most detailed record of the case publicly available is the 130-page ruling of the British Court of Appeals, I will draw from that presentation of the case. I must state from outset that, in many points, my account follows the ruling almost *verbatim*; indeed. Always when the sentences are in quotation marks. At the time of the ruling, the text was to be found on the Internet at www.courtservice.gov.uk. Unfortunately, once the media focus on the case faded away, that useful reference disappeared also.

only because a common artery enables her sister, who is stronger, to circulate life sustaining oxygenated blood for both of them. Separation would require the clamping and then severing of the common artery. Within minute of doing so, Mary will die. Yet if the operation does not take place, both will die within three to six months, or perhaps a little longer because Jodie's heart will eventually fail. The doctors are convinced they can carry out the operation so as to give Jodie a life which will be worthwhile.

The parents, however, cannot bring themselves to consent to the operation. The twins are equal in their eyes and they cannot agree «to kill one even to save the other.» As devout Roman Catholics they sincerely believe that «it is God's will that their children are afflicted as they are and they must be left in God's hands.» Because of the parent's resistance, the hospital sought a declaration that the operation may be lawfully carried out and on August 25, a high court judge ordered the hospital to separate the twins against the parents' wishes. Meanwhile, the parents' position was supported, both implicitly and explicitly, by public declarations of Catholic church officials. On August 28 the Vatican offered «a safe heaven to the twins and their family in an indefinite and complete free pledge of medical services as an ethical alternative to the decision of the judge.» Cardinal Ersilio Tonini of Ravenna, in North-East Italy, told the Sunday Times in London that «the family would be offered safe harbor in Italy in one of two specialized centers.»

The initial Catholic position on the case was summarized by Archbishop of Westminster, Cormac Murphy O'Connor: «The refusal of the parents of Jodie and Mary to consent to surgery to separate them involves no injustice towards either of the children and is indeed wholly reasonable on the grounds that they have advanced...The good end would not justify the means. It would set a very dangerous precedent to enshrine in English case law that it was ever lawful to kill, or to commit a deliberate lethal assault on an innocent person that good may come of it, even to preserve the life of another.»[8]

---

8 For the complete documentation of Archbishop Murphy O'Connor's statement see «The Conjoined Twins Mary and Jodie: Ethical Analysis of Their Case,» in *Origins*, December 2000, 269-272.

Eventually an appeal was lodged in the Court of Appeal by solicitors acting on behalf of the parents. On September 22, the British Court of Appeals dismissed the appeal upholding the previously High Court ruling that surgeons must operate on Baby Jodie and Baby Mary and try to separate them. At this point, the parents decided not to appeal further to the House of Lords or to the European Court of Human Rights. In doing so, they cleared the way for the separation surgery which took place in November 2000 and ended with Mary's death.

2. The Details

The Parents

The father is forty-four years old; his wife is ten years younger. They have been married for two years and have no other children. Life is hard for them on the island of Gozo where they come from, just off Malta. There is simply no work for the husband. He has been unwillingly unemployed for eight years. The mother was more fortunate but her work terminated during her pregnancy. They have, somewhat, managed to accumulate very modest savings and were in process of building a home for their expected family.

When about four months pregnant, an ultrasound scan revealed that the mother was carrying twins and that they were conjoined. A doctor at the hospital had trained at St. Mary's Hospital, Manchester, and knew of its expertise and excellence. Therefore, he advised that they should seek treatment there. Through long established links between the government of Malta and England, a number of patients can be sent to England to be treated under the British National Health Service. Indeed, an assessment panel in Malta had judged that the case could not be managed by the local resources. Thus, the government of Malta paid for the mother to travel to Manchester in mid-May for treatment during her pregnancy and the father managed to join her there. Further scans were taken and an MRI scan was undertaken at Sheffield. As a result of these scans, it became clearer during the latter stages of the pregnancy that the difficulties with the twins were more than originally suspected. For a number of weeks toward the latter end of the pregnancy, the clear indication was given by the treating doctors that, sadly, the smaller of the two twins would probably not survive. However, the parents never considered the option of terminating the pregnancy. As a result of their desire for non-intervention, the obstetrician took the unusual step of allowing the twin pregnancy to continue until the mother went spontaneously into labor at 42 weeks. Normally, obstetricians consider delivery before that time because of a concern as to whether the placenta can adequately nourish both fetuses. As agreed by the parents, the obstetrician delivered the twins by Cesarian section at the last possible moment in labor. «This was to meet the parents' desire that the pregnancy was as non-interventionist as possible.»

The Birth

The twins were born on August 8, 2000. Their combined birth weight was 6 kg. They were immediately taken to the resuscitation venue. While Jodie was easily intubated and made spontaneous breathing effort, Marie was unable to ventilate, presenting no chest movement or breath sound. The anesthetist found that although he could pass the end of the clear tube into Mary's main airway, he was not able to detect any gas way at all, nor when he put a monitor into the ventilator to track the excretion of carbon dioxide did he detect any carbon dioxide, which should be exhaled. In the anesthetist's words, «we never had any evidence that she has breathed for herself at all.»

The Conjoined Twins

They are *ischiopagus* (i.e. joined at the ischium) *tetrapus* (i.e., having four lower limbs) conjoined twins. The ischium is the lower bone which forms the lower and hinder part of the pelvis – the part which bears the weight of the body sitting. The lower ends of the spines are fused and the spinal cords joined. There is a continuation of the coverings of the spinal cord between one twin and another. The bodies are fused from the umbilicus to the sacrum. Each perineum is rotated through ninety degrees and points literally. Jodie's head seems normal but Mary's is obviously enlarged, for she has a swelling at the back of the head and neck, she is facially dysmorphic and blue because she is centrally cyanosed. Between these two heads is a single torso about forty centimeters long with a shared umbilicus in the middle. Two legs, Mary's right and Jodie's left, protrude at an acute angle to the spine at the center of the torso, lying flat on the cot but bending to form a diamond shape. The external genitalia appear on the side of the body. Internally each twin has her own brain, heart, lungs, livers and kidneys and the only shared organ is a large bladder which lies predominantly in Jodie's abdomen but which empties spontaneously and freely through two separate urethras.

The most crucial anatomic fact, however, is that Jodie's aorta feeds into Mary's aorta and the arterial circulation runs from Jodie to Mary. The venous return passes from Mary to Jodie through a united inferior vena cava and other venous channels in the united soft tissues.

The Twins' Condition

Jodie has an anatomically normal brain, heart, lungs and liver. Her bowel is also normal and appears to be totally separate from her twin Mary. There is an abnormal vertebra in the lower thoracic area of the spine. She has two kidneys and a full spinal cord. She has two normal lower limbs, which move normally but are widely spaced because of the pelvic diastasis. The hip joints are both normal but the sacroiliac joins are dislocated and externally rotated causing the lower limbs to lie at right angles to the spine. Neurologically, she has various neonatal response which appear to be normal including a Moro response, plantar grasp and palmar grasp responses. There is normal routing response and a glabellar tap. In the cranial nerves, the optic fundi are normal and she has normal external ocular movements. Facial movements are normal and she is capable of sucking and swallowing. In the limbs, appearance, tone, movements, and muscle developments of the chest wall and the diaphragmatic movements are satisfactory. No obvious abnormality was seen in the cervical, dorsal and lumbar spine. The bladder is shared with her co-twin Mary. The pictures of the ultrasound brain scan showed no obvious abnormality and the findings suggest Jodie may have normal brain development. In fact, at three weeks of age she showed «normal reactions and normal development as expected for a child of her age and gestation.» Of particular concern, however, is the capacity of her heart to sustain life for herself and her sister. At three weeks, «Jodie's heart remains stable and appears to be coping well with the circumstances. The results of blood gas analysis are below normal indicating a degree of oxygen deprivation for both twins. Despite this presently Jodie does not show any clinical signs of concern.» A later report from the hospital's cardiologist dated August 31 states that «Jodie was comfortable breathing air, alert, and hungry. She was feeding from a bottle without distress. She is on demand oral feeds...Oxygen saturation was 100% ...Arterial pulse were palpable in all limbs. There was a good peripheral perfusion. Leg blood pressure was record 80/50 mmHG. The precordial impulse was not overactive. Heart sounds were normal. I could hear a heart murmur.»

There are some complications in that there is only one external opening which communicates with the urinary bladder and vagina and there is no opening of the anus. More importantly, the twins caloric intake is insufficient to allow growth. It is a feature of conjoined twins, even when both are neurologically normal, that one is less active and feeds less than the other. Conversely, the active, feeding twin is thinner than the other. Here Mary does very little and Jodie does all the work. She is not receiving enough calories to grow normally and this is not a favorable situation for her in the long term. The fact that Mary is drawing nutrition from Jodie and growing at her expenses may have implications for the timing of a possible operation.

Mary is severely abnormal in three key respects. First, she has a very poorly developed «primitive» brain. The brain scan showed various abnormalities including reduced cortical development, ventricular enlargement, partial agenesis of the corpus callosum and Dandy Walker type malfunction of the hindbrain. A neuronal migration defect may have occurred. These are the result of a major malformation which was probably present early in fetal life. Similar brain malformations are not compatible with normal development in post-natal life. The child could possible develop hydrocephalus and corpus callosum in later childhood is associated with disorders/ epileptic fits. The second problem is with the heart which is very enlarged, almost fitting the chest with a complex cardiac abnormality and abnormalities of the great vessels. Thirdly, there is a virtual absence of functional lung tissue (severe pulmonary hypoplasia).

The Available Options and the Doctors' Views

From a medical point of view, there are three ways of treating this disturbing situation:

(A) Permanent union with the potentially normal Jodie carrying her very abnormal sister. The looming problem for Jodie would be high cardiac output failure, probably in three to six months. If Jodie were to deteriorate to the point where Mary was to die because of Jodie's heart being compromised, then both twins would die simultaneously.

(B) Elective separation will lead to Mary's death, but will give Jodie the opportunity for a separate life with a high probability of a normal health. There are concerns regarding the possibility of acute heart failure for Jodie at the time of the separation. Jodie may have bladder and anorectal control problems and is likely to require additional operative intervention over time. She may have musculoskeletal anomalies, which may also require surgical correction. It is expected, however, that separation will give Jodie the option of a long-term good quality life. She should be able to walk unaided and relatively normally. Separation should allow Jodie to participate in normal life activities as appropriate to her age and development.

(C) Semi-urgent/urgent separation will need to be considered in the event of an acute catastrophe such as Mary's death, the development of progressive heart failure for Jodie, or the development of a life-threatening condition. The prognosis for Jodie would be markedly reduced and mortality highly likely, particularly following the death of Mary. For Jodie the prospects of a successful separation (60% mortality) compare poorly with those of a planned elective separation (6% mortality). Clearly, for Mary, separation will always mean death.

3. The Nature of the Proposed Operation to Separate the Twins

The operation will be in separate parts. After a basic anatomical exploration, surgeons will need to determine «from which parts each bit of each organ is being supplied so as to know which bit to give to whom.» They will then proceed with the separation of the bladders and will look at the anatomy of the anal rectum. Once it is established which bits are going to whom, the actual separation then starts by separating the pelvic bones, the linings of the spinal cords and the terminal ends of the spinal cord. At that point, surgeons should be left with possible some muscle union at the pelvic floor which too will need to be divided. Finally, the surgeons will have to deal with the separation of a major blood vessel, which is a continuation of Jodie's aorta bringing blood across to Mary, and similarly, the vena cava returning blood from Mary to Jodie. That is the most dramatic moment for at that point Mary will most likely die. The rest of the operation for Jodie would then be essentially a reconstructive operation.

All along, the operation will be a major procedure. It will take many hours and involve various teams of surgeons. Three observations on the operation itself seem of additional importance for an ethical judgment. The first deals with the issue of Mary's bodily integrity. Separation of the twins would necessarily involve exploration of the internal abdominal and pelvic organs for both twins and particularly the united bladder. However, it is expected that each twin would have all its own body structures and organs. It is not anticipated or expected to take any structure or organ from either twin to donate to the other. The second observation refers to the skin element. In order to ensure that surgeons would close the surgical wound with Jodie, it would seem a more prudent course of action to favor Jodie. Yet this is not required, i.e., no part of Mary would be given to Jodie. The final observation points to the site where the clamping of the aorta would occur. According to the surgeons, interruption of the blood supply from Jodie to Mary would occur at the level of the united sacrococcygeal vertebrae. This site could be biased toward Jodie.

4. *The Prognosis for the Twins*

Surgery would probably be a low risk procedure for *Jodie*. The operation itself and possibility of later complications would probably carry an overall risk of death of perhaps 1-2%. According to the surgeons, evidence from the literature suggests that the separation is usually well accepted minimal psychological effects on the survivor. Moreover, she should be able to stand and walk on her own without support. In the worst scenario, it is possible that she will never walk, she may need a wheelchair, or an appliance in the form of a crutch or brace, although this is not what is expected. Of course, Jodie will require further surgery. Since her large bowel seems normal, she is expected to have normal bowel control. However, given that attachment of the muscle in the pelvis will be absent or at least tenuous on one side, one cannot be absolutely certain that bowel control will be normal. Further operations will be required to provide a functioning vagina. Jodie does have a hemi vertebra at the lower end of her thoracic spine. It is possible that she would need scoliosis surgery should a curvature of the spine develop. At present, the need for surgery cannot be predicted and one would need to await further spinal growth.

If the operation to separate the twins is carried out, *Mary* will be anaesthetized against all pain and death will be mercifully quick. If, on the other hand, the twins remain fused, the evidence is that Mary will have a 75% or more chance of developing hydrocephalus. This, in turn, would be very difficult to treat because usually the end of the shunt system would either go into the abdomen cavity which is abnormal in her case or into the heart which is also not possible in her case. The effect of untreated hydrocephalus will be increase brain damage. As for the issue of pain, it is difficult to draw final conclusions. The child's responses are extremely primitive and more like mass movements to a stimulus, be it pleasurable or painful. There is a response to stroking and a response to pinprick, but they are the same.

Once the common aorta has been severed, theoretically Mary could be kept alive if she were immediately attached to a heart lung machine.

5. Ethical Assessment

Methodological Observations

As already mentioned, at the time of the Court decision, the case was still unfolding. Although the details of the case seemed quite clear, there still was room for unexpected developments. In particular, the prognosis for the twins, were they left to be untreated, remained difficult to assess with precision. Quick judgments and clear-cut conclusions had to be kept in check. This, of course, applies to my considerations as well. I will proceed in two steps. First, I will look at the case from the perspective of the agents involved, in particular, the parents and the health care team. In doing so, I hope to gain a clear sense of the intentions and motivations underlying their decisions. In particular, I will confront the issue of whether parents and physicians at the hospital are justified in their contrasting requests to, respectively, let both twins die or try to save at least one of them.

Secondly, I will step back from the agents perspective and look at the case from the outside, so to speak. My intention in this moment of the ethical analysis is to clarify some conceptual issues whose meaning, although raised by the case, go well beyond its particularity. In particular, the meaning of the distinction between killing and letting die and the conditions for the applicability of the theorem of double effect will be analyzed.

The Perspective of the Agents

It is clear from the reconstruction of the case that *the parents* have, at least subjectively, the best interests of their babies in mind. From the moment the twins' condition was discovered, the parents made every effort to best cope with an obviously terrible situation. Their decision to go to England, which carried considerable economic and emotional burdens for them, was clearly motivated by their willingness to offer the twins the best available health care. Moreover, they ruled out terminating the pregnancy on the basis of their religious convictions.

All along, they have manifested a *non-interventional* attitude which is further reflected in their understanding of the twins' predicament and the nature of the medical intervention. In reference to the former, they are concerned with the imposition of what they see as a terrible burden upon Jodie: The separation itself appears to them a highly experimental procedure with an unclear therapeutic outcome; they see a long series of surgeries awaiting their baby and wonder how they will cope with them emotionally or psychologically, not ultimately from an economic point of view. Furthermore, if their baby survives the operation, they would not be able to take her back to their own country. They rightly feel that the medical system in their own country is unable to provide, technically and financially, for such a case in the long run; by the same token, they cannot accept the idea of separating themselves from their baby so that she can be cured.

The non-interventionist attitude of the parents is also reflected in their understanding of the nature of the surgery. They immediately see the separation as a «sacrifice surgery»i.e., an intervention in which one twin will have to be «sacrificed» so that the other may live. It is obvious that this is more than just a *factual* rendition of the situation: Perhaps it is their way of making sense of a biological condition they feel should not be acted upon because it is «God's will.» Moreover, it expresses their fundamental belief that the twins are equally valuable and, for this reason, can never become mere variables in an instrumental equation. Both twins are their babies and as such, they cannot become the object of a choice in which one is pitted against the other.

Without entering into the merit of the parents' judgment, it may be worthy to point out the following: first, it is surprising that only one of the twins, Mary, is seen as the one to be «sacrificed» by the intervention. One could conversely claim that, if not separated, Jodie becomes the victim of Mary's «dragging» her to death. Second, since Jodie's death can be prevented, the decision to «let her die» represents as much a choice as the one to separate her from Mary. Third, the idea that a separation in this case amounts to sacrificing one twin for the benefit of the other overlooks the clear differences in the actual conditions of the babies.

The *medical team* finds itself in a very agonizing situation whose features, although similar to those faced by the parents, differ from theirs in many important respects. The moral challenges of the case are for them not only those entailed by the twins' biological condition, but also those stemming from the parents' decision to let them both die. Therefore, they have to look at the technical feasibility of the separation not as if it were takin place in a vacuum, but rather as it is perceived and understood by those who are in charge of the twins' condition as legitimate and valid decision makers.

As a medical team, specializing in the treatment of conjoined twins, the physicians feel that they have the expertise necessary to carry out the operation. Their focus, however, is inevitably different from that of the parents: They look at the medical condition of the twins as one which needs to be optimized by the best overall outcome. Confronted with the dilemma of saving one life or losing both, they rather favor the first option. The difference between the two babies does not carry for them the existential weight it has for the parents: It is a purely clinical difference defined by Jodie's good statistical chances to enjoy a decent life versus Mary's lack of any meaningful hope for survival. Still, the doctors must face the central question concerning the morality of the means they employ to achieve the good end, on the presupposition that they would save both babies if this were possible. The moral judgment concerns in particular the action of clamping the aorta that unites the twins. Although this action will cause the death of Mary, it will also free Jodie of the burdensome condition that would eventually kill her. It is important to take into account the fact that the life of Jodie is saved without compromising the bodily integrity of Mary. Thus, from a factual point of view, the physicians cannot share, and rightly so, in the parents' interpretation of the surgery as a sacrifice of one life for the sake of the other.

There are additional elements defining the intentionality of the physicians. One may expect that, at a medical facility specializing in highly sophisticated surgical interventions, the medical team will be more inclined to pursue the separation. In this sense, their pre-supposition is exactly the opposite of the parents; one. Theirs is, by default, an *interventionist* attitude in which the success of the separation and of further interventions to secure Jodie's medical condition in the future will possibly be regarded with pride as a stunning surgical achievement. However, as already pointed out above, the possible success of the surgery is only one face of the moral challenge confronting the medical team in the case. The other relates to their ability to interact with the parents' firm decision not to pursue the separation. The parents relate to their children and their predicament in a way which is entirely different from the medical team. Whereas the latter see the operation as an intervention to save one life, the former look at it as a form of violence against both their children. The right thing to do is for them not just the result of neutral theoretical insight, but a synthetic judgment defined by an existential predicament of frustration, fear, insecurity, cultural diversity and anxiety toward the future. All these elements define their care toward the twins they equally love.

Are the physicians justified in their pursuing their case against the parents' position all the way to the court? Whereas I happen to share the physicians' desire in this case to save one life rather than lose both, I cannot make sense of their decision to force the operation upon the parents. To start with, one wonders about the quality of the interaction between the parties involved. Could it be that the radically different understanding of the «moral object» of the action is due to a lack of communication? Furthermore, even if differences of opinion concerning the very nature of the intervention remain, the medical team should recognize, in the end, whether the parents have reasons to defend their own position.

As a matter of fact, respect for the parents' autonomy – itself grounded in the particular nature of the parents' relation to their children – would not require physicians to agree with the parent's reasoning entirely; it would require them to recognize the reasonableness of the parents' position even if they disagreed with it. Indeed, however problematic it may appear from the doctors' perspective, the parents' position is not without reason and, thus, not immoral. The ground for their decision to refuse the separation is defined by two factors: a judgment on the proportionality of the treatment and a judgment on the *meaning* of the surgical procedure. Both these factors shape their understanding of the case. One could object that the parents' pre-comprehension is not entirely justified by the «objective» features of the case; yet how could that objectively be gained? Why is looking at the action as an attempt «to save, at least, one baby rather than losing two» *more objective* than seeing it as a sacrifice surgery in which one is killed for the sake of the other?

Conceptual Clarifications

The evaluation of the moral object of the action represents in my opinion the central theoretical problem of the case. More specifically, that evaluation depends entirely upon the implicit understanding of the nature of the killing(s) involved.

One could presume to settle the situation by drawing immediately upon the distinction between «killing» and «letting die» on the presupposition that, whereas the former can be considered always and under ever circumstance immoral, the latter would be permissible only under certain conditions. This seems indeed to be the parents' position which they defend on the ground of their Catholic belief. In their view, the intervention to severe the aorta that unites the babies amounts to killing Mary, whereas to forgo the separation simply means allowing Jodie to die. Moreover, «it is God's will that their children are afflicted as they are and they must be left in God's hand,» or, as the Archbishop of Westminster put it, the parents' refusal to consent to surgery «involves no injustice towards either of them.»

In reality, both the parents and the Archbishop beg the question by relying upon a distinction whose plausibility presupposes the very moral judgment they still need to make for the distinction to apply in the first place. Indeed, Jodie's death can be construed as allowing her to die only if there are moral reasons to justify forgoing the intervention that could save her. Otherwise, her death is as much an act of killing as the presumed killing of Mary. Conversely, killing Mary can be understood as allowing her to die only if there are moral reasons to justify her death. The tautological nature of the recourse to the aforementioned distinction between killing and allowing to die ultimately forces moral reasoning into a vicious circle: To take as a premise of an argument the very conclusion that needs to be demonstrated clearly represents a *petitio principii*. Once the distinction between killing and allowing to die has been introduced to discriminate between different types of action, there still remains the problem of finding moral reasons warranting the applicability of the distinction.

The same problem of a possible semantic fallacy applies to the other distinction invoked in Catholic moral theology to justify the taking of human, the distinction between direct and indirect killing. The advantage of this second distinction, however, is the fact the action causing death is clearly recognized in its *ontic* meaning as an act of killing; yet that act remains open to further *ontological* specification by the introduction of additional elements which, taken together, provide the unequivocal ground for a moral judgment to be made. Mary's death occurring as a result of the separation from Jodie, as well as the latter's death subsequent to the decision to forgo the procedure are both, however paradoxically, the result of acts of killing. With that, I mean two things: first, they physically cause one or the other twin, respectively, to die;

and secondly, they do so even if in the case of Mary the killing appears to be an action, whereas in the case of Jodie an omission. At a purely factual level, however, the two actions are morally the same; in fact they still are, from a moral point of view, indifferent. The moral difference must then be provided by the qualifier direct-indirect. If moral reasoning proves that the act of separating Mary is only an indirect killing, then the moral object defining that action cannot be understood as «sacrificing Mary for the sake of Jodie,» but rather as an attempt to prevent the latter from needlessly following into the same destiny of the former, or, to put it more positively, as «an attempt to save her.»

The moral reasoning just referred to finds in the so called principle of double effect the practical guidelines that would in the end decide whether separating Mary is direct or indirect killing. The conditions for the specification of the principle of double effect to the situation at stake are four. The surgeons' goal, i.e., to save Jodie, is by its very nature, good. Secondly, they intend *only* that goal because they are prevented from saving the life of Mary; Mary's death, which they accidentally cause, is *praeter intentionem*, to use the expression of the manuals, because there would be no possibility of saving her even if they wanted. In that sense, and this is the third condition, Mary's death is not the means to achieve the good end of saving Jodie's life. Even from a physiological point of view, the bodily integrity of Mary is not compromised in the course of the surgical procedure for the sake of Jodie. Finally, there is a proportionate reason, or a balance of good over bad, warranting the action in the first place. In the end, one could say that since the theorem of double effect obtains, Mary's killing is indirect and thus morally justified.

This conclusion is, of course, only a *probable* interpretation of the Catholic tradition, and I mean this in the sense of the moral system known as «probabilism.» The function of moral systems is to clarify a doubt of conscience and thus provide, at least subjectively, a valid reason for a decision to be made. In the end, the morality of an action is defined by the intentionality of the agent. Of course, such an intentionality does not exist in a vacuum, but only as interpretation of an objective reality whose meaning remains flexible. To recognize the central place of subjectivity in ethics does not amount to subjectivism. When seen from the perspective of the agent, moral reasoning appears more clearly for what it is by its very nature, a form of *practical* reasoning. This, in turn, entails a particular way of looking at what represents the «moral object of the action» which is also the object of moral judgment. In so far as such as object represents a change in reality effected by someone's decision, it cannot be equated to a fixed state of affair ready to be grasped in an objectively detached fashion. The moral object of an action is not just an essence-out-there-in-the-world, but the correlate of the subject's engagement with reality. And since the correlation between the agent and the action can best be defined by the notion of intentionality, the moral object of an action already entails a subjective component. Thus, the objectivity pursued by moral reasoning will not be claimed against the subjective elements of the case, but rather, gained reflectively through an interpretation of their very meaning.

The central place of intentionality in determining the morality of an action becomes apparent also in reference to the other argument used by the parents against the operation, that the separation would be a disproportionate burden for *all* the subjects involved in the case. It would be an unacceptable burden for Jodie, for it would impose upon her a series of interminable, or at least not immediately predictable, operations. It would be an impossible burden for the medical system of the country where the family comes from because, that system cannot even technically cope with Jodie's surgical needs in the future. Finally, it would be an unbearable burden for the parents for reasons that should be clear at this point.

Of course, the judgment of the parents concerning the disproportionate nature of the treatment is rooted in objective facts. In this sense, it is not purely subjective. Yet, the parents' judgment represents a plausible inference from the facts because those facts have a certain meaning for the parents. To leave their daughter Jodie back in England so that someone (perhaps the state) will be able to care for her medical needs is for the parents a fact that carries a certain meaning. The same fact has another meaning for the doctors that would like to keep operating upon Jodie.

In conclusion, I think the parents are justified in opposing the operation on their own moral ground. However, to let Jodie die with Mary is the right thing to do not because of the *intrinsic* immorality of the surgical intervention. From a different perspective and under different circumstances, the parents could come to recognize that intervention as an indirect killing. As a matter of fact, it is precisely the predicament of their own perspective and the reasonableness of their own (implicit) argument what allow them to justify the death of Jodie as an indirect killing as well, or as they put it, letting her die. Their position is indeed coherent with the basic tenets of Catholic moral theology. Thus the Church's backing of their decision was commendable. However, is the Church aware of the fact that, in so doing, she was implicitly supporting a more nuanced notion of moral objectivity?

CHAPTER 7
BETWEEN RESISTANCE AND SURRENDER
In Search of an Anthropological Horizon
for an Ethics of Dying Well

It is not easy to add entirely new perspectives to the vast bioethics literature that has emerged, over the years, on the ethics of dying well.[1] As for the U.S. experience, the passing of «aid-in-dying» laws in several states, whether defined by statutes or as a result of popular referenda, simply stokes a fire that was never really extinguished, only kept alive under the embers of previously defined conceptual systematizations by new publicized cases of requests for «assistance» in dying.[2] The latest count finds ten American states with laws on physician assisted suicide, but the story is not over yet, and an extension to the rest of the country seems easy to predict for the near future.[3] What is at stake in the conversation is the problem of articulating the conditions for a *good death* – and more specifically, for a good death when faced with the vulnerability of

---

[1] Tom Beauchamp offers a philosophical articulation of the main bioethical issues. For an example, the distinction between assisted suicide and euthanasia, the problem of causation and the moral relevance of the distinction between killing and letting die, intention and foresight, and the entire question of the principle of double effect. See T. L. Beauchamp (ed.), *Intending Death: The Ethics of Assisted Suicide and Euthanasia*, Prentice Hall, Upper Saddle River, NJ 1995. Recently, with particular reference to the European experience, D. Jones – C. Gastmans – C. MacKellar, *Euthanasia and Assisted Suicide: Lessons from Belgium*, Cambridge University Press, Cambridge 2017.

[2] For a recent overview of the legal landscape, J. Keown, *Legal Issues at the End of Life*, in I. Carrasco De Paula – R. Pegoraro (eds.), *Ageing and Disability*, Pontifical Academy for Life, Rome 2014), 203-213. More broadly on theoretical questions concerning public policy, especially with respect to the legal principle of the inviolability (sanctity) of human life, J. Keown, *The Law and Ethics of Medicine*, Oxford University Press, Oxford 2012, and Idem, *Euthanasia, Ethics and Public Policy,* 2nd ed., Cambridge University Press, Cambridge 2016.

[3] I speak of «physician assisted suicide» rather than «aid in dying» for reasons of semantic clarity. The attempt to fudge the issue by recourse to a more palatable language fails to convey the reality at hand.

old age, terminal disease, and unbearable suffering. Such a task remains quite formidable both in relation to its philosophical foundations, as well as with reference to the analysis of specific ethical quandaries.[4]

The Ethics of Dying Well and the Suspension of Meaning

The discussion about the ethics of dying centers on the resources of a restricted language game, defined by the conditions for a control of death.[5] At the heart of the reflection is the use of medical technology, of the medical power to prolong life. Such power has pushed the limits of our technical possibilities, creating an imbalance between what can and what should be done, thus leading to the question of the quality of life endorsed or maintained by a particular treatment. To use the distinction made famous by philosopher James Rachels, there is a difference between «being alive» and «having a life».[6]

But how to decide on the boundary between the two? If «having a life» depends on a perspective of value regarding what renders life livable or worth living, do we end up in the trap of a subjectivism that dispenses with all criteria? Moreover, how to reconstruct an objectivity that obtains for the medical act itself, beyond two equally false alternatives: either turning medicine into a function of patient's individual preferences, or reducing it to maintenance of purely biological mechanisms. In the former case, the *lex suprema*

---

[4] The work of Marie-Jo Thiel on ageing and disability is especially sensitive to the larger spectrum of anthropological and ethical issues that define the vulnerability of the dying. See, among others, M.-J. Thiel (ed.), *Ethical Challenges of Ageing*, Royal Society of Medicine Press, London 2012. Also in I. Carrasco De Paula – R. Pegoraro (eds.), *Ageing and Disability*, and Idem, *Assisting the Elderly and Palliative Care*, Pontifical Academy for Life, Rome 2015.

[5] For a historical analysis, which highlights the particularity of our contemporary attitude toward death, see the classic works of Ph. Ariés, *Western Attitudes toward Death: From the Middle Ages to the Present*, Johns Hopkins University Press, Baltimore 1974, and *Images of Man and Death*, Harvard University Press, Cambridge, MA 1985.

[6] J. Rachels, *The End of Life: Euthanasia and Morality*, Oxford University Press, New York 1986.

of the medical act is conflated with what the patient wants (*voluntas aegroti*); in the latter, the well-being of the patient (*salus aegroti*) is confused with stubborn insistence on the biological functioning of individual organs. In both cases medicine loses sight of its ultimate end, namely, the *integral* good of the patient.[7]

The debate on the good death – and thus on the normativity of dying – seems paradoxical, in that it unfolds on the premise of a suspension, a bracketing placed on the meaning of death. One often speaks of the ethics of dying, of «dying well,» but without always knowing in relation to what. Bioethics claims to provide normative criteria. It does so, however, on the presupposition of suspending any symbolic horizon capable of saying what death is, what it represents for the person. Of course, one cannot but be pleasantly impressed by the formal elegance and analytical consistency with which the moral principles of a presumed «common morality,» logically shared by all rational agents, are put into play, when faced with the most complicated ethical conflicts.[8] So called «principlism,» with its mantra of beneficence, non-maleficence maleficence, autonomy, and justice, in turn specified by the rules of proportionality, informed consent, etc., might constitute a helpful point of reference when tackling the many conflicts faced by health professionals, patients and their families, in the different clinical settings. Yet, when ethical «principles» and «rules» are employed mechanically, as if in a kind of a priori framework, such elegant bioethical theory looks more like a game without any grasp on re-

---

7   At stake in the discussion is the viability of the foundational principle of Hippocratic medical ethics, i.e., beneficence, in its relation to the patient's autonomy. For a classical and still very valuable articulation of the question, see E.D. Pellegrino – D.C. Thomasma, *For the Patient's Good: The Restoration of Beneficence in Health Care*, Oxford University Press, New York 1988.

8   T.L. Beauchamp – J.F. Childress, *Principles of Biomedical Ethics*, 8th edition, Oxford University Press, New York 2018. The approach in question, known as «principlism,» has been the subject of sharp criticism in the debate over the method of bioethics, both in the United States as well as in Europe. See E.R. Du Bose, R. Hamel, L. O'Connell (eds.), *A Matter of Principles? Ferments in U.S. Bioethics*, Trinity Press International, Valley Forge 1994; H. Ten Have, *Approcci europei all'etica della medicina clinica*, in C. Viafora (ed.), *Comitati etici. Una proposta bioetica per il mondo sanitario*, Gregoriana, Padova 1995, 91-118; G. Khushf (ed.), *Handbook of Bioethics: Taking Stock of the Field from a Philosopcial Perspective*, Kluwer Academic Publishers, Dordrecht 2004.

ality. In this version of ethics, «substantive» rationality gives way to «formal» rationality and reflection on ethical content dissolves in sheer proceduralism.[9]

How could such an approach provide recommendations for a truly good and dignified death? Could it be that ethical formalism betrays, in the end, a lack of any points of reference, if not resentment, as Nietzsche might suggest, when faced with the void of sense? Morality, in this case the morality of a good death, would only be smoke and mirrors, a nihilistic enchantment.

I ask: Can we leave unaddressed the existential aspects of death and dying considered in their experiential value, that is, as dimensions of our journeying (*experior*) toward the end? Likewise, can we pass over in silence those dimensions that speak to the trial (*peiros*) entailed by the agony of passing, the physical pain, the loneliness of suffering?

A *good death* can only be envisaged as the fulfillment of a *good life*, and this with reference to a life that will inevitably age. What I mean is that death can be considered good when it succeeds in expressing the meaning of living, understood as living well. That living well must end, at times even tragically, and why – all these are questions that do not belong to the discipline of ethics per se. Yet ethics cannot even begin to reflect on its proper criteria if not because it lets itself be challenged by the existential perplexity such questions entail. I believe such questions have a metaphysical quality to them: they interrogate our attitude toward being as such and the meaning of things.[10]

---

9 Sociologist John Evans has criticized bioethics, especially in the American version, for being a purely *formal* endeavor, autonomous with respect to premises defined by substantive moral content. See J.H. Evans, *Playing God? Human Genetic Engineering and the Rationalization of Public Bioethical Debate*, Chicago University Press, Chicago, IL 2002. The analysis of Evans, which turns on Max Weber's distinction mentioned above between «formal» and «substantive» rationality, refers to the discussion on genetics and genetic research. Nevertheless, in its basic meaning, it obtains for the entire field of bioethics, as it has developed in the United States. For a more generous account of the history of bioethics as public discourse, see A.R. Jonsen, *The Birth of Bioethics*, Oxford University Press, New York 1998, especially 352-376.

10 Dietrich von Hildebrand, commenting on Pascal (*Pensées*, VI, frag. 347), writes: «And so (Pascal) alludes to in a singular manner the contradictory

I find the distinction between *conatus* and *passio* helpful in this context.[11] One might identify in it something like a heuristic path toward the articulation of two different paradigms for an ethics of dying, whose basic features point in the direction of alternative existential modalities. The retrieval of deeper human postures in facing death belong to a phenomenological moment that precedes normative discourse. To an extent, it grounds it, for it brings into relief those very premises about the meaning of death which contemporary ethical discussion tends to bracket.

Resisting Death and the Effort to Be: Conatus Essendi

Conatus essendi is a way of standing before things defined by the endeavor, the effort to be. We do not choose such a posture. We are already endowed with it by virtue of our relation to the world, though such posture may take up a certain primacy, given our emphasis on doing or acting. The *conatus* is the defining posture of modernity, in whose larger narrative the «effort to be» makes the appearing of things conditional upon a subjectivity that posits and determines. Being *is* insofar as it responds to the (transcendental) forms of its apperception by a subject that measures and rules any phenomenic presence.[12]

Such posture has obvious epistemological importance, which Descartes and Kant will fully unpack. Given the ambiguity that

---

nature of the metaphysical situation of mankind, in part due to the fallenness of life, yet also to the ineffable survival of his personal condition after death.» D. von Hildebrand, Über den Tod, EOS Verlag, St. Ottilien 1980, 33.

11 The distinction is central in thought of Leuven Philosopher William Desmond. See W. Desmond, *The Intimate Universal: The Hidden Porosity Among Religion, Art, Philosophy, and Politics*, Columbia University Press, New York 2016. Also W. Desmond, *Pluralism, Truthfulness, and the Patience of Being*, in C. Taylor – R. Dell'Oro (eds.), *Health and Human Flourishing: Religion, Medicine, and Moral Anthropology*, Georgetown University Press, Washington, D.C. 2006, 53-68. The distinction between *conatus* and *passio* is one of ontological order, before it applies to ethics, and can be properly understood in light of Desmond's complete work. For a study of Desmond's thought, see T. Kelly (ed.), *Between System and Poetics: William Desmond and Philosophy after Dialectic*, Ashgate, Burlington 2007.
12 W. Desmond, *Ethics and the Between*, SUNY, Albany 2001, 17-47.

marks the appearing of things – thus the doubt about them, the *cogito* tries to recover an irrefutable certainty, starting no longer from the promise of meaning that inhabits reality, but from the subjective certainty that defines the very act of thinking. Such act must necessarily presuppose – and beyond all doubts no less – the existence of the *cogito* who thinks. In this way, however, one sees a shift, a «Copernican revolution» as Kant will define it, in the relation of subject and object, conditioning the way in which one understands the priority of the former over the latter.

This «anthropological turning point» entails important cultural consequences, and frames our way of thinking about nature. We no longer «undergo» nature, so to speak, but actively shape it according to heuristic models, which, reducing nature's complexity to mathematical univocity, enables us to describe and empirically verify it. For sure, the «book of nature» has much to say still, but will do so because the «spectator scientist» now sets the conditions to prevent its hiding, thus forcing nature to yield its secrets, as Galileo suggests in *Il Saggiatore*.[13]

All this presupposes a neutralization of being to *mathesis universalis*. One thinks of the meaning of such neutralization with respect to the distinction between primary and secondary qualities, or the Cartesian reduction of the human body to *res extensa*, now become a mechanism separate from the mind. If true reality can only be rendered in mathematical terms, then it is imperative to bring the unverifiable *pathos* of things back to the dianoetic precision of scientific *formulae*. This holds true also for the subject, whose emotional complexity will have to be reduced, now, to the act of «thinking clearly and distinctly.» In the words of Spinoza: «*non ridere, non lugere, neque detestari, sed intelligere.*»[14]

---

13 «Nature loves to hide» said Heraclitus. See J. Barnes, *Early Greek Philosophy*, Penguin Books, London 1987, 122. But the modern gaze is more akin to an act of unveiling. The forcing of nature also signals the end of teleology: «*Naturam finem nullum sibi praefixum habere, et omnes causas finales nihil nisi humana esse figment*» («Nature has no fixed goal and all final causes are but figments of the human imagination,» B. Spinoza, *Ethics*, trans. Samuel Shirley, Hackett Publishing Company, Indianapolis, IN 1992, 59.

14 B. Spinoza, *Theological-Political Treatise*, 2nd ed., trans. Samuel Shirley, Hackett Publishing Company, Indianapolis, IN 2001, 1,4. Nietzsche interprets somewhat sarcastically the spirit of Spinoza's quotation: «What does knowing mean? *Non ridere, non lugere, neque detestari, sed inteligere!* Says

So much for the epistemological significance of the *conatus*, whose implications, however, extend to ethics, and, more specifically, to the ethics of dying. I would say the following: the subject who fashions reality also *grants value* to it. This is so because the neutralization of being, with respect to the object, entails something like a rebound effect, a kind of «contraction of value» – especially with Kant – in favor of the subject. Only the person possesses an intrinsic value: as a good in-itself, she is never to be treated as a means, only as an end.

Unlike nature, understood now as phenomenal field open to endless manipulation, the person is not neutral; rather, being the source of absolute meaning, it becomes the condition for the very possibility of meaning's attribution. But, as Desmond points out:

> Here is the sting. The subject cannot live with this devaluation of otherness, and even less with the devaluing of its own valuing. It will not be passive to this. It will be active. The subjectification of value inevitably leads to the primacy of self-activity that impresses itself on the other... We witness the recoil of the subject on itself out of the hiding of neutrality it had schemed for itself. There is no escape from itself, but now when it awakens again to itself, it has been transformed into a more *radically self-assertive subjectivity*.[15]

In this paradigm, the good death is the *humanized* death, death lived as chosen, not as undergone or endured. «Choosing death» is to determine it, the way our choice determines the theoretical models that grant access to reality as such. To be «the measure of all things» is to be greater than death. Thus, the latter will be neutralized, if not in its inevitability of fact that inexorably happens, at least, in its dramatic quality of experience to be resisted.

This is also the final outcome of Heidegger's reflection in *Being and Time*, possibly the most subtle reflection on the significance of death as existential possibility for *Dasein*, and a measure of the latter's call to authenticity. For Heidegger, death must be recognized

---

Spinoza, so simply and sublimely, as is his wont. Nevertheless, what else is this *intelligere* ultimately, but just the form in which the three other things become perceptible to us all at once?» *The Gay Science*, Book IV, 333 («What does Knowing Mean?»), 261.

15 W. Desmond, *Ethics and the Between*, 29 (emphasis in the original).

in its inevitability of fact that occurs; on the other hand, there is in him a tendency to overcome death «trans-subjectively»:

> For Heidegger's *Being and Time*, death as the end of earthly existence no longer constitutes a philosophical term in and of itself, rather the ultimate attitude toward an inevitable fact. In this sense, death cannot be completely absorbed by subjectivity. It is precisely in death that the limits of existence and human action become clear. Nonetheless, to the extent in which death is no longer viewed as the «ultimate enemy» of human life [...], rather, it is interpreted as the highest point of human maturity and perfection, it becomes necessary to understand the facticity of death «trans-subjectively».[16]

The effort to be, the *conatus essendi*, is a struggle against death, the attempt to indefinitely postpone it, or else, to anticipate it «rationally,» as in the case of euthanasia or assisted suicide. Absolute passivity is not worthy of man.

This paradigm is not without important emphases. The efforts to humanize death, above all through the contributions of medical and scientific research that aim at the treatment and management of pain, are essential part of our modern way of relating to death. An inhumane and dehumanizing death cannot be good.

Yet, this paradigm contains also the seeds of a possible degeneration. It risks thinking of a «good death» according to the logic of scientific-technological control and neutralizing planning, which, in the paradigm of modernity, renders a life worth living.

But the truth is another: we can prolong life, eliminating from it all pain and suffering, but will never succeed in «managing death.» Death will always come, an unexpected surprise and an expression of the heteronomy of nature, even more striking now, because it seems to contradict the autonomy by which we attempt to completely define ourselves.

---

16 L. Scheffczyk, *Die Phänomenologie des Todes bei Dietrich von Hildebrand und die neuere Eschatologie*, in *Truth and Value: The Philosophy of Dietrich von Hildebrand*, Peter Lang, Bern 1992, 271. Heidegger's effectual history, however, has been important with respect to theological reinterpretations. For an example, K. Rahner, *On the Theology of Death*, trans. C.H. Henksey, Herder and Herder, New York 1961, and L. Boros, *The Moment of Truth: Mysterium Mortis*, trans. Gregory Bainbridge, Burns and Oats, London 1969.

The separation, the dualism of person and nature, constitutes the condition of possibility for controlling death, but it can also lead to conflicting results: a technological effort that de-personalizes nature; or a will to power that de-naturalizes the person, reducing it to self-determining rationality. The *epoche'* on any search for the meaning of death is the inevitable result of both these developments, stemming from the same root.

Surrender to Death and Letting Go: Passio Essendi

I ask: what if something else companioned, more originally (*co-natus*), the vector of intentionality that drives our own effort and strive for control?[17] What if a more radical openness, perhaps even intimacy, to reality sustained our ταυμάζειν, our astonishment at the fact that being *is*, when it could also *not* be?

In wonder, we take up our residence in the between, attuned to the saturation of meaning that dwells in things, in their value, and in whose hospitality we build a world for ourselves. The *conatus essendi* can only be a derivation, of course possible and legitimate, of a more original *passio essendi*, of an undergoing (*passio*) that also becomes a «passion» for being.

With respect to the previous paradigm, the *passio essendi* bears with it the recognition that we are not the origin of meaning. Only because originated, can we attribute meaning to things, and do so on the condition of a previous attunement (Heidegger's *Stimmung*) to the promise of meaning that already inhabits things. In this paradigm, there is no separation between being and value, fact and meaning, for being is, intrinsically, promising and valid, good and beautiful. Of course, we produce and make, search and fashion, yet do so without bracketing the charged sensuousness inhabiting the world in which we dwell. Our activity perfects nature, acknowledging in being a reserve of meaning to make our own and bring to fulfillment.

---

17   I say «more originally» because the very etymology of *co-natus* presupposes a communion, an intimacy (*cum*) with being that already companions our effort to be. My point, following Desmond, is that such «porosity» to being, what I call *passio*, precedes the *conatus*, keeping it true to its deeper ontological constitution.

The receptivity at work in this is clearly not a form of passivity either, for it is to a consciousness and to its active intentionality that the meaning of things discloses itself. And yet, the activity of consciousness rests on the inexhaustible mystery of things (with a bow to Gabriel Marcel), on their endless and never to be reduced profundity, which makes itself known, because it opens itself up, because it reveals itself.[18]

In this perspective, our relation to being is a relation of trust, rather than doubt, of promising proximity, rather than distancing suspicion. With respect to freedom, we come to recognize that its task is indeed to do and build, to fashion the world, but only because, prior to this, the world was «let be.» Thus, the task of freedom is essentially «responsorial,» in fact, a responsibility, beyond the autonomy that finds fulfilment in will to power, seized at another's expense: «I am summoned as someone irreplaceable. I exist through the other and for the other, but without this being an alienation.»[19] Rather, it is a freedom that lets things be in the generosity of love and giving.[20]

---

18 The turn to *givenness* is, of course, central to phenomenologically inspired thinking. This goes beyond differences among phenomenological schools, whether «realistic» or «transcendentally» defined. For an example of the former, see the beautiful book of E. Kohak, *The Embers and the Stars: A Philosophical Inquiry into the Moral Sense of Nature*, University of Chicago Press, Chicago and London 1984. Marion's position is an attempt to overcome the dichotomy in question by reversing the transcendental position itself, and affirming the excess of intuition over intentionality: before coming to itself as self-determining, the subject is already called into reciprocity, already «appealed to.» See Jean Luc Marion's essay on *The Saturated Phenomenon*, in *The Visible and the Revealed*, trans. C.M. Gschwandtner and others, Fordham University Press, New York 2008, 18-48.
19 E. Levinas, *Otherwise Than Being Or Beyond Essence*, trans. Alphonso Lingis, Pittsburgh University Press, Pittsburgh, PA 1998, 114.
20 This inevitably entails a de-mystification of the modern ideal of autonomy, a recognition that, in the long run, Kantian autonomy degenerates into «will to power,» as in Nietzsche. On this reading, see already R. Guardini, *Das Ende der Neuzeit. Ein Versuch zur Orientierung*, Echter Verlag, Würzburg 1951, and H. De Lubac, *Le drame de l'humanisme athée*, Éditions Spes, Paris 1945. For a different, more positive interpretation of modernity, with respect to the ideal of autonomy and its possible Christian reinterpretation, see J. Schwartländer (ed.), *Modernes Freiheitsethos und christlicher Glaube*, Kaiser Verlag, München 1981, especially the contributions of Schwartländer, Honecker, Kasper and Böckle.

The relation to death unfolds within this context, and according to the same logic. The humanization of death will be possible on the condition that death be accepted, not suppressed or censured. Death is, after all, part of the human experience, an event whose significance cannot be anticipated; it is a disclosure, in fact, a total revelation of meaning, both promising and significant. From a Christian point of view, this acceptance concerns both suffering and death, and yet not in the sense of a masochistic passivity. Klaus Demmer writes:

> In the end, the Christian faith is anything but an ideology of suffering. Even for the Christian, suffering does not possess value in and of itself, and therefore it is never sought for its own sake. Rather, one accepts it, almost as an anticipation of death, which, too, must be accepted.[21]

As Eberhard Jüngel suggests, the «passivity» implied by death is, therefore, an expression of the more general receptivity of life:

> There is a passivity without which man could not be man. Part of the reason for this is the fact we were born, that we were given-birth-to. Here there follows the fact that we are loved. So, too, is the fact that we die.[22]

The point is also made by Levinas, deserving of broader exegetical attention and interpretive articulation, well beyond the limits of a single quotation. Still, here is one, as a *donne à penser:*

> ...The subjectivity of the subject, its very psyche, (is) a possibility of inspiration. It is the possibility of being the author of what has been breathed in unbeknownst to me, of having received, one knows not from where, that of which I am the author. In the responsibility for the other we are at the heart of the ambiguity of inspiration.[23]

---

21 K. Demmer, *Leben in Menschenhand: Grundlagen des bioethischen Gesprächs*, Herder Verlag, Freiburg 1987, 146. On the topic of suffering, see the reflections of M. Scheler, *The Meaning of Suffering*, in *Max Scheler (1874-1928): Centennial Essays,* ed. M.S. Frings, Martinus Nijhof, The Hague 1974, 121-163. From a theological perspective, see D. Sölle, *Suffering,* Fortress Press, Philadelphia 1975.
22 E. Jüngel, *Death: The Riddle and the Mystery,* Westminster Press, Philadelphia 1975, 85.
23 E. Levinas, *Otherwise Than Being Or Beyond Essence*, 148-149. «Inspiration» is existing «through the other and for the other, but without this being alienation,» *Ibidem*, 114-115.

The acceptance of death is still bound to an act of preparation on our part, one that opens up for us a space of creativity. We all die, yet we face death *differently*. In the same way in which life requires its own special art, accomplished daily in the cultivation of virtues, so, too, does death require a kind of art, the *ars moriendi*. Death is a threshold toward which we journey together, as if in pilgrimage, comforted by prayer.[24]

In a Christian framework, death is, at bottom, an eschatological event, one which belongs to the personal narrative of each and every human being. Death also points to a trans-historical fulfillment, to definitive communion with God beyond the limits of history: «Birth and death are thresholds and transitions, and as the radical transition of birth is creation, the radical transition of death may not be nothing, but resurrection.»[25]

Some Ethical Implications

My reflections, at this point, would have to become more attentive to the particularity of cases and situations, thus letting anthropological considerations merge more smoothly into the ethical. The passage from the anthropological to the ethical is not without difficulties, the logical pitfall implicated by the so-called «naturalistic fallacy» being only one of the potential missteps.

A more daunting task consists, in my opinion, in the difficulty to articulate a nimble casuistry, which takes into account nuances of contexts and diversity of clinical situations. Though a universal experience, death reserves for each of us a more intimate invitation. To prepare for the ultimate journey, we must face *the fact* that we die alone,[26] even when surrounded by others, given over, in the most radical way, to the mystery of our own singularity.[27]

---

24 On prayer as a dimension of the *ars moriendi*, see W. Reich, *L'arte del prendersi cura del morente*, in *Itinerarium* 4 (1996) 31-43.
25 W. Desmond, *The Intimate Universal*, 363.
26 Is this something analogous to a Kantian *Faktum der Vernuft?* In which case, the facticity of death would be the ultimate test confronting our freedom, now, no longer a freedom of choice, only a freedom to let go.
27 There is also an *ethical* singularity thrown into relief by the *ontological* singularity in question. On this, already, K. Rahner, *On the Question of a For-*

For example, the condition of patients who lost their autonomy to cognitive disability will impose ethical challenges that are different from those of patients who never possessed such discretionary autonomy. Likewise, we will have to distinguish criteria defined by substituted judgment standards from those based on rules of beneficence or non-maleficence, as in the case of best interest assessments by surrogate decision makers.

Such a detailed casuistry will have to find more adequate treatment elsewhere.[28] What is relevant, in this context, is the realization that the ethical quality of one's death depends on embracing, rather than rejecting, the inevitable passivity entailed by life's ontological condition. Such condition is not chosen, but given, all the accomplishments of our own making notwithstanding.

Consider the anticipation of treatment decisions in advance directives. In the materiality of the «letter», one such document may betray a different attitude of «spirit»: the acceptance of death, in the logic of the *passio essendi*, or the other, more desperate effort, indeed the *conatus*, to deny it. In articulating our personal preferences for this or that treatment, we might only exorcise our fears, reassuring ourselves that in managing its terminal phases, we will grant the mute alterity of death a face we can at least recognize. Let there be silence after that: if something more is to be had, it will be according to our own measure, beyond all feelings of dependence, and without surrendering to any kind of heteronomy, such as the one imposed by a treatment we do not want.[29]

I am not rejecting the importance of advance directives, only the potential abstractness that might accompany their drafting, when such *preparatio mortis* bespeaks an exercise of freedom that stub-

---

    mal Existential Ethics, in *Theological Investigations*, vol. II, Helicon Press, Baltimore, MD 1961, 217-234.

28   One might consult for this the interesting *Report* produced by the U.S. President's Council on Bioethics, *Taking Care: Ethical Caregiving in our Aging Society*. The *Report* can be retrieved at https://bioethicsarchive.georgetown.edu/pcbe/reports/taking_care/index.hml

29   The literature on advance directives is quite extensive. A first introduction to the topic in P. McCarrick (ed.), *Living Wills and Durable Power of Attorney: Advance Directive Legislation and Issues,* in *Scope Note 2*, National Reference Center for Bioethics Literature, Georgetown University, Washington, DC 1992.

bornly decides and plans. Here one finds a *resistance* that is still unable to *surrender*.[30] Still, death will come to us in a future that is *ad-ventus*, both indeterminate and indeterminable.

In trusting abandonment, death must be let happen, for we cannot escape it. In this light, the attempt to control death can become a paradox, especially when we fail to see how the scrupulous articulation of our directives, for an example, in relation to treatment decisions, points more to the radical alterity of death, than to our presumed capacity to domesticate it. Of course, we can clearly state what we want, even with a document that possesses the power of binding others to our wishes; yet, it will always be *others* who are charged with the task of respecting our desires and carrying them out: *their* decision will be, eventually, beyond *our* control.

I think of this paradox with regard to the already mentioned «aid in dying» normative provisions. The request for assisted suicide, and even more for euthanasia, will be heralded as an instance of self-determination, and in the name of «death with dignity.» And yet, while asserting their own autonomy for the last time, patients who choose to die must still abandon themselves to someone else, who, providing the lethal cocktail or the lethal injection, does for them what they can no longer do on their own.[31]

It is clear that, if the language of dying – even in the ethically exemplary case of advance directives, is articulated by the patient against the backdrop of an *epoche'* on the existential meaning of

---

30 *Wiederstand* and *Aufhebung* are two modes of being analogous to what I called *conatus* and *passio*. With reference to the distinction of Dietrich Bonhöffer, see P. Cattorini, *Tra resistenza e accettazione*; *indicazioni etiche per superare accanimento vitalistico ed euthanasia*, in P. Benciolini – C. Viafora (ed.), *Etica e cure palliative. La fase terminale*, CIC, Roma 1998, 77-87. For a physician's narrative of care of patients at the end of life, specifically from the perspective of palliative care, see I. Byock, *The Best Care Possible: A Physician's Quest to Transform Care Through the End of Life*, Avery, New York 2012.

31 Michael Maret defines euthanasia as «the paradoxical figure of autonomy.» See M. Maret, *L'euthanasie. Alternative sociale et enjeux pour l'ethique chrétienne*, Editions Saint-Augustin, Paris 2000, 71-100. For an articulation, both anthropological and theological, of the dialectic of activity and passivity in dying see K. Demmer, *Handeln als Einüben des Sterbens. Ein Kapital theologischer Anthropologie*, in A. Holderegger (ed.), *Das medizinisch assistierte Sterben. Zur Sterbehilfe aus medizinischer, ethischer, juristicher und theologischer Sicht*, Herder Verlag, Freiburg 1999, 175-191.

death, the doctor and the healthcare personnel, in turn, cannot but share this «conspiracy of silence,» in which that which cannot be spoken about, will inevitably be passed over in silence.[32]

Of course, doctors will have their own reasons for resisting talking about death with their patients, and will express their will to power in their very unique way. For an example, by hiding themselves behind the technological imperative to fight death until the end, by opting for an aggressive treatment that has become futile, or by stirring up for patients and their families an impossible hope of recovery, which is but a mask of fear-the mask of the doctors, no less![33] The bracketing of the reality of death, as well as the privatization of the criteria for dying, renders any solidarity with the dying impossible: the last word in this predicament can only be the loneliness of the dying person.

---

[32] J. Katz, *The Silent World of Doctor and Patient*, Johns Hopkins University Press, Baltimore, 2002.

[33] For an empirical analysis of the relation between doctor and patient with respect to death, see the instructive study of K.K. Curtis – M.G. McGee, *An Overview of Physician Attitudes toward Death and Dying: History, Factors, and Implications for Medical Education*, in «Illness, Crisis, and Loss» 8 (2000) 341-349.

CHAPTER 8
CONFRONTING THE DOBBS DECISION
On the *Legality* of Abortion

Introduction

The purpose of this chapter is to addresses a number of questions, at the boundary of morality and law: what moral premises have informed the American legal discussion on abortion? What premises ought to do so?

I speak of *moral* rather than legal *premises*. Though it is necessary to distinguish morality and law, it is also impossible to separate them. I stand by the presupposition that legal developments do reflect maturations (and sometimes declines) in the ethos of society, in the perceptions of right and wrong, good and bad that substantiate our living together.

Moral premises about abortion then, whether supporting a Constitutional «right to choose» (as in *Roe* and *Casey*) or, conversely, denying such a right (as in *Dobbs*), stand in the background of the norms articulated by the law. They define recessive premises, informed with philosophical and ethical presuppositions that might be less evident in their meaning than the legal provisions they overtly convey. Still, we cannot have a full picture of the legal landscape without taking into account such moral premises. The *letter* of the law becomes fully clear only in light of the *spirit* that pervades it.

I offer three considerations. The first concerns the *Dobbs* decision. The second looks back at the framework overridden by *Dobbs*. The third outlines a personal position on the legal matter of abortion in our country.

The Decision

The first consideration focuses on the moral premises of the *Dobbs* decision.[1] I find the decision morally inadequate for the following reason: *Dobbs* fails to provide a full account of the moral agency of women as autonomous beings, endowed with full equality and the freedom to make a substantial choice over, perhaps, their most personal and consequential of all life decisions.

It is true that the failure in question concerns *per se* only the recognition of a *constitutional* right to abortion. It does not deny the *freedom of the states* to enshrine such right in their laws – through statutes, legislations, or referenda.

However, it also leaves the door open for the states to do the opposite. In the potential conflict between a woman's claim to autonomy and a state's right to determine the future of her pregnancy, the *Dobbs* decision sides with the latter over the former, rejecting any space of «personal liberty» for women, even in cases of rape or incest.

Why so? I find it interesting that the argument for the position adopted by the Supreme Court in *Dobbs* is not grounded in the concern to protect prenatal life. Nor it is based on the affirmation that the right to life of a developing human being overrides a woman's right to choose.

The state's interest in protecting prenatal life plays no part in the majority's analyses. In fact, the majority takes pride in not expressing a view «about the status of the fetus.»

Their argument relies purely on the following question: does the woman's decision to end a pregnancy involves any *Fourteenth Amendment* «liberty interest» or not? Their answer is that it does not.

The legal reasoning justifying such denial of a woman's liberty interest is not my immediate concern, at this point of our conversation. I leave it to legal scholars to decide whether and why *Dobbs* sides with a so-called «originalist» interpretation of the Constitution, or whether its decision to abandon *stare decisis* as a principle central to the rule of law is justified.

---

1   Supreme Court of the United States, *Dobbs v. Jackson Women's Health Organization* (2022).

I want to call attention to something deeper, and that is the apparent decision's blindness to developments in our society that have contributed to the affirmation of women's moral agency.

According to *Dobbs*, a woman's decision to end a pregnancy does not involve any *Fourteenth Amendment* «liberty interest,» for the simple fact that *the law did not intend to do so when the amendment in question became law*. In other words, since the *liberty* in question *did not extend to the protection of a woman's choice in 1868*, it cannot serve as a basis for its later justification either.

I am quite puzzled by such argument. As the dissenting opinion points out, there are several other things the law did not protect, back then. In fact, it failed to do so for decades to come. It did not protect the right to same-sex intimacy and marriage. It did not protect the right to marry across racial lines. It did not protect the right to contraceptive use or the right not to be sterilized without consent.

Dobbs fails to recognize, then, that our social and cultural perceptions of women's autonomy, in all its dimensions, including the power of control over their bodies, have dramatically changed since 1868.

I do not see such changes in perception as a sliding into some kind of moral relativism. I also do not subscribe to the notion that, then, «everything is up for grabs,» unless we hold firm to an «a-historical» interpretation of the *Fourteenth amendment*.

The changes I am referring to are the results of social struggles by women to affirm their place within our American society. I see them as contributing to a maturation in our moral sensibility, in finally coming to recognize women as full moral agents.

Such moral agency gives women the freedom to decide whether and when to have children. It determines how they live their lives and how they contribute to the society around them.

Furthermore, since full moral agency defines the status of citizens within a secular, democratic polity, the failure to recognize dimensions that are essential to women's freedom risks curtailing the requirements of full democratic participation for more than half of our society.

To impose a choice on women over matters that belong to their most intimate sphere threatens to compromise their integrity, bodily and otherwise, as persons.

It also undermines basic requirements of tolerance toward the pluralism of moral perspectives within society. *In matters of personal life*, a democracy differs from a totalitarian regime because it maximizes, rather than restricting, a space of personal freedom for *all citizens*, including women.

Jurisprudential Precedents to the Dobbs Decision

I come to my second consideration. I want to unpack now the underlying premises of the framework overridden by *Dobbs*, and submit that the latitude of a woman's right to choose in *Roe* and *Casey*, extending all the way to *viability*, goes too far.

My argument relies upon two premises: a notion of the human being defined by a *relational, rather than absolute*, autonomy; and a *positive* rather than negative understanding of the purpose of the law.

I stand by a notion of the human being – a *philosophical anthropology*, to use a technical term, in which freedom of choice is not absolute. We are embodied beings, not isolated monads. Since we are dependent upon one another, we are also responsible for each other's vulnerability, which we share in our common human condition.

I believe society has a responsibility to support and care for both a woman facing pregnancy and the developing human being in her. We need to overcome an anthropology for which «each person is an island,» and a notion of society that is simply the sum total of individuals living side by side, caring for their own interests only, trying not to stamp on each other's feet.

What I am saying here does not contradict at all what I articulated in my first consideration. A right to a fully free decision toward one's pregnancy stands within a *wider matrix of additional rights* the law ought to defend. They include the right to prenatal care, work leaves, family financial support, and a slew of social goods necessary to the flourishing of women (and children) as human beings.

The agency of women, which I fully support, can find the conditions of its practical sustainability only in the relativity and support of others.

In light of this, the anthropology that underlies the framework of *Roe* and *Casey* seems defective to me. It rightly recognizes the centrality of women's autonomy in matters central to their existence. At the same time, it fails to pay heed to the *responsibility generated by the relation of mother and child*. If at all, it does so too late.

For *Roe* and *Casey*, such responsibility, and the subsequent interest of the state, sets in *only at the point of viability*, when, presumably, the developing human being is *able to survive on its own*, becoming now a self-sufficient being, potentially capable of autonomy and self-determination.

With respect to the *unborn life before viability*, such notion has little to offer: the unborn is not autonomous or self-determining. Thus – so goes the argument, if the fetus is not autonomous, abortion harms no one. If autonomous, then the harm cannot be justified.

As I said before, at stake here is also a way of thinking about *the purpose of the law*. We tend to see the basic *purpose of the law as essentially negative*. In this view, the «business» of the law is strictly to set boundaries and impose limits. It restrains individuals from harming other self-standing, autonomous individuals seen in isolation.

For my part, I see the law as transcending a purely negative purpose. *The law has also a positive function.* It can help foster a sense of care for the bond of reciprocity between human beings, including the bond of mother and child, and the bond a pregnant woman shares with the community to which she belongs.

In America, it is this positive understanding of the law what has shifted the way we look at issues like civil rights, medical leave, disability, etc. I indulge, for the sake of argument, on the latter.[2] All the way up to mid-1970s, many American cities, including Chicago, Columbus, and Omaha, prohibited persons with disability to have a normal public presence. The Chicago Municipal Code of the time had a provision, known as «ugly law,» which required the following:

> No person who is diseased, maimed, mutilated, or in any way deformed, so as to be an unsightly or disgusting object or improper person, is to be allowed in or on the public ways or other public places in this city, or shall therein or thereon expose himself to public view, under a penalty of not less than one dollar nor more than fifty dollars for each offence.

---

2   For what follows, I borrow from the excellent volume of Cathleen Kaveny, *Law's Virtues: Fostering Autonomy and Solidarity in American Society* (Washington, D.C.: Georgetown University Press, 2012)

The historical discrimination against persons with disability may appear hardly believable, when not simply dumbfounding, to our contemporary sensibility. Yet it puts us in the condition to appreciate the cultural maturation we have been able to achieve when we introduced, in 1990, a new legal framework, the *American with Disabilities Act* (ADA).

Such framework and its measures have had a profound impact on our social sensibility toward the disabled. They have produced changes not only in our personal attitudes toward them, but also in the legal recognition of their individual rights. In the *American with Disabilities Act*, the law goes beyond the task of setting boundaries among individuals, carving out a space of isolation for those who cannot claim full autonomy and self-determination.

It *positively* calls for the integration of the disabled. Furthermore, it engages citizens to *come to terms with the reality (I am tempted to say, with the «visibility») of persons with disability*, to recognize their potential contribution, to assure equality of opportunity and full participation within society.

This model of the law offers a different understanding of freedom because it sees autonomy as a dimension that is integral, rather than alternative, to social solidarity. It also shows the potential effect of a different anthropological ideal, which recognizes the embodied condition of the person, her vulnerability and dependence within social and historical ties.

Abortion and the Law: A Proposal

I have come to my final point, in which I try to bring together the considerations developed in my previous two sections. I offer suggestions for what I consider a potential compromise on the *legality* of abortion, a compromise in which women's autonomy and social solidarity toward prenatal life might come to mutual recognition.

I said before that an anthropology of dependence and vulnerability ought to provide the context within which to frame the *extension of autonomy*. The nature of freedom, for women as for men, is not reducible to a «freedom from,» but only grows in relativity to the other, in becoming a «freedom to.» It is a freedom that finds limits, but also meaning, in serving the «face» of the other. As the Jewish philosopher Emmanuel Levinas reminds us, we are truly free only when we become responsible.[3]

Freedom is mysterious. It gives us the choice to exploit the vulnerability of the other who is dependent on us, and turn such dependence into an opportunity for the exercise of power. In this case, freedom becomes an opportunity for exploitation. But freedom can also give us the courage to pay heed to the vulnerability of the other, allowing into full visibility the injunction inscribed in the other's face, the command that says, «do not kill me!» Even when it does not speak, even when it is incapable of speaking, the face of a human being summons to respect and reverence. It does so especially when it suffers and grimaces in pain. This is for me the threshold beyond which a freedom to choose cannot legally go. This threshold we cross substantially earlier than viability.

I grant that the life of a developing human being does not exist in a vacuum: it is embodied in the flesh of a woman. In this sense, it is «entrusted» to her choice. The latter however, extends only so far. I submit that the ability of the fetus to feel pain represents the limit at which the legal right to choose ought to end.

It is not entirely easy to come to scientific consensus around the moment such emergence of pain occurs in fetal development. Evidence seems to point to somewhere between 15 and 20 weeks, but doubt with respect to extinguishing human life imposes prudential restraint, rather than uncertain daring.

This is why the law should limit a woman's freedom to choose at the point when the possibility of inflicting pain upon a developing human being sets in. In fact, for prudential reasons, it should do so before that possibility becomes actual reality. Beyond that point, I believe abortion represents a form of cruel intervention on the fetus and our sense of communal belonging falters. This we owe, minimally, to the vulnerability and dependence of a developing human being. For most of European countries, the space of freedom to choose is limited, with certain exceptions, to the first trimester of

---

3   The point, with respect to the metaphysical meaning of the *face*, in Emanuel Levinas, *Totality and Infinity: An Essay on Exteriority*, trans. Alphonso Lingis (Pittsburgh, PA: Duquesne University Press, 1969)

pregnancy. I stand by this threshold for our state laws as well, on the premise that such space is sufficient for a woman to make an informed choice, as it should be for all medical decisions.

Conclusion

I would hope that the change of heart entailed by hospitality to the disabled, of which I spoke before, might bring about a legal landscape for the unborn that is consistent with the compromise I articulated. It is a *minimum morale*, but it might be a starting point toward a more robust «culture of life.»

The decision toward a developing human life belongs in principle to the woman who bears it, to her responsibility and choice. This, as I said, within limits.

From a moral point of view, and for reasons that belong to a different essay, I would hope for that decision to be *in favor of life* and to honor the mystery of a new human being *emerging* in a woman's body. The reason for this new human life «being there,» a body in the flesh of another, transcends the biological mechanisms that brought it into existence.

The *moral response to life* cannot be, in my opinion, anything less that *life-affirming*. One can easily recognize that «being alive,» for each one of us, owes first to our mother accepting *our* emergence in *her* flesh.

I do not assume that for *all* women such acceptance was entirely free. There are forms of constriction upon a woman more subtle, and perhaps more violent, than those imposed by a state. Still, I hope for a legal framework in which a woman's decision toward her pregnancy is the response of a moral agent endowed with dignity and, for that, capable of recognizing the dignity of the life that grows in her as well.

PART II
PHILOSOPHICAL PROBINGS

CHAPTER 9
INTERPETING CLINICAL JUDGMENT
Epistemological Notes on the Praxis of Medicine

Clinical judgment represents a topic of such complexity and importance that to pose the epistemological question in a direct fashion seems an act of either intellectual hubris or methodological naiveté. This is the case for at least two reasons. There is, first of all, the problem of defining the appropriate field for the topic. The vast body of literature dealing with issues of «philosophy and medicine» is still searching for methodological consistency and for the coherence of a systematic frame.[1] Sharing in the current post-modern cultural mood,[2] philosophers of medicine have become increasingly weary of comprehensive systems[3]. Indeed, the «deconstruction» goes so far as to cast doubts on the plausibility of philosophy

---

1   I see two different issues here. The first concerns the need and the plausibility of an overarching frame in a foundationalist fashion. The second question pertains to the epistemological status of philosophy of medicine as a separate discipline. See Edmund Pellegrino, «Philosophy of Medicine: Towards a Definition», *The Journal of Medicine and Philosophy* 11 (1986): 9-16, and more recently, idem, «What the Philosophy of Medicine Is», *Theoretical Medicine and Bioethics* 19 (1988): 315-336.
2   For an interpretation of postmodernism as mood, or *Stimmung* – the term is, of course, Heideggerian in tone – see Richard J. Berstein, *The New Constellation: The Ethical-Political Horizon of Modernity/Postmodernity* (Cambridge, Mass.: The MIT Press, 1992).
3   According to Stephen Toulmin: «This movement shared, at least, the conviction that all earlier quests for a comprehensive system of knowledge, based on permanent, universal systems of overarching principles, were misguided from the start, and are by now discredited. Claims to philosophical universality and permanence can be ignored: their only interest lays in the ways that they could serve as a «cover» for the collective interests of the nations, social groups, or genders with which their authors – novelists, philosophers, or play writers – were affiliated, «The Primacy of Practice: Medicine and Postmodernism» in Ronald A. Carson and Chester R. Burns ed., *Philosophy of Medicine and Bioethics: A Twenty Years Retrospective and Critical Appraisal* (Dordrecht: Kluwer Academic Publishers, 1977), 41-42.

of medicine itself.[4] According to Henk Ten Have the emphasis on bioethical issues has practically reduced the threefold spectrum of philosophical traditions in medicine – the epistemological, the anthropological, and the ethical – to the last one. As a result, «it seems that philosophy of medicine has come to an end, or that it has been transformed into bioethics.»[5]

With that I mind I come to the second, no less powerful difficulty in delimiting the topic, namely, the plurality of scientific approaches of formal analysis to clinical judgment, and the problem of assessing the claim to objectivity exhibited by each one of them[6].

The differences among competing theories – from those based on syllogistic procedures to various probability theories, judicial algorithms, and Bayesian analysis – run so deep as to imply a particular understanding of the meaning of medicine as a scientific enterprise and an implicit or explicit theory of human knowledge. I submit that an epistemology of clinical judgment should not be reduced within the limits of a purely *formal* enterprise. Skeptical of such a reductive epistemology, I will fame the problem of an epistemology of clinical judgment within the broader context of the *praxis* of medicine, ultimately defined by the reality of the clinical encounter.[7]

My notes unfold along the line provided by three fundamental reflections. First, I lay out the essential features of clinical judgment as they appear through broad phenomenological analysis. Secondly, I mention the interpretive nature of clinical judgment by focusing on the experience of illness and the physician-patient relationship. Finally, I address clinical decision making with a hermeneutic model of application.

---

4   See Arthur L. Caplan, «Does Philosophy of Medicine Exist», in *Theoretical Medicine* 13 (1992): 66-77.

5   Henk ten Have, «From Synthesis and System to Morals and Procedure: The Development of Philosophy of Medicine» in R.A. Carson and C.R. Burns, ed., *Philosophy of Medicine and Bioethics, op. cit.*, 107.

6   Alvan R. Feinstein, *Clinical Judgment* (Baltimor, MD: Williams Wilkins, 1967). Also H. Tristram Engelhardt, Stuart F. Spicker, and Bernard Towers, *Clinical Judgment: A Critical Appraisal* (Dordrecht: D Reidel Publishing Company, 1979).

7   On this very point Marx Wartofsky, «What Can the Epistemology Learn from the Endocrinologists? Or is the Philosophy of Medicine Based on a Mistake?» in Carson and Burns, ed., *Philosophy of Medicine and Bioethics, op. cit.*, 55.

Phenomenology of Clinical Judgment

The thrust of my analysis is descriptive or phenomenological *latu senso*, in that it takes description of experience as the first step in any philosophical approach. The task at hand consists in describing the phenomenon of medical judgment, so as to grasp the meaning of the experience that leads the physician, through a process of argumentation and reasoning, to a particular therapeutic action on behalf of the patient.[8] The presumption underlying this approach is that phenomena can speak to us if only we allow them to fully appear for what they are.[9] Thus, describing is more than an empirical transcript, a pure recording of reality. Indeed, phenomenologists underline the *a priori* nature of «phenomenological experience,» so that everything and anything that is given rests on experience. As for the *empirical* nature of this experience, consider the following observations of Max Scheler:

> He who wishes to call this empiricism may do so. The philosophy which has phenomenology as its foundation *is* empiricism in that sense. It is based on facts, and facts alone, not on construction of an arbitrary «understanding.» All judgments must conform to *facts*, and methods are *purposeful* only insofar as they lead to prepositions conforming to facts.[10]

The goal of phenomenological description is not to provide a copy of things, but to uncover those fundamental structures that make things what they are. The act of phenomenological description is, in the words of Dietrich von Hildebrand, *a prise de conscience*, an intellectual insight that transcends the realm of empirically verifiable factuality.

---

8   I am aware of the fact that my analysis presupposes a particular type of clinical encounter as paradigmatic, one in which the ill person seeks diagnosis and treatment through a single health care professional. The physician-patient relationship may, of course, assume other forms that are as plausible and valid.
9   This may sound like a contradiction considering the claim to *Voraussetzungslosigkeit* or «suspension of presuppositions» invoked by phenomenologists. Indeed, the decision to suspend judgment already involves a fundamental ontological judgment on the formal intelligibility of reality.
10 Max Scheler, *Formalism in Ethics and Non-Formal Ethics of Values* (Evanston, Northwestern University Press, 1973), 51-52.

Clinical Judgment as Action

I take the essential phenomenological feature of clinical judgment to be its dynamic, process-like nature. Whatever formal method is chosen to analyse the logic that leads to a particular medical judgment, one can relate the particularity of a single operation to a series of other operations. Indeed, clinical judgment must be considered within a larger contextual frame, spanning from the collection of medical data, to diagnostic and prognosis assessments, to the selection of a specific treatment; it is a continuous action distended along time. Of course, it is possible to isolate various moments in the process, and to analyse them in their own individual meaning. Yet, those individual moments constitute segments of a broader context that only as a whole can account for the reality of the clinical action. Within this broader context, each moment relates to the next; moreover, each moment necessarily moves in a *crescendo* that culminates in the medical decision. The latter represents the focal point, the situation of maximum ontic density toward which the other points of the clinical action converge. The broader phenomenological characteristic of the clinical phenomenon considered as a process is its intrinsic *teleology*: all the elements of medical judgment acquire their meaning in light of the final point, or *telos*, toward which they move.

Metaphysical difficulties historically raised against teleology impose a more nuanced reflection. In particular, the teleology underlying clinical action cannot be described as an *ideal* trajectory of unrelated points whose connection is being projected onto the action itself by an act of «synthetic apperception.» Such a connection would not define the action in its ontological consistency, but only a function of the observer's mind. In Kantian terms, it would be only a transcendental condition on the side of the subject.

On the other hand, the presumption of phenomenological description is one of *ontic* realism. Something is seen as the objective correlate of an intentional relation, even if no conclusions are immediately drawn on the *ontological* nature of such intentional correlate. Yet, one might say that the process-like nature of clinical judgment *is* the reality of an action understood in relation to a situation of illness. The clinical decision represents the closure of a process in which all the elements are taken into account and synthesized around the final question, «What should be done for this particular patient?»

In his study *The Anatomy of Clinical Judgment*, Edmund Pellegrino takes this meaning of teleology much further, signifying not only the culmination of clinical judgment as a process, but the fulfilment of the medical encounter and, ultimately, of medicine *tout court*:

> The end of the medical encounter, and the process of clinical judgment through which it is achieved... is restoration and healing – some corrective, remedial or preventive action is directed at what the doctor and the patient's wholeness, each in his/her own fashion. The end is not diagnosis, a scientific truth, testing an hypothesis or evaluating a treatment, though the knowledge derived therefrom enters into several states in making the decision to act.[11]

Perhaps Pellegrino's use of teleological language commands a deeper articulation of different levels of discourse at work here. Suffice it to say that the reference to a decision that functions as the fulfilment of the clinical process is important: it clears the ground for understanding clinical rationality as *practical* rationality, one in which the explanatory model implied by the scientific understanding of disease is not suspended or superseded, but rather integrated as a dimension, however essential and important, of the larger *praxis* of medicine.

Scientific Reasoning and the Practice of Medicine

If the ontic specificity of clinical judgment is dynamic progression toward a *telos*, «a being which is becoming», in the words of Aristotle, then one must exercise a methodological epochè, a suspension of belief on any approach that could potentially compro-

---

11 Pellegrino, «The Anatomy of Clinical Judgments. Some Notes on Right Reason and Right Action», in Engelhardt, Spicker, and Towers, *Clinical Judgment, op. cit.*, 172. Also Italian philosopher of medicine Paolo Cattorini, «Sulla Natura della Bioetica: una nota epistemologica sull'applicazione dell'etica alla scienza e alla clinica», in Elio Sgreccia, Vincenza Mele, and Gonzalo Mirando, ed., *Le Radici della Bioetica* (Milano: Vita e Pensiero, 1996), vol. 1, 77-83.

mise the specific nature of the phenomenon in question. The relation of intentionality itself requires such a suspension as it presupposes a structural homogeneity between subject and object, consciousness and experience, *noesis* and *noema*.

Phenomenological epochè has, first, a *negative* and critical meaning. Its function, rather than indicating the approach to take in order to grasp the essence of a particular realm of experience, is to clear the very ground of the subject's consciousness from any theoretical presuppositions that could potentially compromise the full affirmation of the phenomenon's *eidos*. In particular, the *epoche'* functions as a critical *caveat* against positivistic approaches to medicine and to scientific inquiry, approaches that ultimately define the general character of modern natural sciences.

In *The Enigma of Health*, Hans Georg Gadamer poignantly describes modern scientific knowledge as a capacity to produce effects: the mathematical-quantitative isolation of laws in the natural order provides human action with the identification of specific contexts of cause and effect; also, it empowers human action with new possibilities for intervention.[12] Gadamer's hermeneutic of the modern scientific enterprise echoes, in many ways, another critical account of the scientific idealization of experience. In his famous *Die Krisis der europäischen Wissenschaften*, Edmund Husserl provides a «genealogy» of experience that, insofar as it represents an experience of *Lebenswelt*, or life-world, precedes its being idealized by science.[13] Using a concept consciously formulated in contrast to a concept of the world that includes what can be made objective by science, Husserl calls «life–world» the world in which we are immersed in the natural attitude (*natürliche Einstellung*) that never becomes an object as such for us, for it represents the pre-given basis of all experience.[14]

---

12   H.G. Gadamer, *The Enigma of Health: The Art of Healing in a Scientific Age* (Stanford, CA: Stanford University Press, 1996), 35.
13   See in particular Edmund Husserl, *The Crisis of European Sciences and Transcendental Phenomenology: An Introduction to Phenomenological Philosophy*, (Evanston: Northwestern University Press, 1970). For a general overview, Paul Janssen, *Edmund Husserl: Einführung in seine Phänomenologie* (Freiburg: Verlag Karl Alber, 1976).
14   For a general hermeneutics of the notion of *Lebenswelt* in Husserl see Paul Janssen, *Edmund Husserl, op. cit.*, 135-145, Maurice Natanson, «The Leb-

The positivistic abstraction underlying the concept of technology in the modern scientific ideal acquires new and specific possibilities in the field of medicine and its healing procedures.[15] In relation to clinical judgment, such an idealization could be understood, in the first place, as a negation of the particular nature of the phenomenon, as a tendency to reduce the contextual *praxis* within which the clinical judgment takes place to the objectivity of theoretical knowledge; moreover, to interpret the healing process itself as a production of effects. Of course, the application of scientific reasoning to clinical judgment is not being questioned. In trying to determine what is wrong with the patient, in attempting to identify and explain the cause of symptoms, the physician does indeed deploy probabilistic laws and rules, theories and principles, of the biomedical sciences. Concepts of normal and abnormal, for an example, are statistically derived concepts based on scientifically validated norms of human biological functions. In the attempt to classify the patient's symptoms as manifestation of a particular disease entity, the physician relies upon the intrinsic possibilities of hypothetic-deductive reasoning.

Modes of scientific reasoning also define the therapeutic question. In trying to determine what can be done to remove or alleviate the cause of the patient's suffering, the physician appeals to prognostic knowledge about the course of diagnosed disease and about the efficacy and toxicity of relevant therapeutic possibilities.

Yet, *clinical* reasoning cannot be entirely equated to *scientific* reasoning. The goal of the former is not to relate different segments of scientific explanations to a unified theory. Rather, it is to bring together – in a synthetic action that is theoretical and practical at

---

enswelt», in Erwin Straus, ed., *Phenomenology: Pure and Applied*, (Pittsburgh: Duquesne University Press, 1964), 75-104. Also Hans Georg Gadamer, *Truth and Method.*, 2nd edition (New York, The Continuum Publishing Company, 1994), 242-254; idem, «The science of the Life-world,» in Anna-Teresa Tymienicka, ed., *Analecta Husserliana*, vol. II (Dordrecht: D Reidel Publishing Company, 1972), 173-185.

15 For an historical account, see S.J. Reiser, *Medicine and the Reign of Technology* (Cambridge: Cambridge University Press, 1978). From a more theoretical perspective, Salvatore Natoli «La costituzione dello sguardo medico: dal gesto terapeutico alla scientificità della medicina», in *Nuovi Saggi di Medicina e Scienze Umane* (Milano: Instituto Scientifico San Raffaele, 1985), 31-70.

the same time – an understanding of illness with a specific medical decision on behalf of the patient. Unlike the patho-physiology of disease, the phenomenon of illness cannot be observed, analysed, and explained in itself. It must be understood as a part of the life-world of the subject in whom it manifests itself.[16] This is the reason why medicine represents a unity of theoretical and practical knowledge within the domain of the modern sciences, «a peculiar kind of practical science for which modern thought not longer possesses an adequate concept».[17]

Retrieving the Subject: From Action to the Agent

A further fundamental step in a phenomenological inquiry of clinical judgment leads to the question of the *subject* (or the subjects) involved in the clinical action. Originally influenced by positivistic ideals of science, analytic philosophy has defended the possibility of a philosophy of action in which the question of the subject is never uttered and, ultimately, completely ignored.[18] The difficulty of this position is dramatically illustrated by the development of analytic philosophy itself, for which the «semantic of action» is pulled back to its inter-subjective condition of possibility, in a «pragmatic of action».[19] As Toulmin himself points out, this development can be seen in a wide range of analytic philosophers, from Wittgenstein, to Austin, and Searle.[20]

---

16 See Edmund D. Pellegrino, «The Lived World of Doctor and Patient: A Phenomenological Perspective on Medical Ethics», lecture at Yale University, April 11, 1996, *Bioethics and Public Policy Symposium*.
17 H.G. Gadamer, *The Enigma of Health*, op. cit., 39.
18 See for instance, A.I. Melden *Free Action* (Routledge and Kegan Paul, 1961) and Stuart T. Hampshire, *Thought and Action* (Notre Dame, IN: University of Notre Dame Press, 1983). Also Arthur Danto, *Analytic Philosophy of Action* (Cambridge: Cambridge University Press, 1973(.
19 On this question, see the observations of Paul Ricoeur, *Oneself as Another*, translated by Kathleen Blamey (Chicago: The University of Chicago Press, 1992), 27-55.
20 «Wittgenstein's focus in the *Tractatus Logico-Philosophicus* on «propositions» (*Sätze*) shifted, in the *Philosophical Investigations*, to *Sprachspielen* («language games») and ultimately to *Lebensformen* («forms of life») as the occasion of language games. What this shift made clear is that words, sentences, and other lexical items are not connected to their occasions of

Yet, positivistic prejudices and factual idealizations keep lurking behind the analytic skills of an «agent-less» theory of action. Within the general constraints of such a paradigm, it matters not who the subject of the action is: the action is treated simply as a sub-class of impersonal events. In a series of articles collected in the volume *Actions and Events*,[21] Donald Davidson presents a theory of action in which the distinctive teleological character of action is subordinated to a *causal* conception of explanation. Causal explanation serves, in its turn, to place actions within a general ontology in which events are understood as incidental occurrences, as irreducible entities placed on the same level of substances as fixed objects. This ontology of «impersonal events» ends up structuring the entire gravitational sphere of the theory of action, preventing an explicit, thematic treatment of the relation between action and agent.

In light of an impersonal ontology of events, clinical judgment could very well be interpreted as the result of a computer based operation that depends exclusively on the completeness and accuracy of the information being submitted.[22] One can see that the increasing specialization of medicine as a discipline, the anonymity of hospital procedures, and the powerful influences of economical forces operating behind the health care industry contribute to slowly, but surely, concealing the reality of a personal agent in medicine. And yet, the question of the subject cannot be ignored in a phenomenological analysis of experience. Here, the symmetric polarity of subject and object must be interpreted – by the necessity of an essential

---

use by formal, logically necessary relations... J.L. Austin reminded us how many utterances operate more as *performances* than as representations of facts (*Bilder der Tatsachen*). John Searle similarly argued that they gain a meaning not in the way mathematical formulae do – by formal definition – but as other human gestures and songs, that is to say, only in being «speech» acts rather than acts of other kinds», Stephen Toulmin, «The Primacy of Practice», *op. cit.*, 45.

21 Donald Davidson, *Essays on Actions and Events* (Oxford: Clarendon Press, 1980).
22 A computer-assisted diagnosis is part of a larger trend in medicine leading to what Wartofsky calls the «technologization of the medical subject». See «What Can the Epistemologists Learn», *op. cit.*, 61. Also Luigi Stella and Antonella Crescenti, «Principi di Informatica clinica», In *Nuovi saggi di Medicina e Scienze Umane* (Milano: Instituto Scientifico H San Raffaele, 1985), 293-360 and Elliott Sober, «The Art and Science of Clinical judgment: An Informal Approach», in *Clinical Judgment, op. cit.*, 29-44.

connection – within the frame of an intentional relation. How does the question of the subject become relevant in the specific case of the action being at issue here, namely the clinical judgment?

Intentionality is a term that refers originally to the theory of knowledge.[23] It underlines the fact that the consciousness of the knowing subject does not exist in-itself, prior to its relations to an object, but it is always object-oriented, i.e., consciousness *of* something. Correlatively, the object never exists as an object-in-itself, but always as a correlate of a consciousness. Along the same line, philosophers working in phenomenology have uncovered the intentional dimension of other operations in the subject. For an example, Max Scheler points to the intentional meaning of emotions and feelings, bringing forth their deeply personal and spiritual dimensions against deterministic and materialistic hermeneutics of instinct.[24]

The application of the notion of intentionality to the realm of action, however, represents a kind of extension in the phenomenological theory. In fact, it provides a framework for the interpretation of clinical judgment within the context of a personal relation; better, in terms of a *personal* encounter. The intentional dimension of action can be described retrospectively. It implies going up-stream against the intentional flux, in order to reach out for the agent, or the agents, involved in the action. In this light, an action always proceeds from somebody and is directed at somebody else. Looking at the action intentionally means, therefore, overcoming an objectivistic attitude that rests, *de facto*, upon the separation between subject and object.

An action is never just an impersonal state of affairs out-there-in-the-world. There is something ambiguous about the language of bioethics when reference is being made to the «puzzle» posed by

---

23   In the fifth of his *Logical Investigations,* Husserl elaborates the nature of intentionality. Even if the concept of intentionality is given a definite elaboration in *Ideas*, here already consciousness is understood not as an «object», but as an essential coordination. This constitutes an important starting part in overcoming objectivism. See the observations of Gadamer in *Truth and Method*, 244.

24   Max Scheler, *The Nature of Sympathy*, translated by Peter Heath (London: Routledge and Kegan Paul, 1954). Also Paul Ricoeur, *Freedom and Nature: The Voluntary and Involuntary*, translated by Erazim V. Kohàk (Evanston: Northwestern University Press, 1996).

clinical cases.²⁵ Taken by itself, a case represents an abstraction, an objectification that extrapolates from the intentional context, or the life-world of meaning and experience within which it is always embedded. In the words of Gadamer:

> the concept of life-world is the antithesis of all objectivism. It is an essentially historical concept...(it) means the whole in which we live as historical creatures... It is clear that the life-world is always at the same time a communal world that involves being with other people as well. It is a world of persons, and in the natural attitude, the validity of this personal world is always assumed.²⁶

For this reason, the context of clinical judgment is always a personal context. Which is to say that a case is always somebody's case: somebody's life, but also somebody's responsibility and conscience in dealing with the complexity of the situation are more than accidental variables in the circumstantial texture of the case. They are the very stuff of which the case is made. In the same fashion, a particular decision on behalf of the patient is not just a strategic solution, or a technical fix to the complexity of an anonymous incident. From a phenomenological point of view, that action represents, first of all, the *actualization* of a subject, a person's practical involvement whose effect is more than a change in reality. Indeed, it represents a change in the subject's experience, a modification of the subject's *being-in-the-world*.

The change pertains primarily to the subject who is a *patient*. The modification brought about by the physician in restoring health cannot be adequately described simply as the production of a biological state of affairs; rather, it represents the re-composition of a natural equilibrium whose essential features extend to the life-world of the patient.²⁷ Just as sickness represents a situa-

---

25 For the notion of «puzzle» see Thomas Kuhn, *The Structure of Scientific Revolution*, 2ⁿᵈ edition (Chicago, University of Chicago Press, 1970), 36.
26 H.G. Gadamer, *Truth and Method, op. cit.*, 27.
27 For Gadamer the understanding of human health in terms of the natural condition of equilibrium (*Gleichgewicht*) implies a more fundamental understanding of nature as equilibrium: «If we presuppose this idea of nature, then medical intervention must be understood as an attempt to restore an equilibrium that has been disturbed», Hans G. Gadamer, *The Enigma of Health, op. cit.*, 36.

tion of *dis-ease*, a rupture and break in the position of the human individual within the totality of being, so restoration of health predisposes the ground for a new personal synthesis, the re-unification of a life-world previously shattered or compromised by the event of illness.

Yet, the change is no less radical for the *physician* than the patient. Precisely because the physician's action is more than a production of effects, it demands from him/her not just the application of technical skills; the physician's life-world is put in question as well. This is, in my opinion, the radical meaning of the notion of care and the very ground upon which an ethics of care rests.[28] Of course, the good physician is one who cares for the patient in a *moral* sense, in so far as he/she empathizes with the patient's situation and frames his/her action in relation to this situation. Yet, care pertains to the quality of the action only because it defines the ontological condition of an acting person. It is the reality of a personal actualisation continuously challenged and kept in motion by the intrinsic demands of a relationship what gives meaning to the clinical experience and to its different dimensions.

Clinical Judgment as Interpretation

The phenomenology of clinical judgment ends with an attestation of subjectivity. Even if the epistemological importance of such attestation needs to be further determined, the reference to an inter-subjective polarity at the heart of clinical judgment seems to lead in the direction of interpretation. Indeed, the process of clinical judgment is defined by interpretation from the beginning to the end.

---

[28] I am aware of my not so implicit reliance upon Heidegger's hermeneutics of *Sorge* as ontological-existential category, i.e., as defining the being of Dasein. Such reliance does not imply, however, a negation of the specific ethical meaning of care. For an overview of the discussion on the ethics of care see the accurate reconstruction of Warren T, Reich in the *Encyclopedia of Bioethics*, 2nd edition (New York: Simon and Schuster MacMillan, 1995) vol. 1, 319-343.

The Patient's Experience of Illness

This can be seen, first of all, in relation to the way in which a patient understands illness and seeks the help of a physician. As any other human interaction, the clinical encounter rests on the freedom of the persons who enter in relation with one another. Ideally, a patient partakes of a particular healing relationship without being forced into it. This holds true independently of the concrete availability or choice of physicians.

Correspondingly, the physician accepts the reality of the patient on the basis of a personal commitment, a promise to help, itself predicated upon freedom. Clinical medicine is the inter-subjective exchange in which an individual in need of healing entrusts himself/herself to another individual who professes and promises to heal on the basis of acquired knowledge, skill, and experience.[29]

There are, at the same time, deterministic dimensions to this relation that are rooted in the very reality of illness. There are very good reasons for referring to the *fact* of illness as the first, constitutive element of medicine, if illness is indeed something that *happens* to a person. Even when a particular disease can be traced back, etiologically, to a certain life-style, one never chooses the particular suffering that comes with the ailment. By definition suffering entails an element of passivity (*pati*), which is absolutely personal. In the words of Paul Ricoeur, «*La souffrance est, avec la jouissance, la retraite ultime de la singularité*».[30]

We may suffer of the same disease, yet we undergo the experience of suffering in different ways, radically left to our own individuality. Even as empirically reducible entity, the fact of illness exists only as interpreted fact, experienced and recounted by a particular patient. Consider how, from the moment in which medical data are collected, the subjective perception of facts takes central place. Clinicians comment often on patients' tendency to inadvertently shift their language from the pure enumeration of symptoms to a kind of

---

29 E.D. Pellegrino, «The Healing Relationship: The Architectonics of Clinical Medicine» in Earl Shelp, ed., *The Clinical Encounter: The Moral Fabric of the Physician-Patient Relationship* (Dordrecht: D. Reidel Publishing Company, 1983), 153-172.

30 Paul Ricoeur, «le trois niveaux du jugement medical » *Esprit*, 12 (1996): 21-33, at 22.

self-inferred diagnosis. The phenomenon might be explained in a variety of ways; not ultimately, the psychological need to control and define what one experiences. Insofar as it represents a spontaneous tendency, however, it throws into relief our need to interpret illness, to «reduce» its brute facticity to meaning.[31]

The subjective dimension of illness, including the transcendental function of interpretation, leads to two additional observations. Illness might not be associated with demonstrable pathology. Likewise, pathology might be present even when the patient does not experience himself/herself as sick. These observations presuppose understanding the distinction between illness and disease: the former referring to the subjective experience, and the latter to the objective construct.

Even if the distinction between illness and disease is basic to the critique of a model in which medical objectivity is seen as a «flight from interpretation»,[32] it is no less important to avoid separating the two notions as completely foreign to one another. Indeed, the separation would simply legitimise an approach to illness, as well as an understanding of medicine, totally unrelated to their subjective and, therefore, interpretative variables. Interpretation represents the proper epistemological mode for understanding illness because our very access to reality is structurally mediated by an act of interpretation. In this sense, human understanding is hermeneutic at root.[33]

---

[31] Indeed, the patient's narrative has not only a diagnostic, but even a therapeutic significance. According to Drew Leder the very ability to bring the disease to the level of labguage «counteracts two primary features of illness that give rise to suffering: senselessness and isolation... The fact of translating disease into a language begins to overcome this twofold alienation. What was a private pain is now made public, what was senseless and random is woven into a meaningful tale. The narrative context itself can have healing force,» Drew Leder, «Clinical Interpretation: The Hermeneutics of Medicine,» *Theoretical Medicine* 11 (1990): 9-24, at 13.

[32] See Drew Leder, «Clinical Interpretation», *op. cit.*

[33] Martin Heidegger, *Being and Time*, trans. By John Macquarrie and Edward Robinson (New York: Harper and Row, 1962), 182-203. On the importance of Heidegger's and Gadamer's philosophy in understanding this claim see Richard J. Berstein, «From Hermeneutics to Praxis», in *Philosophical Profiles: Essays in a Pragmatic Mode*» (Philadelphia: University of Pennsylvania Press, 1986), 94-114: «Implicit in Heidegger and explicit in Gadamer are two central claims: the ontological primacy pf hermeneutics and its universality. We are thrown into the world as beings who understand; and

By looking at the patient, I have shown how the intentional dimension of clinical judgment shapes the subjective perception and definition of the experience of illness. The interpretative nature of this experience becomes evident in the language used by the patient, in the emotional mood underlying his/her narrative of symptoms and pain. It is important to stress that narratives of illness are never purely descriptive: insofar as they are embedded in the lifeworld of patients – bespeaking their beliefs, fears, uncertainties – they are already value-laden. In their narratives, patients are already trying to understand; in the process of understanding, they are also interpreting.

Physician's Attitude: Listening and Caring

At the other side of the relation, the physician listens to the patient's story. There is something very profound about this attitude of listening that directly affects the clinical judgment. Indeed, hearing what the patient has to say conditions the clinician's ability to understand the phenomenon of illness.[34] More than just a professional virtue, listening represents for the health care professional an ontological specification, or, in the words of Martin Heidegger, a fundamental existential possibility:

> If we have not heard «aright,» it is by no accident that we say we have not «understood.» Hearing is constitutive for discourse. Listening to is Dasein's existential way of Being-open as Being-with for Others... Dasein hears, because it understands.[35]

For the patient, as for the physician, illness becomes the intentional correlate of an act of interpretation. For this reason, in the diagnostic process, the physician needs to rely on the patient's story. Disease is never just a thing-in-itself, ready to be grasped in a scientific act of detached objectification, predicated upon the «sus-

---

understanding itself is not one type of activity of a subject, but may properly be said to underlie all activities,» at 96.
34   See Richard Baron, «I Can't Hear While I Am Listening,» *Annals of Internal Medicine* 103 (1985): 606-611.
35   Heidegger, *Being and Time*, 206.

pension» of the patient's narrative, experience, and history. On the contrary, the objectification of disease cannot be separated from the subjective account of illness. Indeed, it is *through* the interpretative nature of the patient's story that the physician will eventually explain the objective nature of illness. Just as disease is embedded in the experience of illness, so scientific judgments of its objective nature represent derivative modes of interpretation, grounded in the reciprocal understanding made possible by the clinical encounter. With a term borrowed from hermeneutics, one could re-express the kind of understanding that occurs between the physician and the patient in terms of a fusion of horizons (*Horizontsverschmälzung*). The horizon represents the pre-comprehension structuring – from different perspectives – the understanding of illness[36].

The category of fusion of horizons stands dialectically against the notion of the clinical gaze as an act of objectification[37]. Of course, the ultimate goal of the clinical encounter is to isolate the cause of illness, by objectifying the etiology of the disease, not to establish a personal relation. In this sense, the «fusion» just referred to is more a means toward an end, than an end in itself.

Yet, the scientific objective of medicine can be achieved only through the intrinsic possibilities and difficulties of dialogue. For this reason, the diagnostic process does not represent a moment apart from the contextual reasoning that makes the communication possible. Such reasoning is *practical*, for it takes place within the *praxis* of communication defining the physician-patient relationship, and it is functional to achieving the *telos* of that relation. The inter-subjective nature of the physician-patient relationship does not add a new element, a purely external one, to the properly «scientific» side of the clinical judgment, as if the latter could stand independently of the relation within which it takes shape. Rather, the opposite is true: the relation itself intrinsically structures and defines the scientific side of the process in its very meaning.

---

36   In the process of understanding, a real fusion of horizons occurs – which means that as the historical horizon is projected, it is simultaneously superseded. To bring about this fusion in a regulated way is the task of… historically effected consciousness.» H.G. Gadamer, *Truth and Method*, 307.
37   M. Foucault, *The Birth of the Clinic: An Archeology of Medical Perception* (New York: Vintage Book, 1975), 107-123.

Phronesis *and Application in Clinical Reasoning*

With the question, «What should be done for this patient?» clinical reasoning comes to closure. In determining a particular course of action for a particular patient, the clinician brings the process of interpretation of illness, grown out of the clinical encounter, to fulfilment. The understanding of illness as subjective experience finally translates into therapeutic intervention. The interplay of theoretical and practical dimensions, of objectivity and subjectivity structuring clinical judgment and, with it, the process of clinical reasoning, provides the framework for understanding the meaning and the importance of the clinical decision. I have argued that the need to contextualize the patho-physiology of disease within the patient's life-world is based on the expectation that the physician will *act* upon the patient as subject, rather than intervening on the biological entity as an object-in-itself.

Unlike the pure scientist, the clinician's job ends with the practical application of knowledge. The clinician's interest in the disease is ultimately oriented toward making the *right* decision on behalf of the patient. The epistemological significance of this last statement depends entirely on our understanding of application.[38]

Two competing models – which I will call scientific and hermeneutic – are available. In the *scientific* model, theory and praxis describe two entirely separated levels of reality. Indeed, separation represents the condition whereby theory is established in its universality, gaining validity through progressive abstraction from any particular case; theory stands independently of praxis. Application, understood as a function of verification or falsification, makes possible the passage from the general to the particular, from the objective to the subjective. Yet, such a passage functions only insofar as particular cases fit the hypothetical model of general explanatory principles. From this perspective, the rightness of a clinical decision is formally defined by its conformity to pre-existing parameters. What the clinician requires is purely the skill of the craftsman who learns to apply scientific knowledge and discoveries with the purpose of restoring health.

---

38 My observations are grounded in Gadamer's analysis of application in *Truth and Method*, especially 307-311.

The *hermeneutic* model, on the other hand, recognizes the structural interplay of theory and praxis at the heart of clinical reasoning. Here the theoretical and the practical are not separated levels of reality, but polar dimensions whose validity is reciprocally co-determined. Far from being a purely mechanical function, application represents a true *mediation* between the universal and the particular, the theoretical and the practical. In this light, the theoretical understanding of illness is not gained independently of its practical manifestations and its subjective interpretation, but rather through them. Indeed, the rightness of a particular clinical decision will represent a synthetic mediation of universal and particular, whereby the general laws of diagnostic procedures are understood and interpreted in relation to the specific situation of illness affecting a specific patient.

It is impossible to miss the analogy between the hermeneutic model of application and the Aristotelian notion of *phronesis*.[39] In Book VI of the *Nicomachean Ethics*, Aristotle distinguishes *phronesis* from the «intellectual virtues» of *episteme* and *techne*. In the same way here the notion of application as mediation is gained dialectically, namely, by playing out its hermeneutic meaning in contrast with the scientific model of application. The analogy reveals a form of reasoning appropriate to the specific praxis of medicine.[40]

In particular, Aristotle's *phronesis* seems to account for two important features of clinical judgment that were thrown into relief by the previous phenomenological description. The first concerns the *practical* nature of clinical judgment, the mediation it establishes by means of deliberation and choice between the universality of scientifically validated laws concerning the patho-physiology of disease and the particularity of concrete, i.e., personalized phenomena of illness. The second feature is the *intentional* dimension of clinical judgment, its necessary correlation to a subject who actual-

---

39   The hermeneutic relevance of Aristotle is, of course, recognized by Gadamer himself in *Truth and Method*, 312-324. For a commentary, see Richard J. Bernstein «From Hermeneutics to Praxis,» *op. cit.* and Paul Schuchman, «Aristotle's Phronesis and Gadamer's Hermeneutics,» *Philosophy Today* 23 (1979): 41-50.
40   For an understanding of *phronesis* as a paradigm of clinical rationality see the study of Daniel Davis, «Phronesis, Reasoning, and Pellegrino's Philosophy of Medicine,» *Theoretical Medicine* 18 (1997): 173-19.

ises himself/herself in the very process of judging. *Phronesis* is not to be confused with *episteme*, with an «objective knowledge» that is detached from one's own being and becoming. Just as *phronesis* (prudence) determines what the *phronimos* (the prudent person) becomes, so the application of clinical judgment progressively shapes the clinician into a moral agent. Conversely, it is the *phronimos* who – by relying upon experience and practice – is most likely to make the right decision and to fully account for the particular features of the case at hand.

Conclusion

These last observations seem to indicate a direction rather than a conclusion. My notes would probably have to abandon, at this point, their general epistemological concern and become more clearly ethical in tone. But this seems required by *die Sache selbst*, the very nature of medicine as praxis of healing. In the end, to interpret clinical judgment means to recognize the practical nature of medicine. In my analysis, I have shown that such recognition depends upon the intrinsic teleology of clinical judgment, the interpretive character of the physician-patient relationship, and the «phronetic» application of general principles of diagnostic intervention to particular instances of illness.

CHAPTER 10
WHY CLINICAL ETHICS?
Experience, Discernment, and the *Anamnesis* of Meaning at the Bedside

In this chapter, I reflect on the function of clinical ethics in medicine as oriented to the retrieval of meaning. In a somewhat Platonic vein, I will term such a task, always poised between forgetfulness and remembrance, the *anamnesis* of meaning.[1]

The task calls for preliminary clarifications. On account of its closeness to the professionals and their practice, clinical ethics can be seen as a form of mindfulness that impels the practice of medicine towards its own *telos*, i.e., the ends proper to medicine. At the same time, because it articulates the ends of medicine in the context of a communal ethos, with its needs, values, and priorities, clinical ethics may be better understood as a function of critical analysis that borrows from the anthropological *milieux* in which it operates. The *telos* of medical action cannot be found independently of the context it is supposed to serve.[2]

With the philosopher of medicine Edmund Pellegrino, I do share in the belief that such *telos* can only be pursuit in an attitude of faithfulness to the *internal morality* of medicine. And yet, unlike him, I would refrain from postulating a notion of internal morality that reduces it to a univocal concept, one that borrowing from the resources of *naïve* realism, leads to a sort of abstract definition. In this view, the internal morality of medicine becomes something like

---

1   In the Italian context, Corrado Viafora has dedicated his reflection to the elaboration of a philosophically grounded version of clinical ethics. His latest accomplishment is the fruit of such reflection. See Corrado Viafora, *La cura e il rispetto. Il senso della bioetica clinica* (Milano: Franco Angeli, 2023). From a more international perspective, but always with a preoccupation for foundational questions, see Corrado Viafora, ed., *Clinical Bioethics: A Search for the Foundations* (Dordrecht: Springer, 2005).
2   For the debate on the goals of medicine, see Mark J. Hanson and Daniel Callahan, *The Goals of Medicine: The Forgotten Issue in Health Care Reform* (Washington, D.C.: Georgetown University Press, 1999).

an *eidos*, or an idea, grasped once and for all in an intuitive insight, untouched by time and accidental circumstances.³

Against the assumptions of such an uncritical epistemology, one might see the internal morality of medicine more as a dynamic process, unfolding through the concrete intermediation with particular ideologies of human fulfillment. This latter movement, unlike the fixed essentialism portrayed above, commands an appreciation for the disclosure of meaning in history, and for the truth of the *humanum* that inhabits the social context in which medicine operates.

I see the function of clinical ethics in medicine as articulating a twofold commitment to the search for meaning, a search that has been hindered, in the medical context, by the limited vision of positivist natural sciences, and in ethics, by an excessive preoccupation with normative dimensions. The former is a recurring temptation of medicine, most visible, of late, in the discussions on matters of genetics and research. As for the latter, search for meaning entails much more than simply re-arranging the «internal coherence» of a «content-thin» ethical strategy, each time awaiting for the next edition of the *Principles of Biomedical Ethics* by Beauchamp and Childress.⁴ It is not enough to keep the system open to the latest normative integration, in an endless exercise of «reflective equilibrium,» if such a system fails to address «deepest matters of our humanity,» to quote Leon Kass. Brilliant moral theories might come too late, when ethics has already lost its soul.⁵

---

3 For a clarification on the presuppositions of Pellegrino's philosophy of medicine, one can see the collection of essays edited by Roger Bulger and John McGovern, *Physician Philosopher: The Philosophical Foundation of Medicine. Essays by Dr. Pellegrino* (Charlottesville, VA: Carden Jennings Publishing, 2001).

4 Tom Beachamp and James Childress, *Principles of Biomedical Ethics* (New York: Oxford University Press, 2013). The book, first published in 1979, is now in its 8th edition. The changes in the evolution of the book testify to the authors' attention for the unfolding of the methodological debate in bioethics. On the other hand, their commitment to a principle-based approach remains unshakable, in spite of mounting external criticism.

5 On the condition of contemporary bioethics, relative to a lack of questioning about moral meaning see Leon Kass, *Life, Liberty and the Defense of Dignity: The Challenge for Bioethics* (San Francisco, CA: Encounter Books, 2002), especially 55-76; also Gilbert C. Meilaender, *Body, Soul, and Bioethics* (Notre Dame: University of Notre Dame, 1995).

Perhaps the central claim of these reflections is to encourage a dramatic shift in paradigms that turns, first and foremost, to the interpretation of experience – in this case, the experience of clinical practice with all its complexity and nuances – as the central task of clinical ethics. In this perspective, the ethical methodology specific to clinical ethics cannot be defined *a priori* to the challenges of clinical practice itself; rather, as the articulation of an ethical gesture that already pervades such practice in a quest for intelligibility.

Mindful of the latter suggestion, I intend to convey the following thesis: the contribution of clinical ethics to the practice of medicine at the bedside – I take this restriction of the *material* object of ethical analysis to define the specific task of clinical ethics vs. bioethics more in general – consists in a twofold retrieval of meaning: relative to medicine, first; and, secondly, relative to ethics. In more synthetic terms, I would identify the function of clinical ethics in what I call the «anamnesis of meaning».

1. *The Separation of Principles and Meaning in Medicine*

The problem of the search for meaning in medical ethics might be illustrated by the metaphor of the stethoscope. Richard Baron, in a famous article for the *Annals of Internal Medicine*, tells the story: «It happened the other morning on rounds, as it often does, that while I was carefully auscultating a patient's chest, he began to ask me a question. "Quiet" I said. I can't hear you while I'm listening.»[6]

The stethoscope metaphor is emblematic of the inattention to meaning («not hearing») brought about by the reductionist focus (the mode of restricted «listening») in the methodologies of both modern scientific medicine and contemporary ethical theory.

To start with, the mind-set created by modern scientific medicine has required for medicine *to be inattentive*, i.e., not to hear, the sick person's experience of illness. Influenced by a positivist framework, 19th century medical scientists popularized the notion

---

6   Richard Baron, «An Introduction to Medical Phenomenology: I Can't Hear You While I'm Listening,» *Annals of Internal Medicine* 103 (1985): 606-611, at 606. In an analogous phenomenological vein, see Richard Zaner, *Ethics and the Clinical Encounter* (Englewood Cliffs, N.J.: Prentice Hall, 1988).

that *practical* clinical medicine should be viewed as a form of applied *theoretical* medicine. In the United States, the reformation of medical studies introduced by the medical educator Abraham Flexner, in the first part of the 20th century, completed the picture. Moreover, this happened as a result of modernity's understanding of scientific knowledge, which Hans Georg Gadamer poignantly describes as a capacity to produce effects. In the *modern* version of scientific knowledge, the mathematical-quantitative isolation of laws of the natural order provides human action with the identification of specific contexts of cause and effects, together with new possibilities for intervention.[7] In relation to clinical medicine, such an idealization entails a tendency to reduce the *praxis* of medicine, with its matrix of subjective components and contextual features, to the detached «objectivity» of theoretical knowledge, and to interpret the healing process itself as a production of effects.[8]

Of course, one cannot in principle question the application of scientific reasoning to medicine. In trying to identify and explain the cause of symptoms, medicine employs probabilistic laws and rules, theories and principles, of the biomedical sciences. Concepts of normal and abnormal, for an example, are statistically derived concepts, based on scientifically validated norms of human biological functioning. In the attempt to classify symptoms as the manifestation of particular disease entities, medicine relies upon hypothetic-deductive and inductive reasoning. Moreover, in order to determine what can be done to remove or alleviate the cause of particular diseases, medicine appeals to prognostic knowledge about the course of the diagnosed disease, as well as efficacy and toxicity of relevant therapeutic possibilities.

And yet, in spite of its undisputable scientific basis, medicine cannot be entirely equated with science. The goal of medicine is not

---

7   Hans Georg Gadamer, *The Enigma of Health: The Art of Healing in a Scientific Age* (Stanford, CA: Stanford University Press, 1996), 35.
8   Marx W. Wartofsky, «What Can the Epistemologists Learn from the Endocrinologists? Or Is the Philosophy of Medicine Based on a Mistake?» in Ronald A. Carson and Chester R. Burns, ed., *Philosophy of Medicine and Bioethics: A Twenty Years Retrospective and Critical Appraisal* (Dordrecht: Kluwer Academic Publishers, 1997), 55-68. As for Weber's value-free principle, see his «Science as a Vocation,» in H.H. Gerth and C. Wright Mills, ed., *From Max Weber* (New York: Oxford University Press, 1946), 155.

to reduce different segments of scientific explanations into a unified theory; rather, the specific goal of medicine consists in bringing together, in a synthetic action, which is theoretical *and* practical at the same time, an understanding of illness with a specific medical decision on behalf of the patient.[9] Unlike the patho-physiology of disease, the phenomenon of illness cannot be observed, analyzed and explained *noumenically*, i.e., «in itself.» As Gadamer suggests, it can be fully understood only *hermeneutically*, i.e., through an act of interpretation that takes place within the sociological, cultural, and ideological matrix of a defined life-world. For this reason, medicine represents a peculiar unity of theoretical *and* practical knowledge within the domain of the modern sciences, «a peculiar kind of practical science for which modern thought no longer possesses an adequate concept.»[10]

My point here should not be misconstrued. Careful scientific attention to the patho-physiology of disease, together with ever more extensive bio-technological applications, has certainly yielded marvelous advances in modern medicine.[11] Yet, its positivist reduction has also created a mind-set that brackets questions of meaning, themselves highly significant to human well-being and to the ethical aspects of medicine.

Consider the case recently publicized in the news concerning the FDA discussion for approval of an *in vitro* fertilization technique which, in an attempt to prevent certain illnesses, like muscular dystrophy and respiratory problems, uses DNA from three people. [12] Most commentators, especially scientists and doctors, welcome the advent of yet another technological fix to a congenital predisposi-

---

9   Such a perspective has been forcefully maintained by Edmund Pellegrino, «The Anatomy of Clinical Judgment: Some Notes on Right Reason and Right Action,» in Tristram Engelhardt, Stuart F. Spiker, and B. Towers, ed., *Clinical Judgment: A Critical Appraisal* (Dordrecht: D. Reidel Publishing Company, 1979), 169-194; idem, «The Healing Relationship: The Architectonics of Clinical Medicine,» in Earl Shelp, ed., *The Clinical Encounter: The Moral Fabric of the Patient-Physician Relationship* (Dordrecht: D. Reidel Publishing Company, 1983), 153-172.

10   Gadamer, *op. cit.*, 39.

11   Leon Kass, «The Problem of Technology and Liberal Democracy,» in *Life, Liberty, and the Defense of Dignity*, op. cit, 29-53.

12   See Kim Tingley, «The Brave New World of Three-Parent I.V.F.,» in *The New York Times*, June 27, 2014.

tion with an attitude of unquestionable awe. On the other hand, the more critically minded, among them ethicists, are willing to grant that some moral problems for this «three parent baby» solution do exist after all: doubts about safety are raised, together with the fear of unforeseen eugenic slippery slopes. Strangely passed over in silence, though, remains the most obvious question, «whose child will this baby be?» Of course, experts are quick to rebut this preoccupation as scientifically naïve, if not totally unfounded: they reassure the concerned public that because the female donor of healthy mitochondrial DNA to the defective biological mother provides, in the end, a very negligible genetic contribution, she cannot be described appropriately as «a parent.» However, when considered from another angle, namely, that of the *personal* identity of a child thus produced, the question «whose child will this baby be?» comes to the fore as actually very serious. This is so because personal identity is now imperiled by what I would call «an ambiguity of belonging,» in which the embodied matrix of traceable biological debts represents for the child in question more an opportunity for doubt, than a condition for self-identification. The lack of evidence about one's *distinct* genetic lineage turns the trust in the source that gives to be, under normal circumstances the syngamy of two genomes, into puzzlement about one's *own* origin and identity.[13]

The ethical judgment on the technology in question is not the point here. I am not concerned with the ethics of artificial reproductive technologies *per se*, but with the discussion on the more recessive premises about the body, embodiment, and the «embodied self,» premises that drive these technologies in the first place and, more in general, our understanding of reproductive medicine's goals. I ask: how important is it to unpack what remains tacit in the public discussion about a case such as this, and why? What are the philosophical models of embodiment presupposed by medicine, and, consequently, by medical ethics today? How to articulate, in the concrete clinical context, an anthropology that speaks to the na-

---

13 For a stimulating analysis of the way in which biotechnology redefines embodiment, see Marie-Jo Thiel, «La corporéité face à la maladie et la mort,» in *Exploring the Boundaries of Bodiliness: Theological and Interdisciplinary Approaches to the Human Condition*, eds. Sigrid Müller *et al.* (Göttingen: Vienna University Press, 2013), 1-13.

ture of the body as gift, the person as a «unified totality,» and the inter-subjective quality of the body as a medium of relationality. How is one to make *philosophical* sense of those categories, unequivocally rich, yet also culturally opaque?

Perhaps, the judgment of Edmund Husserl in his *Crisis of the European Sciences*, while summarizing the development of modern sciences, offers at the same time a prophetic anticipation of the predicament of contemporary medicine:

> The exclusiveness with which the total world-view of modern man lets itself be determined by the positive sciences and be blinded by the «prosperity» they produced, meant an indifferent turning away from the questions which are decisive for genuine humanity. Fact-minded science excludes in principle precisely the questions which man finds the most burning: questions of the meaning or meaninglessness of the whole of human existence.[14]

With the latter quotation, I come to my first conclusion. The first task of clinical ethics is to foster a search for the meaning of the very questions medicine seems to suspend: the significance of illness and disease, of our human condition as embodied, of birth, suffering and death, and of the service to the ethos of generosity that sustains the healing professions.

2. *Ethics and the Primacy of Experience*

The stethoscope metaphor, symbolizes also the mind-set of the moral philosophy that has dominated and shaped much of our ethical inquiry in medical ethics. In the critical judgment of many, the field has concentrated on a very restricted version of moral language, the language of biomedical quandaries, as well as principles and rules that sustain the rational argumentation for the «solution» (the language here is telling!) of concrete cases.[15]

---

14 Edmund Husserl, *The Crisis of European Sciences*, translated by D. Carr (Evanston, IL: Northwestern University Press, 1970), 5-6.
15 The literature on the methodological debate in bioethics is very extensive. For a thorough examination of the potentials and problems of a principle-based approach, see Raanan Gillon, ed., *Principles of Health Care Ethics* (Chiches-

Such a normative preoccupation with problem solving, however, strongly fosters an attitude of *inattentiveness* – the word recurs again, here – to the moral components and voices that do not communicate in the language of quandary, do not create a challenge for ethical argument, or do not speak with the precision and articulation required in our intellectual culture to attract the attention of «serious» ethical argumentation.

I will delve a bit into the «etiology» of what I would call «the suspension of meaning» in medical ethics. In addition to a critical integration of positivistic attitudes in medicine and the reduction of moral discourse to the normative, one must mention the basic presumption of a cultural situation, which, in the name of post-modernity, raises serious doubts against the possibility of engaging in questions of meaning across moral boundaries.

2.1. The «Suspension» of Meaning in Medical Ethics

A look at the relatively brief history of epistemological developments in medical ethics shows a methodological shift in the fundamental preoccupation of ethicists. The scholars who originally shaped the field of bioethics,[16] did indeed seek a horizon of meaning capable of sustaining ethical discourse that aimed to address the value implications of technological developments in medicine and the life-sciences. Such a horizon of meaning had a pluralistic character: it inspired moral anthropological interpretations in a theological fashion, as well as generally humanistic, when not explicitly non-religious, hermeneutics.

At the end of 1970s and the beginning of the 80s, however, a major shift occurred. Under the increasing influence of contempo-

---

ter: John Wiley & Sons, 1994). For a critical assessment, Edwin R. DuBose, Ronald Hamel, and Laurence O'Connell, ed., *A Matter of Principles? Ferments in U.S. Bioethics* (Valley Forge, PA: Trinity Press International, 1994) and Henk Ten Have, «Approcci europei all'etica nella medicina clinica,» in Corrado Viafora, ed., *Comitati etici. Una proposta bioetica per il mondo sanitario* (Padova: Gregoriana, 1995). Most recently on the methodological debate George Khushf, ed., *Handbook of Bioethics: Taking Stock of the Field from a Philosophical Perspective* (Dordrecht: Kluwer Academic Publishers, 2004).

16 Albert Jonsen, *The Birth of Bioethics* (New York: Oxford University Press, 1998). One might think of people like Paul Ramsey, Josef Fletcher, Hans Jonas, Daniel Callahan, and Warren Reich.

rary Anglo-American moral philosophy, medical ethics developed a preoccupation with the elaboration of normative criteria (so called *principles* of respect of person, beneficence, non-maleficence and justice) that drew their justification from the perspective of a restrictive cluster of concepts in political philosophy.[17] This moral philosophical approach sought to create a consensus based on shared arguments that were divorced from the horizon of meaning and the meaningful narratives that initially inspired them. Under the strong influence of the need to provide a consistent ethical basis for public policy formation, moral philosophy built for medical ethics an area of autonomous reflection centered on the use of principles and rules, together with the ethical theories that articulate them through utilitarian or deontological strategies.[18]

Leon Kass comments critically on the inherent value, or lack thereof, of these principles, when applied to particular cases: they translate mainly into concerns to avoid bodily harms and do bodily good, to respect patient autonomy and secure informed consent, to promote equal access to health care and provide equal protection against biohazards. So long as nobody is hurt, no one's will is violated, and no one is excluded or discriminated against, there is little to worry about. The possibility of willing dehumanization is out of sight and out of mind.[19]

---

17 Classic here remains the work of Tom Beauchamp and James Childress, *op. cit.* and, most recently, Tom Beauchamp and David De Grazia, «Principles and Principlism,» in George Khushf, ed., *Handbook of Bioethics: Taking Stock of the Field from a Philosophical Perspective* (Dordrecht: Kluwer Academic Publishers, 2004), 55-74. In my opinion, however, the commitment to a principle-based approach in bioethics extends beyond the work of the authors mentioned above, for it represents a larger theoretical gesture defining mainstream Anglo-American bioethics. See, for an example, Robert Veatch, *A Theory of Medical Ethics* (New York: Basic Books, 1981) and H. Tristram Engelhardt, *The Foundations of Bioethics*, 2nd edition (New York: Oxford University Press, 1996).

18 The term «principlism» was eventually used, in the wake of critical remarks by philosophers Dan Clouser and Bernard Gert, to designate this approach. See K. Danner Clouser and Bernard Gert, «A Critique of Principlism,» *The Journal of Medicine and Philosophy* 15 (1990): 219-236. Most recently Bernard Gert and Danner Clauser, *Bioethics: A Return to Fundamentals* (New York: Oxford University Press, 1997).

19 Leon Kass, *op. cit.* As an example, one can look at the 1999 document on stem cell research by the National Bioethics Advisory Commission. See *Ethical Issues in Human Stem Cell Research*, vol. 1/ Report and Recommenda-

2.2. The Challenge of Post-modernity

The difficulty of a moral reflection that deals with serious questions of meaning is also blamed on the complexity, both epistemic and moral, defining our «post-modern condition.»[20] Postmodernity entails the definitive overcoming of the modern philosophical and scientific agenda characterized by the optimism of reason; also, the recognition of a structural fragmentation that, forcing us to the inevitability of contextual interpretations, defies any illusion of totality and, with it, the very pursuit of truth as meaningful.

The theoretical indeterminacy of postmodernism as a philosophical label contrasts with the clear dimensions of the problems it creates in practice. Two are particularly important and worthy of reflection. First, the problem of bringing together the plurality of lived moralities, what we call moral pluralism, under the common denominator of a shared ethos, or a «common morality» in bioethical jargon. Secondly, the difficulty of finding a level of discourse that engages differences among moral traditions on questions of substance. Whereas the former problem concerns the moral climate that structures all practical spheres of reality, the latter pertains, more specifically, to the possibility of a theoretical reconstruction of such moral climate, both in terms of ethical discourse and public policy.

Relying upon an analysis of different typologies of moral argumentation, Alasdair MacIntyre observes: «debate between fundamentally opposed standpoints does occur; but it is inevitably inconclusive. Each warring position characteristically appears irrefutable to its own adherents; indeed in its own terms and by its own standards of arguments it *is* in practice irrefutable. But each warring position equally seems to its opponents to be insufficiently warranted by rational arguments.»[21]

A way of solving this predicament is to bridge the gap of cultural fragmentation and the unconvincing nature of arguments between

---

      tions of the National Bioethics Advisory Commission (Rockville, MD: September 1999). The authors of the document can, ultimately, agree on *safety* as the only moral constraint against the practice of reproductive cloning.
20   Richard Bernstein, *The New Constellation: The Ethical-Political Horizon of Modernity/Postmodernity* (Cambridge, MA: MIT Press, 1992), 59.
21   Alaisdair MacIntyre, *Three Rival Versions of Moral Enquiry: Encyclopedia, Genealogy, and Tradition* (Notre Dame: University of Notre Dame Press, 1990), 7.

moral agents by surreptitiously reducing ethics to a purely regulatory task, thus progressively diluting the distinction between the legal and the moral. The tendency to sublate ethics under the law rests upon the assumption that dialogue on moral convictions separates people; only the law, now invested with a kind of soteriological meaning, can bring moral differences under the banner of unifying social rules.

I believe such a notion of ethics not only discourages meaningful exchange across different traditions; it actually entails, in the long run, a neutralizing effect upon the *content* of moral conversation as such. An ethical discourse capable of laying out a territory of discussion, where differences can meet and confront each other, will be expunged from the theoretical agenda of ethics. The latter will, at best, provide a grammar of procedural conditions upon which differences among moral traditions may co-exist, without ever coming into contact with one another. Rather than focusing on questions of intrinsic value, moral discussion is expected to articulate, at best, rules of reciprocal engagement – the *a priori* of the communication – that will allow each moral participant to remain in a safely protected, yet totally separated, moral universe.

In order to overcome the problems posed by our postmodern condition, it seems imperative to rethink the meaning and purpose of ethical dialogue across different traditions and within the public realm of «secular» society. One must move here between the Scylla and Charybdis of a twofold dead end: the reduction of ethical rationality to a purely procedural function of political regulation, and the intellectual impotence toward an incommensurable pluralism that legitimizes the relativity of different points of view.

3. Retrieving Meaning

Moral reflection, especially in the existentially charged realm of clinical ethics, does not begin with the application of normative principles, nor can it be sustained by an attitude of resignation toward the pursuit of the good. It begins, rather, with a free and open confrontation with the meaning of the experience we face.[22]

---

22 For a paradigmatic application of this concept to the field of bioethics see Warren T. Reich, «Ein Neues Paradigma: Erfahrung als Quelle der Bioethik,»

Experience is not merely an objectively described empirical entity, though empirical analysis might have an important part in it. Already at the level of its etymological meaning, «experience» entails a reference to subjective intermediation: experience speaks of the predicament of *peiros*, of the passing or living through a situation of crisis, and of the personal growth effected by such existential challenge. We are summoned by meaning in an integral fashion, and the radicalness of such call can only be answered by a synthetic act of reciprocity, a response to an intrinsic source of value (*Wert-antwort*) to borrow from the phenomenological tradition, which we confront with that most intimate and all-encompassing definition of the self we identify with the notion of *con-science*.[23]

I hear the objection of clinicians: questions of meaning can only have a secondary importance, when tough decisions need to be made, in the *hic et nunc*, the here and now of concrete clinical challenges. In this perspective, «gazing into the meaning of things» can be, at best, an interesting theoretical exercise; at worst, a useless distraction that utterly fails to address the call of the moment. It does indeed make good sense to put meaning in a secondary place and give primacy, instead, to one's immediate reality, when confronting the premature cry for survival in the neonatal intensive care unit, or the puzzlement over the competence of a surrogate decision maker, acting on behalf of an elderly patient now mentally incapacitated.

At the same time, when the larger world of wellness, suffering, being struck with affliction, being sick, dying and so on, does not find its proper way into the decision-making process of clinical ethics; when, instead, clinical ethics relies, in a rather mechanical fashion, on an algorithmic approach to problem-solving, with

---

in Dietmar Mieth and Klaus Steigleder, ed., *Ethik in den Wissenschaften: Ariadfaden im technischen Labyrinth?* (Tübingen: Attempto Verlag, 1990), 270-292.

23 For the notion of *Wert-antwort* (value response) see Dietrich von Hildebrand, *Ethik*, 2nd edition (Stuttgart: Kohlhammer, 1973) [Gesammelte Werke, Bd. II]. A careful study of the notion can be found in Josef Seifert, «Dietrich von Hildebrans philosophische Entdeckung der "Wertantwort" und die Grundlegung der Ethik,» in *Truth and Value: The Philosophy of Dietrich von Hildebrand* (Bern: Peter Lang, 1992), 34-58. Also Bernard Lonergan, *Method in Theology* (New York: The Seabury Press, 1979), 27-57.

its plethora of predefined categories – advance directives, consent forms, values inventory, etc., we end up creating obstacles to good habits of moral reasoning, hindering the disclosure of moral meaning while, quite paradoxically, producing the «right» answer for the quandary at stake.

Attending to the moral meaning of concrete situations entails recognizing that formal modes of logical argumentation are only *derivative* functions of the moral language. Prudential or practical reasoning unfold as dimensions of a more original form of mindfulness, a synthetic act of *discernment*, which includes elements of detecting, sensing, sifting, discriminating, comparing, connecting, and, ultimately, deciding (compare Pascal's *esprit de finesse* against *esprit de geometrie*).

Richard Zaner puts the matter brilliantly, when analogizing such phenomenological probing with the work of a detective: «(One must) deliberately be alert to the multiple ways in which participants interrelate and variously experience and interpret one another and, with that relationship, the relationship itself. Even a brief moment reveals a number of interrelated voices, each with its own emotional, volitional, valuational, and cognitive tonality... The ethicist's involvement is thus *a work of circumstantial understanding*.»[24]

This mode of moral reasoning is certainly relevant to all settings, but it becomes particularly important when questions of meaning need to be addressed beyond the application of normative strategies for «solving» moral problems. In fact, relying upon these strategies might precisely be a way to by-pass larger questions of meaning, questions for which ethicists have long since declared their incompetence, and therefore, gladly pass on to the «care» of alternative agencies, spiritual care personnel, psychologists, etc.

---

24   Richard Zaner, «Experience and Moral Life: A Phenomenological Approach to Bioethics,» in A Matter of Principles? *op. cit.*, 211-239. The same emphasis on the particularity of moral reasoning, especially in medicine, can be found in casuistic and in hermeneutic approaches to bioethics. For the former see Albert Jonsen and Stephen Toulmin, *The Abuse of Casuistry: A History of Moral Reasoning* (Berkeley: University of California Press, 1988); for the latter Fredrik Svenaeus, *The Hermeneutics of Medicine and the Phenomenology of Health: Steps Towards a Philosophy of Medical Practice* (Dordrecht: Kluwer Academic Publishers, 2000).

Let me draw my second conclusion, then. Clinical ethics functions as the anamnesis of meaning, not only for medicine as a practice, but, more importantly, for ethics itself. For sure, the search for meaning does not end in a kind of bracketing of the ideological presuppositions that generate ethical discourse. As Paul Ricoeur has suggested, meaning cannot be reached from a position of neutrality (*Voraussetzungslosigkeit*) that fails to objectify the ideological prejudices already operative in the archeology of meaning.[25] On the contrary, because it puts questions of meaning at the center of its attention, clinical ethics becomes better equipped at unmasking all sorts of ideological mystifications. Consider, for an example, the notion of medicine that feeds into a mode of thinking defined by the presumption to «fix everything.» It is an insidious presumption affecting modern medicine, with disastrous consequences for the motivational and intentional agency of the physician. A medicine with no appreciation for the deepest matters of our humanity will not be able to see how *caring* can still be part of the definition of medicine, when *curing* is no longer possible.[26]

Indeed, when treatment options cease to offer a *meaningful* hope of recovery, there appears to be no patience for the unsuspected disclosure of *another* meaning, one that escape *production* of any kind because it can only be *received* in the openness of attentive receptivity.[27] It is the call to meaning generally entailed by situations of «vulnerability»: the genetically defective fetus, the handicapped child, the elderly patient. In these cases, one comes to a dead end: nothing more can be done, or so we think.

---

[25] Paul Ricoeur, *The Conflict of Interpretations* (Evanston: Northwestern University Press, 1974). Also Fredrik Svenaeus, *op. cit.*, 140-146.

[26] For a thorough investigation of the historical roots and contemporary applications of an «ethics of care» see Ludwig Haas, *Für kranke Menschen sorgen: Die Bedeutung der «Cura» für ethisches Handeln im Gesundheitswesen* (Münster: LIT Verlag, 2000)

[27] What is at stake here is a kind of phenomenological *epoche'* on a naturalistic attitude still closed to the deepest intelligibility of reality. I speak of «attention,» with reference to the Simone Weil, in Roberto Dell'Oro, «La dimensione ermeneutica nell'esperienza della cura. Una ricostruzione fenomenologica,» in Corrado Viafora and Silvia Mocellin, ed., *L'argomentazione del giudizio bioetico: teorie a confronto* (Milano: Franco Angeli, 2006), 183-198, at 197.

Conclusion

I have pleaded for a notion of clinical ethics defined by a twofold retrieval of meaning. Relative to medicine, clinical ethics functions as a reminder of what defines medicine as a human practice, the nature of its action, and its ultimate purpose. In reminding medicine of the moral sources that nourish its doing, clinical ethics also functions as reminder of the ultimate nature of ethics in medicine: to be an interpretation of moral experience as the condition for the articulation of moral principles and norms. In the *anamnesis* of meaning that always inhabits experience, whether of health care professionals or patients, clinical ethics finds its ultimate purpose and scope, as well as the condition of its own significance in the clinical world.

CHAPTER 11
THE PHYSICIAN PATIENT RELATIONSHIP
A Philosophical Perspective on Medical Professionalism and Virtues

In this chapter, I address the problem of the relationship between virtues and professionalism. This is a broad theme, and, to a certain extent, endless, both in relation to the subject of professionalism and the nature of virtue. Yet because what matters here is the notion of their relationship, it is necessary to make certain preliminary choices in how to approach the topic, both in content and in the method.

As for *content*, I will focus on *medical professionalism*. With respect to the *methodological* aspect, I would say that the plausibility of a virtue ethics in medicine depends on a number of premises: how one understands virtue ethics to begin, but also the broader question of the goals of medicine, and of the relation between physician and patient.

The reference to the goals of medicine constitutes the *ontological* condition for an *ethical* articulation defined by reference to virtues: virtue perfects an intrinsic potency waiting to be actualized, i.e., something that is already given in its phenomenological essentiality. In this sense, it seems legitimate to connect the difficulty of talking about virtue ethics in medicine to the crisis of professionalism in medicine.

I submit that the *crisis of professionalism* in the medical field depends, more deeply, on the progressive loss of the goals of medicine, leading, in turn, to a crisis in the understanding of the specificity, both phenomenological and ethical, of the doctor-patient relationship.[1]

In what follows, I want to focus on the physician–patient relationship as the phenomenological *locus* wherein both the ques-

---

1   Edmund Pellegrino, Robert Veatch, and John Langan, ed. *Ethics, Trust, and the Professions: Philosophical and Cultural Aspects* (Washington, D.C.: Georgetown University Press, 1991).

tion of medical professionalism and virtues emerge in their full clarity. Expressed in negative terms, my thesis is this: without a precise understanding of the singularity of the doctor-patient relationship, it is not possible to speak of an *ontology* appropriate for the health care professions, nor of a medical *ethics* based on the concept of virtue.

Philosopher Robert Sokolowski attempts to reconstruct the nature of the professions through an analysis of their essential dimensions. Central to them is a «*relationship of trust*»: «... The natural order in a professional relationship is what allows the relationship to be more than a merely contractual exchange... Both the client and the professional are subject to the nature of the relationship.»[2] The ontological dimension, that is, an understanding of the nature of medicine, subtends the articulation of the ethics of medicine. From the same perspective, Gilbert Meilaender maintains that virtues are intrinsic to any profession: «... Virtues [are] inherent in certain professional commitments...and not simply specifications or applications of a universal common morality.»[3]

If one takes the development of the bioethics literature on the doctor-patient relationship as an indication of the direction of contemporary thinking on the topic at hand, one will not be surprised to find that the issue seems to concern more the legal than the ethics literature.[4]

Ours has become a society evermore complex and more litigious. Yet, at a closer look, the very practice of medicine has turned litigious as well. The doctor-patient relationship represents an increasingly consistent chapter in malpractice law, which seeks to protect the patient from the errors and incompetence of the doc-

---

2   Robert Sokolowski, «The Fiduciary Relationship and the Nature of the Professions,» in Pellegrino, Veatch, Langan, J., *Ethics, Trust, and the Professions*, 23-39).
3   Gilbert Meilaender, «Are Virtues Inherent in a Profession?» *Ibidem*, 139-158, at 145.
4   Paul Ramsey, *The Patient as Person* (New Haven, CT: Yale University Press, 1970); William F. May, *The Physician's Covenant: Images of the Healer in Medical Ethics* (Philadelphia, PA: Westminster Press, 1983); David J. Rothman, *Strangers at the Bedside: A History of How Law and Bioethics Transformed Medical Decision Making* (New York: Basic Books, 1991)

tor, and, vice versa, the doctor from potentially baseless accusations of the patient.[5]

My contribution, however, seeks to be of a different nature. Without disputing the relevance of the legal dimension as central to an *empirical* analysis of the doctor-patient relationship, I want to provide here a *perspective of meaning*, one that might stand as a kind of ultimate horizon within which all the other perspectives fold. In this manner, I hope to remain faithful to my specific philosophical vocation, without any blurring of competencies.

Philosophical reflection in bioethics – the interdisciplinary nature of the field notwithstanding, still carries the burden and responsibility of a *foundational* contribution: I mean «foundational» on account of the truth it elicits and brings to light. The contribution of philosophy does not impose itself – by reducing all contributions to the partiality of a single language. It remains as an offering of meaning that invites to reflection, a free *donne á pénser*.[6]

The Metamorphosis of Medical Ethics

The modern medical practice seems to be characterized by a paradoxical situation, i.e., a kind of disproportion between the complexity of the problems healthcare professionals are called to address and the conceptual tools of an ethical nature at their disposal to resolve them. The explosion of a bioethical reflection, at first glance, represents a response to the emergence of this disproportion in the practice of medicine, and likewise, a

---

5   Ian Kennedy, *Treat Me Right: Essays in Medical Law and Ethics* (New York: Oxford University Press, 1988); G. Timothy Johnson, «Restoring Trust between Doctor and Patient,» *New England Journal of Medicine* 322 (1990): 195-197; Jerry Menikoff, *Law and Bioethics: An Introduction* (Washington, D.C.: Georgetown University Press, 2001).

6   The notion of *donne à pénser*, popularized by French philosopher Paul Ricoeur, refers to the symbol in its function of «giving to think.» It entails an offer of meaning that opens up new perspectives, without any pretension of closure or systematic completeness. I borrow the term to suggest the same with respect to my contribution, which is philosophical, and, thus, «foundational,» in the sense of as an «invitation to think» beyond the boundaries of professional reflection.

systematic attempt to broaden the concerns of traditional medical ethics.[7]

The limits of the latter also bear on the topic at hand. In the classical tradition of medical ethics, the physician–patient relationship relies on classical categories, which are, in their turn, articulated in terms of moral *virtues*: confidentiality, respect for the patient's privacy, faithfulness in the duty to provide treatment, etc.

One might ask how these categories continue to map out the ethical landscape of contemporary quandaries. Consider the explosion of medical information, and the difficulty to keep it under control, the fragmentation and compartmentalization of the medical acts, and the problem of controlling the costs of a healthcare system that are becoming increasingly unmanageable.

Although emerging from within the broader field of medicine, these problems are literally *meta-medical*. Thus, medicine finds itself inserted into a dialectical process of reinterpretation that calls into question the anthropological and moral basis to which the physician feels, almost instinctively, bound. Inasmuch as she or he relies upon a *techne iatrike*[8] oriented toward the restoration of health, the physician pursues an end, a telos, which is not justified in and of itself, on the basis of some abstract scientific value; it can be properly understood only in relation to a good of a nature personal – the good of the patient.

As a personal good, the *good of the patient* cannot be identified simply with the *clinical good*. Consider the meaning of this statement with respect to end-of-life situations. A physician may be inclined to offer yet another treatment, and to prolong life, in a situation the patient deems unsustainable and undignified. If so, the clinical effectiveness of a treatment on the physical condition of the patient might be seen by the latter as incongruous with what she or he perceives as

---

[7] Edmund D. Pellegrino, «The Metamorphosis of Medical Ethics: A 30-Year Retrospective,» *Journal of the American Medical Association* 9 (1993): 1158-1162.

[8] *Techne iatrike* is the term with which the Hippocratic corpus refers to medicine as an «art.» As such, medicine occupies almost like an intermediate space between science (*episteme*) and opinion (*doxa*). It is not science, because it deals with practical, rather than theoretical diemensions of reality. It is not mere opinion, for it relies upon knowledge, if always only within the limits of experiential application.

consistent with his or her good. Such good entails dimensions that surpass purely clinical considerations, to including what the patient considers worth pursuing or not in light of anthropological and even theological ideals. A case in point is the situation of a Jehovah's witness, who refuses a blood transfusion needed to save his or her life, or a person who sees the amputation of a gangrenous limb as contrary to his or her understanding of physical integrity.

The *encounter*, the *dialogue*, the *relation* – these remain the ultimate metaphors of medicine, even in the context of the progressive enhancements of continually new content, or the inevitable shifts of historical paradigms. Hence, it is necessary to return to these metaphors, as one returns to an inexhaustible source of meaning that continues to nourish thought.

The Physician Patient Relationship

In highlighting the particular nature of the physician–patient relationship, I would like to begin with an analysis of the medical secret and the virtue of medical confidentiality, perhaps the most relevant aspect of this unique human relation.[9] The text of the Hippocratic Oath, famously attributed to the fifth century and redacted by the school of Kos, reads:

> And whatsoever I shall see or hear in the course of my profession, as well as outside my profession in my intercourse with men, if it be what should not be published abroad, I will never divulge, holding such things to be holy secrets.

Let us examine the text of the Oath in its symbolic character, more than its literal meaning. An exegetical discussion of the Hippocratic text, relative to the precise meaning of the terms it defines–for example the meaning of «care», «divulge» and «holy secrets», can and must continue. According to eminent historians of medicine, all the notions in question here are questionable and, to some degree, uncertain. Also, the medical secret seems to have had a pragmatic,

---

9   Pierre Loiret, *La Théorie du Secret Médical*. (Paris: Masson, 1988).

rather than moral value: it elicits the language of precautionary conduct, rather than that of duty in the strict sense.

Notwithstanding the historical clarification, the secrecy continues to be part of the Oath by which one enters the medical profession, an integral part of the public declaration (the Latin *profiteor* comes from the Greek *pro-phemi*: it means «to proclaim aloud in front of an audience») whose terms, albeit open to interpretation, are not just some *flatus vocis*.[10]

Yet the question «why the secret?» remains legitimate. What is the point of the *separation* that, as the word's Latin root suggests (*secretum* from *secernere*), sets aside or singles out the doctor-patient relationship from the rest of human interaction? I submit that the separation in question refers to, at least, two elements: the object (1), and the nature of the physician–patient relationship (2).

1. *The Object of the Relationship*

The physician–patient relationship distinguishes itself from all others in virtue of its object. All human communication is communication about something, even when marked by the silence of voiceless allusions.[11] This «something» in the doctor-patient relationship is the request for help with which the sick person implicitly addresses the physician in her or his clinical narrative, through which a history of personal suffering and pain progressively comes to light. Most often, the sick does not know what is it that is causing the problem; in the indeterminate nature of the illness the call for help is also an indication of a person's vulnerability, a sign of the asymmetry that defines the relationship.

---

10   I mean that such terms have a real meaning, rather than a purely nominal character. They are not just names without a corresponding objective reality. The point in question is polemical with respect to the tendency to deconstruct the central commitments of medical professionalism, what Edmund Pellegrino terms the «internal morality of medicine,» in favor of a notion of medical exchange defined by social constructionism. In this latter case, the morality of medicine would be no longer *internal* to the practice of medicine but *external* to it. This in the sense of being defined by the morality of the society within which medicine operates.

11   Pedro Lain-Entralgo, «La palabra y el silencio del medico,» in idem, *Ciencia, tecnica y medecina* (Madrid: Alianza, 1986), 234-247.

The patient waits, in an attitude of trust, for the doctor's verdict, further buttressing the state of passivity (*pati*) already imposed by the illness. The particular pathology exhibited by the patient enhances her or his state of nakedness before the doctor, and this not only in a physical sense. The patient lays bare before the doctor that from which she or he suffers. Since she or he cannot hide her ailment from the clinical eye of the expert, *she or he becomes, in a real sense, her or his illness*.

Paul Ricoeur speaks of suffering as that dimension of intimacy which defines individuality: *la souffrance est, avec la jouissance, la retraite ultime de la singularite*[12] (Ricoeur 1996, 22). Even if the physician has to «bracket» the personal dimensions that accompany the ill person, so as to let the diagnosis focus on what is causing the illness in the patient, she or he cannot totally ignore the subjective responses provoked by any pathology: pathologies have a concrete reality that always appeal to the metaphysical dimension of alterity, or otherness. Only with an attentive disposition and readiness to listen can the physician arrive at a diagnosis that touches the root of what is troubling the patient. The narrative of particulars and circumstances by which the symptoms of the person make themselves manifest is an integral part of the process with which *the dis-ease* – the *objective* dimension of the patient's malaise, comes to light in the patient's *illness*, that is, her *subjective* appropriation.[13]

Furthermore, the patient's narrative acquires therapeutic meaning, at least from a subjective standpoint. The capacity to bring illness to the level of language stands in powerful contrast to the meaninglessness of loneliness, and ultimately to all those components that point to the etiology, or cause, of the particular suffering provoked by illness.

---

12 Paul Ricoeur, «Le trois niveaux du jugement médicale,» *Ésprit* 12 (1996): 21-33. The phrase is pregnant with meaning, and thus, almost impossible to translate. I would say, «suffering, like joy, belongs to the most intimate dimension of the person.» Indeed, suffering, like joy, defines a space of withdrawal (*retraite*), of separation from others and from the world. In this separation the person is herself/himself in the most elemental, singular, way.
13 See Roberto Dell'Oro, «Interpreting Clinical Judgment: Epistemological Notes on the Practice of Medicine,» in Corrado Viafora, ed., *Clinical Bioethics: A Search for the Foundations*. Dordrecht: Springer, 2005), 155-168.

To be able to *talk* about one's illness is, in itself, a cathartic act, in which a *two-fold alienation* – isolation and lack of meaning – begins to be overcome: the pain endured in secret is made public, and the sense of meaninglessness one undergoes is brought to the fore, now confronted with the possibility of a «logical» explanation.[14] In speaking of her or his illness, the patient exceeds the objective dimensions of her disease, alluding, in this «more,» to the existential predicament of her or his personal condition.

The object of the secret, which the physician cannot communicate to others, but must keep jealously to herself or himself, is not just the name of the disease. Rather, it points to the gap, otherwise unbridgeable, between the clinical problem and the world of the patient that undergoes it. Indeed, only at the price of silence is the doctor given access to such a world of intimacy. Thus, the break of the medical confidentiality signified by the secret is something more than just a professional infraction. By reason of its object, such infraction amounts to a form of betrayal, a lack of respect for the personal dignity of the patient. This dignity is maintained only because the world of the ill person, objectified in her or his disease, is kept close to the curiosity of the prying eyes, the indiscrete insolence of vulgar chattering.

One might say that medical confidentiality protects the *difference* between the disease and the world in which the patient lives, between the objectivity of the symptoms the patient suffers and the subjective appropriation of that suffering, signified by the name of the illness. In a parallel vein, medical confidentiality protects the *identity* of the illness and of one's personal world, the identity of the *physical* (Körper) and of the *lived body* (*Leib*).[15]

Illness does not exist in and of itself, independently of the subject that it affects, but always and only as the objectification of an «ill subjectivity.» While the body of the ill person undergoes the auscultation of a stethoscope, the physician will

---

14   On this, see Drew Leder, «Clinical Interpretation: The Hermeneutics of Medicine,» *Theoretical Medicine* 11 (1990): 9-24.
15   The distinction is originally of Edmund Husserl. For two contemporary articulations, see Virgilio Melchiorre, *Corpo e persona* (Genova: Marietti, 1987) and Umberto Galimberti, *Il corpo* (Milano: Feltrinelli, 1987).

continue hearing the voice of the patient who screams: «I *am* my body, this body.»[16]

If this is true, one can conclude that the violation of the patient's dignity in breaking confidentiality expresses a fundamental blindness: that of the physician, but, more generally, the blindness of modern medicine, now become unable to recognize the symbolic character of an ill body, to trace a personal presence in the embodied evidence of the sick body.

2. *The Form of the Relation*

What kind of relation can sustain the weight of such intimate and personal communication? If the medical act represents the answer to a request, articulated in the exercise of technical competency, yet not fully exhausted by it, must it not possess the *form* of the promise, the assurance of fidelity? Will not the confidentiality secured by the medical secret be the necessary condition for protecting such intimacy?

These questions are clearly rhetorical, but they do bring to the fore in an unmistakable manner the fact that the indiscretion and rupture of medical confidentiality constitute a paradigmatic expression of the abuse of power.

Far from being a supererogatory expression of medical ethics, as such optional, the promise of confidentiality speaks to a need for *justice*, *commutative* justice, by which the physician renders the patient what is hers or his: the right to an intimate personal space for the recovery of dignity. After passing through the expropriation of nakedness and shame, of being touched and questioned, the patient returns, at last, to be a subject. Confidentiality grounds the personal character of the relation and sustains the conditions for its *just* articulation: the relation between two persons, one vulnerable, the other with the power to exploit such vulnerability, is rescued from its asymmetrical character, and restored to a relationship of equals.[17]

---

16 Richard Baron, «An Introduction to Medical Phenomenology: I Can't Hear While I'm Listening,» *Annals of Internal Medicine* 103 (1985): 606-611.
17 Richard Zaner, *Conversations on the Edge: Narrative of Ethics and Illness* (Washington, D.C.: Georgetown University Press, 2004) and idem, «A Med-

This need for justice is historically expressed by the restoration of the *patient's autonomy*, thus overcoming the paternalistic shortcomings of traditional medical ethics. The Oath itself, and in general the Hippocratic *corpus*, are silent about the imperative of truth telling, for an example, with respect to the communication of medical diagnosis. Even the secret does not serve a function of redress, thus contributing to the dialogical parity of the relation. After all, it would be anachronistic to read into an ancient text from the fifth century B.C. the traces of modern personalism.

On the contrary, the physician remains the one who defines the space of the secret and the terms of confidentiality. Moreover, as Diego Gracia notes, «in the whole Oath, the patient is treated like an handicapped, not only from the physical point of view, but also from the moral.»[18]

The still relatively recent history of bioethics is, in many ways, the attempt to recast the relevance of justice in the relation between physician and patient, and to articulate all its dimensions in the many areas of medical practice. New categories have emerged, previously unknown to the tradition of medical ethics: in addition to the «rights of the patient,» one finds «informed consent,» «decision-making competence,» «treatment alternatives and prognostic communication,» «advanced directives» for the suspension of treatment in terminally ill patients, etc.

Yet, paradoxically, the push toward autonomy has not brought about a reinterpretation of the physician patient relation in terms of a relation of friendship, which according to the great Spanish medical anthropologist and humanist, Pedro Lain-Entralgo, represents the original metaphor of the human relation, *inasmuch as it is* a relation among equals.

In Book VIII of the *Nicomachean Ethics*, Aristotle poses the question: «Whom might I call a friend? My friend can only be the one who, by nature, is *another myself*.» Referring to the Aristotelian definition of *philia*, Lain-Entralgo comments: «Whom then can I call friend? Indeed, only he who, being by nature equal to me, is

---

itation on Vulnerability and Power,» in Carol Taylor and Roberto Dell'Oro, ed., *Health and Human Flourishing: Religion, Medicine, and Moral Anthropology* (Washington, D.C.: Georgetown University Press, 2006), 141-158.

18   Diego Gracia, *Fundamentos de Bioetica*, 68.

capable of *speaking with* (and not just *at*) me, bring about my personal good, as I, in turn, do for him. From this, one deduces that the exercise of medicine when the doctor aspires to be, from the medical point of view, friend to his or her patients, could be defined along the following lines: affable giving of technical help to another myself who is ill»[19]

American bioethicist Al Jonsen analytically documents, from an historical point of view, the relationship of coherence that binds the patient's movement toward autonomy with the general movement of civil rights in American society in the 1960s.[20] The notion of autonomy clearly throws into relief the need to acknowledge the responsibility of patients, together with a renewed awareness of the moral implications accompanying treatment decisions. It is, nonetheless, a notion rooted in an attitude of suspicion toward the medical establishment, if not the individual physician, and thus the opposite of a trusting relation generated by friendship.

The physician patient relation has evolved in the direction of a *contract-like exchange*, regulated by legal sanctions and clear-cut stipulations of services. The now regained symmetrical, rather than asymmetrical, quality of the relation does not necessarily contribute to the reciprocity of personal encounter, where terms of giving and receiving are, so to speak, functional to actualizing the identity of each subject in the relation. Indeed, a physician does not exist in the abstract, i.e., independently of her or his relation to a patient.[21] She or he becomes a doctor *in* the relation, and *by virtue* of, the relation itself. Hence the student of medicine will *become a doctor* in the public profession of the Hippocratic Oath, i.e., in the act of proclamation (*professio*), when she or he declares her readiness to offer care and healing to those in need of it. The patient is the very *raison*

---

19 Pedro Lain-Entralgo, «La palabra y el silencio del medico,» 239. The Spanish reads: *afable donacion de ayuda tecnica al semejante enfermo*. The statement has an obvious *analogical meaning*, for physicians in different medical disciplines encounter the patient *differently*. The relation to the patient, in the case of a family physician, is certainly different from the one of, say, the radiologist. Still, even the latter encounters the patient, if only in the mediation of an image.
20 Albert Jonsen, *The Birth of Bioethics* (New York: Oxford University Press, 1998)
21 The term profession, as one can easily deduce from its root *profiteer*, refers in fact to the act of «proclaiming aloud and in public.»

d'être of the physician, the good of the patient the driving force of a physician's competence and dedication.[22]

The bilateral nature of a contract-based relation is, on the other hand, a *functionalistic transaction*. In it, the subjects in the exchange remain in the background, so to speak, their identity objectified in the «exchanged product,» in the case of a medical transaction, the production of health. The result is the neutralization of the physician patient relation. Progressively, even the semantics defining the relation gets stripped of its personalistic connotations, replaced by an equivocal, economical semantics: the physician becomes a «health professional,» the patient a client, consumer etc.

The apparent functionality of a similar strategy cannot hide the grave consequences it introduces in the understanding of the physician patient relationship, as well as in the reduction of illness to an observable, objective state of affairs.

Indeed, the success of a *contractualistic* notion of medicine is due in large part to a positivistic view of illness. The more the categories of quantity, technological proficiency, and financial incentives prevail in the definition of medical professionalism, the more easily the terms of the contract between doctor and patient will be measured by quantifiable standards of performance and functionalistic criteria.[23] The most dramatic consequence of this tendency, as already noted by Ivan Illich in his famous book *Medical nemesis*, will be the reduction of healthcare to commercial transaction, of

---

22 See Edmund Pellegrino, «Towards a Reconstruction of Medical Morality: The Primacy of the Act of Profession and the Fact of Illness,» *Journal of Medicine and Philosophy* 4 (1979): 32-56. As for other practices within the capitalist society, also the practice of medicine falls under the criterion of technical performance. The philosophical critique of the Frankfurt School is directed at the reduction of every human relation to *Leistung*.

23 The point can already be found in Theodor Adorno, *Minima Moralia. Reflections from a Damaged Life*, trans. E.F.N. Jephcott (London: New Left Books, 1974). Of course, respect for the principle of autonomy need not imply the adoption of a contractualistic model. Pellegrino and Thomasma offer an attempt to mediate the modern principle of autonomy with the traditional perspective of beneficence in Edmund Pellegrino and David Thomasma, *For the Patient's Good: The Restoration of Beneficence in Health Care* (New York: Oxford University Press, 1988).

medical service to marketable good, and of the patient, with her or his vulnerability, to neutral client.[24]

For sure, at the heart of the now re-defined relation in question there remains a good, but it is no longer be the good of the patient, considered in its integral wholeness (remember that «healing» stands semantically close to «wholeness»). Indeed, in this new definition, the healthcare professional knows little of such integrity, nor of the human dimensions that define it. The very principle of autonomy, when understood in the rigid terms of political liberalism and juridical contractualism, maps out the space of individual decision making as unassailable, one for which the individual only can give justification, based on idiosyncratic reasons. When the physician patient relationship has been stripped of its personal connotations, the medical transaction becomes a neutral exchange. The physician might know that she or he ought to respect the decision of the patient, but without knowing how this might actually be good for her or him, what good it might produce. At worst, the *good* of the patient is understood simply as his or her *right* to be left alone, and, more generally, as freedom to choose what one wants.

Conclusion

I began this essay with a reflection on the notions of medical secret and confidentiality. I then turned to the essential dimensions of the physician-patient relationship with respect to both content and form. Medical professionalism can be understood only in light of the latter, as it is born of an ever renewed consciousness of the phenomenological specificity of medicine as «*pro-fession*», that is, as the responsibility (from *respondere*) of a professional who, in *friendship*, as Pedro Lain-Entralgo suggests, openly declares and promises her availability and commitment to help a vulnerable subject.

Without this consciousness, medical professionalism is reduced to a bureaucratic and impersonal transaction. It becomes a technical exchange, one that, even when defined by criteria of

---

24 Ivan Illich, *Medical Nemesis: The Expropriation of Health* (New York: Random House, 1976).

scientific competency, is still reduced to neutral reciprocity and economic convenience. Such contractual relation is unable to «ground» the ethical importance of personal dedication and the moral imagination sustained by virtuous habits of compassion, trust, and confidentiality.

One might ask whether the phenomenology of the doctor-patient relationship outlined above might re-orient modern medicine, and restore the ontological basis on which ethical virtues rest. An *ontology of medicine* belongs to another chapter, yet it seems necessary in the articulation of a systematic medical ethics. Such ontology will ground both virtues and moral principles, such as autonomy, beneficence, non-maleficence, and justice. Insofar as they are signs of a universally shared ethics (common morality), moral principles stand on the solid ground of medical professionalism, the expression of the social solidarity with which a community constantly integrates the moral experience of individuals through dialogue and communication.

CHAPTER 12
THE MARKET ETHOS AND THE INTEGRITY OF HEALTH CARE

Clarifying the Title

In this contribution, I intend to reflect on a notion I find essential to the ethical discussion on health care in our country, one that seems in danger of losing its meaning: the notion of professional integrity as it applies to health care. Some clarifications on the meaning of such a notion must first be made, then one can argue that integrity in health care, in particular, the integrity of individual professionals and of institutions depends on a large cultural integrity defining society at large. This should provide some clarity to the conditions upon which a plausible retrieval of the notion of professional integrity could be made possible.

What Kind of Integrity?

«Integrity and compliance programs» have proliferated in today's economically stressed healthcare environment. These programs proclaim loudly the commitment of the institutions to the highest standards of morality and articulate the values upon which business relationships among members, customers, employees and stakeholders must be conducted. Yet, in spite of its increasing prevalence, the corporate language of integrity is far from being univocal. Its understanding presupposes familiarity with the corporate reality. For those who do not distinguish the delivery of healthcare from ordinary commerce, integrity and compliance are complementary. The organization that acts to achieve its commercial purpose, i.e., that behaves efficiently, productively, profitably, has integrity. From this perspective, the integrity of healthcare organizations, as with other commercial organizations, will depend upon compliance with the capitalistic vision of the marketplace and with the forces of commercial culture. Indeed, the mission of many health care institutions betrays a notion of integrity which is entirely a function of the maximization of profit. Integrity simply means compliance with the organization's economic interests.

With relation to the health care industry, the promise of the market is to increase competition and to rationalize the system without necessarily altering the fundamentally moral nature of the clinical exchange or undermining the professional standards entailed. Yet, many questions remain, both in relation to the ability of the market to deliver what it promises and in relation to its willingness to save the intrinsic morality of medicine.

Critical thinking carried out by several schools of thought have long since warned against the subtle social and ideological implications of the market and of capitalistic mechanisms in general. Indeed, the ideological spectrum is quite wide, comprising neo-Marxist philosophers of the so called «Frankfurt School,» such as Adorno, Horkheimer and Habermas, as well as the most recent social encyclicals of the Catholic church.[1] This body of critical thinking makes us aware of the social tendencies inherent in the absolutization of the market: the tendency to neutralize non economic values such as compassion, empathy, care, concern for the common good; to reduce interpersonal relations to mechanistic exchange; and to replace the experience of gratuitousness and esthetic appreciation with the concern for the production of material goods.

In fact, economists are also becoming increasingly sensitive to both anthropological presuppositions and broad social consequences of the market. They contend that total reliance upon its presumed self-correcting dynamics is normally accompanied by two related dangers. First, market institutions drive out extra-market institutions. Faced with competitive pressure, non-market institutions such as charity hospitals begin looking and behaving more like for-profit ones. Second, market norms drive out non-market norms. To quote Robert Kuttner, «when everything is for sale, the person

---

1 See Theodor W. Adorno, *Minima Moralia: Reflections From Damaged Life* (1974); The Encylicals of Pope John Paul II, *Laborem Exercens, Centesimus Annus and Sollicitudo Rei Socialist;* See more broadly, Charles E. Curran & Richard A. McCormick, *Readings in Moral Theology No. 5: Official Catholic Social Teaching* (New York: Paulist Press) (1986).

who volunteers time, who helps a stranger, who agrees to work for a modest wage out of commitment to the public good, who forgoes an opportunity to free-ride, begins to feel like a sucker.»[2]

But what are the moral challenges that the increasing commodification of medicine poses to both health care institutions and health care professionals?

Making Sense of Moral Constraints in Today's Health Care Market

Although the literature is already filled with recriminations concerning the bad influences of the market on the practice of medicine, it is difficult to find good arguments explaining and defending those recriminations.

I contend that many of the problems we face in the delivery of healthcare today–in particular, the problem of measuring the influence of the market against the canons of professional integrity-stem from the inability or the unwillingness to look at the new situation created by the increasing institutionalization of medicine. It is a kind of structural nearsightedness which prevents health care professionals especially from seeing the correlation between the practice of medicine and more general trends affecting the rest of society. Healthcare professionals are turning a blind eye to the fact that the corrosive influence of commercial values on their profession is an inevitable implication of its dominance in society as a whole. An ambiguity exists about their rejection of a market within healthcare on the one hand, and their apparently uncritical attitude toward its dominance in the larger culture on the other.

Understanding this hypothesis will help us better understand the nature of the threats to moral integrity faced by health care professionals.

The Institutionalization of Medicine

Sociologists of medicine such as Steven Toulmin and David Rothman point out that since the end of World War II, the focus of medical care in the United States has shifted away from the individual doctor's

---

2   Curran & McCormick, *Readings.*, at 62.

office to hospital clinics and medical centers.[3] This also means that the focus has shifted from the personalized environment of a close relationship between health care professionals and patients to the impersonal milieu of highly capitalized and bureaucratic structures.

This trend and its consequences were foreseen at the beginning of the 20th century by a school of thought which includes, among others, Emile Durkheim in France and Max Weber in Germany. These thinkers came to the conclusion that a leading feature of the growth of modern societies is the increasing differentiation of social functions. The natural evolution in all advanced industrial nations is toward bureaucracy and institutionalization in which all forms of personal exchange lose the immediacy of their origins and become more complex.

Although this thesis should not be accepted uncritically, one of its interesting conclusions is that no institution within a modern society can be seen in isolation. This also applies to medical institutions. A large hospital is a complex institution. Moreover, it is not just a complex institution in itself, it also represents a sub-system within a larger systemic structure – what we call the market – driven by the same logic of depersonalization and neutralization.

To speak about the shortcomings of a modern society defined by greater differentiation of social roles and increased bureaucratization in the operation of institutions, Max Weber used the famous image of the «iron cage.»[4] In such a deterministic system, says Toulmin, «professional callings are displaced by job descriptions; ethical obligations give way to functional imperatives; individual responsibility is replaced by institutional excuses.» This situation is particularly problematic when the claims of professional integrity and institutional survival come into conflict. To the extent that the claims of budgetary survival tend to outweigh those of a moral calling in the operation of a modern hospital, the institution acts like an «iron cage.»

---

3   See Stephen Toulmin, «Medical Institutions and Their Moral Contraints,» in Ruth Ellen Bulger and Stanley Joel Reiser, ed., Integrity in Health Care Institutions: Humane Environments for Teaching, Inquiry, and Healing (Iowa City: University of Iowa Press 1990), 21-32. See also David Rothman, «Medical Professionalism – Focusing on the Real Issues,» in *New England Journal of Medicine* (April 27, 2000).
4   See Max Weber, *The Protestant Ethic and the Spirit of Capitalism*, trans. Talcott Parsons (Scribner, 1958), 181.

Consequences Exemplified

There is no intention here to promote a particular sociological theory. For that reason, the fact that we may agree or disagree with Max Weber's analysis is beyond the point. What is intended is to provide a heuristic hypothesis that could help us develop our own personal conclusions. That said, the thesis makes a lot of sense because it helps explain some of the predicaments which have become the daily experience of different agents within the health care industry.

Weber's metaphor of the «iron cage» clarifies the dilemma administrators face. Their predicament is normally described as one of economic pressure in developing strategies to defend both the budgetary soundness of their institutions as well as their public reputation. Let us assume that one incidental by-product of these administrative procedures is to avoid patients who lack insurance coverage, promote vigorous utilization review, and demand high productivity standards. Although unfortunate and perhaps even ethically wrong, given current reimbursement levels and competition, these actions are entirely consistent with the administrator's ability to act. One could even say that this is precisely what the administrator's job requires!

Such administrative practices and management tactics notoriously have a large impact in health care institutions. Physicians, for example, tend to see these practices as limiting their professional discretion. What is worse, caregivers may begin relating to their patients in a cynical rather than generous manner. When faced with the decision of whether to play by the rules or to fight them, many caregivers may ·choose to protect their privileges rather than serve patient needs. It is not clear, however, that physicians should bear all the blame.

Finally, we come to the patients who receive medical care within such institutions; they also cannot be blamed for the attitude of suspicion they bring to their «encounter» with health care professionals. The romantic image of the doctor-patient relationship the medical profession likes to project upon that encounter will soon be shattered in the patient's awareness of the obvious constraints that affect that encounter.

The Impact On the Notion of Integrity

When we take seriously the reality of sociological shifts within health care institutions and refrain from facile moralism, we come to the recognition that the question of professional integrity needs to be radically re-thought. As long as we refrain from addressing the larger picture, that is the hyper-market culture in which commercial values dominate, it should not surprise us if health care professionals operating in today's environment capitulate to, and ultimately act on, a complementary model of integrity in which their behavior comports with their own and their organization's economic interests. The «iron cage» of the system, not necessarily the lack of personal moral strength, has already imposed upon them this particular model of integrity.

If we think of medical institutions as sub-systems within larger systemic structures, then we cannot avoid seeing the values of health care as a reflection of those permeating the very fabric of our society at large. In other words, we should not be surprised if we experience the values of our health care mimicking those that drive societal relations in general. It is very difficult to understand how health care professionals could, on the one hand, deprecate the consequences of a for-profit mentality affecting health care, while apparently at the same time, feeling perfectly comfortable with a for-profit mentality defining the rest of their social outlook. To the extent that they are so divided in their outlook, it is not surprising they are not on the front line reminding the rest of society that, as a social good of a special nature, health care is not just a commodity like all the others.

Reconstructing Integrity

To address the real challenge, we must return to the Weberian metaphor of the «iron cage.»[5] This image symbolizes the hypothesis that many of the problems seen in contemporary healthcare are the result of society's failure to look at the consequences of the increasing institutionalization of medicine.

---

5 See Charles Taylor, *The Ethics of Authenticity* (Harvard University Press 1992), 93-109.

How are we to reconcile the paradox of individual professionals who want to behave with integrity while suffering the deterministic constraints of a system toward which they feel so powerless? Health care professionals will have to answer that question for themselves. I can do no more than make one suggestion: We need to rethink the meaning of professional integrity. We need to do so because, first, integrity does not necessarily depend upon compliance with the values of the organizations and the health care system within which we operate, and second, because the complexity of those organizations and that system have made the recognition of the fundamental values of health care more obscure.

This, at least, is clear from the Weberian thesis: the increasing differentiation of roles has made it more difficult for professionals to define integrity. Without denying the importance of personal responsibility, it is imperative to start with a definition of integrity in which the *princeps analogatum* is society rather than the individual professional. A society that can bear the thought of having in its midst over forty million people who are uninsured-who therefore have inadequate access to care from the healthcare system-cannot think of itself as moral. Indeed the «wholeness» (*integritas*) entailed by the etymology of integrity contrasts sharply with the reality of so many people living at the margins. The very existence of so many uninsured people is evidence that America still has some distance to go if it is to reach true integration. The *right* to health care is no less important than the right to education or to political participation, struggles that this nation experienced earlier in its history.

Only within the historical framework of a· society engaged in a moral discourse about the integrity of its institutions can we make sense of the notion of integrity as it applies to particular organizations and individual professionals.

CHAPTER 13
EMBODIMENT
AS SATURATED PHENOMENON
Medicine, Theology, and some Metaphysical Premises of Modernity

The question of embodiment hardly surfaces as a relevant topic of discussion in contemporary bioethics. The focus on normative dimensions, further exacerbated by the pragmatic concerns of a consensus-based strategy that is geared to public policy solutions, tend to push to the side premises of a deeper philosophical nature, unquestionably central to any ethical reflection.[1] Consider the case recently publicized in the news concerning the FDA discussion for approval of an *in vitro* fertilization technique which, in an attempt to prevent certain illnesses, like muscular dystrophy and respiratory problems, uses DNA from three people.[2]

Most commentators, especially scientists and doctors, welcome the advent of yet another technological fix to a congenital predisposition with an attitude of unquestionable awe. On the other hand, the more critically minded, among them ethicists, are willing to grant that some moral problems for this «three parent baby» solution do exist after all: doubts about safety are raised, together with the fear of unforeseen eugenic slippery slopes. Strangely passed over in silence, though, remains the most obvious question, «whose child will this baby be?»

Of course, experts are quick to rebut this preoccupation as scientifically naïve, if not totally unfounded: they reassure the concerned public that because the female donor of healthy mitochondrial DNA

---

[1] On the predicament of contemporary bioethics and the need for anthropological integration see Carol Taylor and Roberto Dell'Oro, eds., *Health and Human Flourishing: Religion, Medicine, and Moral Anthropology* (Washington, D.C.: Georgetown University Press, 2006). For a sociological analysis concerning the prevalence of «formal» over «substantive» rationality in bioethics, see the intriguing study of John H. Evans, *Playing God? Human Genetic Engineering and the Rationalization of Public Bioethical Debate* (Chicago: University of Chicago Press, 2002).

[2] See Kim Tingley, «The Brave New World of Three-Parent I.V.F.,» in *The New York Times*, June 27, 2014.

to the defective biological mother provides, in the end, a very negligible genetic contribution, she cannot be described appropriately as «a parent.» However, when considered from another angle, namely, that of the *personal* identity of a child thus produced, the question «whose child will this baby be?» comes to the fore as actually very serious. This is so because personal identity is now imperiled by what I would call «an ambiguity of belonging,» in which the embodied matrix of traceable biological debts represents for the child in question more an opportunity for doubt, than a condition for self-identification. The lack of evidence about one's *distinct* genetic lineage turns the trust in the source that gives to be, under normal circumstances the syngamy of two genomes, into puzzlement about one's *own* origin and identity.[3]

The ethical judgment on the technology in question is not the point here. I am not concerned with the ethics of artificial reproductive technologies *per se*, but with the discussion on the more recessive premises about the body, embodiment, and the «embodied self,» premises that drive these technologies in the first place and, more in general, our understanding of medicine's goals. I ask several questions: how important is it to unpack what remains tacit in the bioethical discussion, and why? What are the philosophical models of embodiment presupposed by medicine and bioethics today? Finally, what are the conditions for the articulation of a «theology of the body» in the Christian framework? A theologically defined anthropology speaks to the nature of the body as gift, the person as a «unified totality,» and the inter-subjective quality of the body as a medium of relationality. How is one to make *philosophical* sense of those categories, unequivocally rich, yet also culturally opaque?

In this chapter, I shall attempt to consider these questions in the following way: first, by approaching the question of embodiment through a simple phenomenological observation; second, by addressing the philosophical roots of our contemporary predicament about the body; and, third, by pleading for a correlation

---

3   For a stimulating analysis of the way in which biotechnology redefines embodiment, see Marie-Jo Thiel, «La corporéité face à la maladie et la mort,» in *Exploring the Boundaries of Bodiliness: Theological and Interdisciplinary Approaches to the Human Condition,* eds. Sigrid Müller *et al.* (Göttingen: Vienna University Press, 2013), 1-13.

between philosophical and theological hermeneutics of embodiment on the basis of an alternative, perhaps even counter-cultural, model of embodiment.

Relying upon insights from the phenomenological tradition, I offer an account of embodiment as a symbol of our being given to be, or more precisely, of our *givenness*, on whose ground rests the ethical imperative to care for the body.

Approaching Embodiment

A simple observation, to begin: the reality of my body is, at the same time, the most obvious and the most elusive. It is the most obvious, because I could not make sense of myself, of my position in the world, without relying upon the body I have. As I write, my fingers pass on to a computer the inputs from my brain, which, in a manner still unexplained by cognitive science, articulate my thinking. In the process, my eyes follow on the screen the unfolding lines that will eventually become a finished paper, focusing my attention entirely on the task at hand. Sight, touch, hearing – all my senses, really, are devoted, for this space of silence that is given to me like a blessing in the noises of the day, to what I am trying to say. The statement «I have a body» sanctions the obvious: my being born anew, every day, to the life I live in the body that is *mine*, so close to me that I not only have a body, I *am* indeed one.[4]

Yet, this sense of perfect continuity between myself and my body can easily be suspended, if not broken. Tiredness sets in and my eyes give out, the dim light of the evening now calling my body back to the rest I deserve, but do not control, perhaps even want: I still have thoughts to convey, things to do, and the allure of the night looks more like a partial death, an unwelcome inter-

---

4 The statement is of Gabriel Marcel, but it pervades, in its meaning, the reflection of phenomenology on the body, from Merleau Ponty, to Levinas, to Ricoeur. Consider the following: «…the problem is never one of relating consciousness (a subject) to the body (an object). The link between consciousness and the body is already functioning and is experienced at the core of my subjectivity and your subjectivity,» Paul Ricoeur, *Freedom and Nature: The Voluntary and the Involuntary* (Northwestern University Press, 1966), p. 88.

ruption in the flow of life. Illness provides another, in fact paradigmatic, example of the dialectical relation in question.[5] While healthy, I feel my body as an unquestionable medium of my presence to the world, almost oblivious to the embodied condition in which I live. When undergoing the *pathos* of disease, on the other hand, I suddenly perceive my body as *other* to myself, as literally «ant-*agonistic*» (a term which entails the Greek root of the word struggle, *agon*).

In my suffering, not only do I become aware of a part of myself I had so far taken for granted – a limb now broken, a physiological function become progressively pathological, or a mechanism of the body that will be chronically compromised. Moreover, in the fragility of my diseased body parts, I reckon with the possibility that illness will turn my *whole* self into somebody else: a patient, a disabled person, an invalid.

Let me then draw a first conclusion from this initial reflection: living in my body, I come to realize that I do so *not* in the form of a perfect identity. At best, I *relate* to my body, on the presupposition of a difference from it, a kind of distance that triggers in me the inevitability of a decision to make, the most basic, yet also the most important in my life: shall I consent to the body I have, or refuse it? Shall I accept my body, or remain forever trapped in a condition of estrangement, perhaps even alienation from it?[6]

Each one of us has a kind of «fundamental option» to make, one that becomes especially urgent in the face of illness. The heroic resistance to a disease, as in the decision «to fight cancer,» for an example, already implies the recognition, if not the acceptance, of one's «new identity»: whether I want it not, I have now become a sick person. While experiencing the shock at the sudden realization of a separation from my body, I aim at regaining a unity with it. Just as illness constitutes an «ontological assault» that plunges me into a situation of *dis-ease*, a rupture and a break within the totality of being, so restoration of health predisposes the ground for a new personal synthesis, the possible re-unification with the life-world that

---

5   On the general question of embodiment and medicine, see Matthias Beck, *Seele und Krankheit* (Paderborn: Schöningh, 2001).
6   On the meaning of this tension, see Paul Ricoeur, *Freedom and Nature, op. cit.*, 444-481.

was shattered or compromised by the event of illness.[7] The dialectical character of the human condition, the tension between identity and difference just mentioned, stands at the heart of the experience of embodiment and, therefore, provides a kind of starting point in thinking more articulately about its meaning.

Paradigms of Embodiment

In the Western tradition, several attempts have been made to interpret and understand embodiment, attempts that, in virtue of their «effectual history,» have left an indelible trace on our own contemporary rendition of the question.

One could say that such efforts are modes of thought, «postures of the mind,» never completely buried in the past, though, for sure, defined by specific thinkers and historical sensibilities. I see them more like paradigmatic attempts, underground currents that continue enriching, for good or bad, our own present thinking. Let me offer a brief overview of such paradigmatic history, apologizing in advance for the inevitable simplifications my account will entail.

The Paradigm of Transcending Unity

I will refer to the first way of thinking about embodiment as the paradigm of *transcending unity*. Such paradigm, which has deep roots in the classical tradition of Greek and Roman philosophy, has influenced to an extent Christian metaphysics through the Middle Ages. It articulates the dialectic of identity and difference from which I began my reflections in terms of a *duality* of body and soul, of the material and the spiritual element in man, a duality, however, that points to a higher, transcending, even transcendent, form of unity. Embodiment here is viewed as a condition of ontological ambiguity, an indeterminacy that needs to be overcome. The tension between body and soul stands at the heart of human existence (*conditio humana*), expressing almost symptomatically its inev-

---

7 See Edmund D. Pellegrino and David C. Thomasma, *A Philosophical Basis of Medical Practice* (New York: Oxford University Press, 1981).

itable flaws: man is poised between nostalgia for a realm entirely spiritual, a world of truth, beauty, and goodness from which he originated, and surrender to a world of imperfect presences. What is real is also impermanent, the precarious show of beings that are either only copies of their true archetypes (Plato and the neo-platonic tradition), or ceaselessly strive toward a *telos*, a perfection that announces itself in them, but always in potency, never fully actualized, except in the end (the Aristotelian tradition).

One of the criticisms of this paradigm points to its latent dualism, with the subsequent undermining, perhaps even hatred, for the body.[8] This is too simple, and even if historically one can certainly find expressions of such an outcome – I am thinking, for an example, of the pervasiveness of Gnosticism and its derivatives upon Western culture all the way into the 20th century, one has to be cautious. For sure, such criticism cannot be leveled against the Aristotelian and, later on, the Thomistic tradition, for which the unity of body and soul is understood as the reciprocity of form and matter in the unity of one dynamic substance.

Here are a couple of additional observations. First, the duality of body and soul throws into relief the relative *independence* of the human body from the spiritual sphere. Though secondary in ontological significance, the human body acquires nonetheless an importance of its own, gaining its own value, so to speak, *vis a vis* the soul. The unity to which man aspires then, is not one of assimilation («sublation») of the body under the soul, but rather one of service of the lower to the higher power. The body, in its own splendor and glory (remember the celebration of the Olympics!) honors the superiority of the soul, and it does so the more it perfects itself, orienting its own aspirations to complement those of the higher part in man. Beauty is the harmony of order, the overcoming of the natural recalcitrance of the irrational to the rational, the peaceful orientation of the material to the spiritual in the imperturbability (the *ataraxia* of the Stoic sage) of a life dedicated to higher things, the activities of the soul.

Embodiment, and this is the second observation, is not so much an experience of the individual human being, but a *metaphysical*

---

8   For a detailed analysis of the problem, see Giovanni Reale, *A History of Ancient Philosophy: II. Plato and Aristotle* (Albany, N.Y.: SUNY Press, 1990)

condition of the cosmos, signifying the intermediate (metaxological) character of the whole: to be is to be in the middle (*metaxu*), between an origin that was lost, and a destination to which we aspire. Although a transitional state, the human condition is grounded by the truth, beauty, and goodness of being (the supreme idea!), endowed with a sensuousness that is certainly fleeting, yet able, at the same time, to become the symbol, the «icon» or image of another world. As Plato reminds us in the Timaeus (29 b), the text that will become the most important for the Middle Ages, «it must be that the cosmos be an image of something else» (*pasa ananke tonde ton kosmon eikona tinos einai*).

Third, embodiment is also a sign of the complex, yet harmonious intermediation of beings within the protective embrace of nature, the ecological symphony of plants, animals, humans, and gods, each with their own distinct part to play, each with their own logos. «To be,» and this also in relation to the specific modalities of being that define «being healthy» or «being sick» is thus to undergo, almost in an attitude of *passivity*, the unfolding life of the natural order, trusting its providence and destination.

What are the consequences of this approach to embodiment for medicine and the «care of the body»? For sure, because of its own relative value, the body becomes also the object of attention, of specific care.

There is a *piety* for the body that drives the care of the sick: the wounded warrior, the infirm athlete, the blind beggar at the doors of the city.[9] Yet, medicine seems to be more like a technical response, in fact a «*techne iatrike*», to a call for the re-composition of the universal order of nature, an attempt to restore an equilibrium that has been disturbed, rather than a service to the concrete vulnerability of the sick person. In some cases, the art of medicine will have no other recourse but to simply obey the necessity of nature, the logic entailed by the inscrutable plot of fate: the disabled child will be left to die, with the blessing of both Plato and

---

[9] See Karl-Heinz Leven, *Geschichte der Medizin: Von der Antike bis zur Gegenwart* (München: Verlag C.H. Beck, 2008), 13-24. Also Ludwig Edelstein, *Ancient Medicine: Selected Papers of Ludwig Edelstein*, ed. Owsei Temkin and C. Lillian Temkin (Baltoimore, MD: Johns Hopkins University Press, 1967).

Aristotle; the elderly, now useless in the hierarchical distribution of services to the community, consign to their inevitable demise, with the blessing of Cicero and Seneca.

I am going too fast, perhaps, and will have to return to this later. But let me anticipate a bit: I find it interesting that in the gospels, the narratives of Jesus' healing are always construed differently, namely, as a response to the call, the cry even, of a *concrete* person, the «face of the other,» to use the beautiful expression of Jewish philosopher Emmanuel Levinas.[10] Jesus is also a healer, and the Patristic literature will refer to him as *Christus medicus*.[11] Yet, Jesus is not a physician by training, belonging to a defined class, as in the Hippocratic tradition. Jesus is not engaged in a lifelong profession dedicated to the universal reversal of disease. He is more like an occasional healer, bent on the concrete wounds of a concrete patient, the outcast, the foreigner, the widow, the child. In the gospels, embodiment has already become the sign of a personal presence, even the gift of a divine manifestation. Care of the body, and medicine with it, turns into agapeic service to the good of the other, an overflow of generosity from one human being to another.

The Paradigm of Dualistic Opposition

Our own understanding of embodiment, however, remains strongly defined by the intellectual and scientific revolution of modernity, which, relative to its antecedent paradigm, can only be seen, to use the expression made famous by Kant, as a «Copernican turning point.»

I come therefore, to the second paradigm, which I call the paradigm of *dualistic opposition*. Elements of continuity between

---

10   The reference is, of course, to his *Totality and Infinity: An Essay on Exteriority* (Pittsburgh, PA: Duquesne University Press, 1969). The focus on the concrete person is beautifully conveyed, in the Christian tradition, by the notion of *cura infirmorum*. In his study, Ludwig Haas refers to such a concept as «the lead concept of the Western/Christian tradition.» See Ludwig Haas, *Für kranke Menschen sorgen* (Münster: LIT Verlag, 2000).
11   On Christianity as a «religion of healing,» see Gary B. Ferngren, *Medicine and Health Care in Early Christianity* (Baltimore, MD: The Johns Hopkins University Press, 2009).

this and the previous paradigm are undeniable, but so is the rupture that welcomes the advent of a fresh attitude, a novel daring toward the old order of things, now obsolete and doubtful. Descartes, in the 16th century, sets the terms of this new «posture of the mind,» in which the received duality of body and soul is being transformed into dualistic opposition. This, in turn, on the premise of a radical questioning about God, reality, and even our embodied self, whose phenomenological immediacy can no longer be taken for granted, but must be demonstrated on the basis of a more original evidence, an act of intellectual self-determination free from the assaults of doubt. If what is immediate is no longer the fleshed presence of my body to myself, but the radical un-questionability of my act of thinking (*ego cogito*), then the whole of reality, including that of my body, becomes somewhat clouded by suspicion. What if an evil genie had tricked me into believing that I have a body?, will ask Descartes in his *Meditations*. And more broadly, what if the show of things were only a play of appearances, without any real grounding in being, without any foundation? We know Descartes will eventually resolve his theoretical qualms and set the course of modern thinking on a new footing: that of a subjectivity asserting itself over a world that now becomes «objective,» in a chasm that sets in opposition the «thinking substance» of the *res cogitans* and the inert, thing-like, matter of *res extensa*. This also will be the outcome of the story relative to the issue of embodiment, for the body will appear to the cogito, to the mind, purely as a lifeless machine, a system of parts organized by the principles of mathematics, ready to be explored, as well as exploited.[12]

Of course, we owe this view much of what is scientific about our own understanding of the body.Significant steps are made, more or less at the same time of Descartes' speculations, by the seminal work on anatomy of Andreas Vesalius, and the important discoveries in physiology of William Harvey. These will be followed, in

---

12   The meaning of the transition inaugurated by Cartesian anthropological dualism, together with the dimensions associated with the «effectual history» of Descartes on contemporary notions of embodiment, are brilliantly articulated, both historically and systematically, by Italian philosopher Umberto Galimberti in: *Il corpo* (Milano: Feltrinelli, 1987).

the 17th and 18th century, by the work of post-Cartesian physicians, such as Boyle, Hoffmann, and Gaub.[13] The importance of such discoveries, together with the renewed commitment to the «relief of the human estate,» will not be questioned. There is genuine love to humankind nurtured by a different attitude toward nature, no longer viewed now as the charged presence of symbolic meaning shrouded in mystery («nature loves to hide,» had once said Heraclitus, and so had repeated with him the monks of the Middle Ages). Rather, nature will be more like the available storage of raw material, open to the exploitation of the scientist. Nature, including the human body, will be forced into speaking the language of empirically verifiable factuality and unleash its secret powers to the hypotheses of the scientist, willingly or unwillingly.[14]

One cannot underscore enough the significance of this attitude for the progress of medicine. In the 17th and 18th century, physicians Giovanni Battista Morgagni (1682-1771) and Xavier Bichat (1771-1802) will put forth a conception of the body as an organic system that is causally determined. Before this time, abundant autopsy reports had been published, but such recorded data had not offered any correlation between clinical and anatomical findings. Medical understanding then was radically altered by the introduction of the so called «clinic-pathological correlation.» For the first time, what was found at autopsy was taken as «explaining» clinical symptoms, observed while the patient was alive. Now disease was no longer associated with a loosely collected set of clinically observed symptoms or with the uncertainty of patients' reports, rather, it took on a highly specific form – the «organic lesion» found inside the body.[15]

Furthermore, in the 19th and 20th century, the work of neurologists, such as John Hughlings Jackson (1834-1911) and clinicians such as William Osler (1849-1919), brought to completion the marriage of clinical medicine to biological science.

---

13   For a historical account see L.J. Rather, *Mind and Body in Eighteenth Century Medicine: A Study Based on Jerome Gaub's De regimine mentis* (Berkeley, CA: University of California Press, 1965). Also, Richard Zaner, *The Problem of Embodiment*, 2nd edition, Phenomenologica, no. 17 (The Hague: Martinus Nijhoff, 1971).
14   Confront the language of Galileo, in his programmatic opus *Il Saggiatore.*
15   See on this the classic work of British historian Charles J. Singer, *A Short History of Anatomy from Greeks to Harvey* (New York: Dover Publications, 1957).

And finally, the educational reforms recommended by Abraham Flexner (1866-1959) in the last century, promoted a medical thinking in which the body is seen as a complex system of interacting structures and mechanisms, governed by multiply interrelated controls seated in the neurological system. With this last reference to neurological conditioning and to the particular version of medical dualism known as «epiphenomenalism,» however, I need to pause for a moment, and go back to a couple of observations about modernity.

Let me highlight the fact that the understanding of embodiment in this paradigm stands within a broader attitude toward being as such. Anthropology reflects a specific view of metaphysics, and the mechanization of the body brought about by modernity will be properly understood only when seen within the horizon of a more general neutralization of reality.[16] This world that has become «objective» stands also empty of meaning before a «subject» that now constitutes the only presence of value (what Kant refers to as *die Würde der Person*, the dignity of the person.) Indeed, the subject is a source of value, first of all, in an epistemological sense – one could say that this is the meaning of the so called «critical» turning point of Kantian and post-Kantian transcendental philosophy. As neutral, the natural order will have no language of its own, no deeper message to convey to an observer willing to see, and this is so because a deep perplexity has now replaced the ancient wonder (*thaumazein*) about the inherent value of being, more, about the inherent goodness of being.

Secondly, the subject becomes the only source of value also in an ethical sense: the good is not «what everyone wants» (*bonum est quod omnes appetunt* – consult Aristotle and Thomas Aquinas); rather, what we want, we call the good (consult Hobbes and his fol-

---

16 For this reconstruction, I rely especially on William Desmond, *Ethics and Between* (Albany, NY: SUNY Press, 2001). In the same vein, the classic work of Catholic theologian Romano Guardini, *The End of the Modern World: A Search for Orientation*, trans. Joseph Theman and Herbert Burke (New York: Sheed and Ward, 1956). A more nuanced and less critical, historical account is offered by the work of German historian of philosophy Hans Blumenberg. See his *The Legitimacy of the Modern Age*, trans. Robert M. Wallace (Cambridge, MA: MIT Press, 1983), and *Säkularisierung und Selbstbehauptung* (Frankfurt: Suhrkamp, 1974).

lowers!) Whether responding to the necessity of a rational ordering of duty, as in the Kantian version of autonomy, or the maximization of value in a network of effective powers, as in the calculative prudence of utilitarian rationality, the moral self of modernity stands before the good as a «radically self-assertive subjectivity.»

Moreover, and this is also very relevant, the moral self stands before the good as a *dis-embodied self*, «auto-nomous» because separated not only from what it sees as the heteronomy of nature, including that of the body, but also from the heteronomy of larger claims to social solidarity, as in the various versions of individual liberalism. What if, rather than with a notion of a self, separated from nature, society, and even its own embodied identity, were we to start with a person that recognizes the variety of debts and historically defined obligations, as McIntyre suggests?[17] What if, rather than deracinated individuals in pursuit of their own self-interest, hoping this will result eventually in the good of society as a whole, were we to recognize that «we are already given to be ourselves, before we give ourselves to ourselves; grown by relations to the others, before we can grow up to be ourselves; already reared and grown in relativity, before being grown up and giving ourselves in relation»[18]

One might wonder what does individual liberalism have to do with embodiment. I ask for the patience that recognizes the connection I am trying to highlight, for the modern self of dualistic opposition described above is also, in the end, the individual that fails to recognize the embodied nature of communal, historically defined ties. I began these reflections with a reference to the latest technique of in vitro fertilization, and so let me bring these reflections on the modern paradigm of dualistic opposition to the end with another reference to it, more specifically to the debate within feminist theories on the ethics of artificial reproductive technologies. I believe such debate suggests something of the tensions intrinsic to the modern understanding of embodiment as neutral, and therefore «constructed.»

---

17 Alasdair MacIntyre, *After Virtue: A Study in Moral Theory* (Notre Dame, IN: University of Notre Dame Press, 1981.)
18 William Desmond, *Being and the Between* (Albany, SUNY Press, 1995), 387.

The spectrum of positions in feminist ethics is a difficult one to summarize, but, in broad strokes, one could say that, when it comes to artificial reproductive technologies, it seems to present a divide between a pro-interventionist and a non-interventionist tendency.[19]

Pro-interventionists tend to welcome developments in reproductive technologies as positive because of their promise to control nature, and therefore to re-define the meaning of gender constructions, relative especially to the distinction between male and female. Because it values self-sufficiency and control, this view praises invasive procedures that break women's links to biology, birth, and maternal nurturing.[20]

On the other hand, non-interventionists see reproductive technologies differently, i.e., as a strengthening of arrogant human control over nature, and thus over women as part of the «nature» that is to be controlled. They see new reproductive technologies as an imposition upon women who look at themselves as failure, if they cannot become pregnant. They insist that technological progress, requiring the invasion and manipulation of women's bodies, must always be critically scrutinized with a kind of «hermeneutics of suspicion,» especially when the market becomes the ultimate mechanism for the exploitation of the body.[21]

Indeed, it is hard to miss the marketing and advertisement strategies associated with fertility clinics and service providers that, understandably, are eager to do what any business does best: sell to prospective customers. But what they are selling is packaged in the

---

19   On the issue, see Paul Lauritzen, *Pursuing Parenthood: Ethical Issues in Assisted Reproduction* (Bloomington: Indiana University Press, 1993).
20   A very recent iteration, with respect to the issue of maternal surrogacy, by Grace Y. Kao, *My Body, Their Baby: A Progressive Christian Vision for Surrogacy* (Stanford, CA: Stanford University Press, 2023). In spite of the «progressive» language, the book really offers a normative account consistent with the traditional, market-based liberal model of justice, with all the anthropological premises involved in the model, beginning with an emphasis on what Bellah and others have termed «expressive individualism.» It is the individual who is the bearer of reproductive rights, and this individual is abstractly defined, it is a dis-embodied, in fact, «unencumbered» self.
21   On this, Barbara Duden, *Disembodying Women: Perspectives on Pregnancy and the Unborn*, trans. By Lee Hoinacki (Cambridge, MA: Harvard University Press, 1993). Also Hille Haker, *Hauptsache gesund? Ethische Fragen der Pränatal-und Präimplantationsdiagnostik* (München: Kösel Verlag, 2011).

language of products and commodity. I ask, is the market-based disembodiment of the self, visible in the exploitations just mentioned, the last station in the trajectory of modernity? Is nihilism, as Nietzsche and Heidegger saw, the necessary destiny of our postmodern condition?

Embodiment and Being Given

In the last part of my essay, I suggest something like a requalification of the meaning of embodiment, and do so by drawing on the philosophical resources of the Christian tradition. I say *philosophical*, because what I am interested in is not so much retrieving specific doctrinal positions on various issues, but throwing into relief a broader intellectual gesture that underpins the tradition, which I understand as living, that is, as organically growing in faithfulness to its origins, rather than as fixed, once and for all. The task of Christian theology is as much defined by discerning engagement with the present, as by vigilance in the protection of received wisdom.[22]

Let me, first, state the obvious. Embodiment is absolutely central to the message of Christianity. The *kerygma*, the good news of Christianity is the «logos become flesh,» and the community that witnesses such event stands before the world as the community of those «who have heard, seen with their own eyes, looked upon and touched with their hands – the life made visible» (1 John). Embodiment is no longer the indeterminate condition of phenomenality, which has either lost or still awaits its ontological perfection, but the over-determined presence of a real manifestation, an epiphany of the true, the good, and the beautiful in the saturated glory of the phenomenon, which is, in this case, a concrete historical event.

---

22 I stay, on this, in the tradition of philosophically minded theologians. Suffice it, for the less inclined, the following quotation of Hans Urs von Balthasar: «In order to be a serious theologian, one must also, indeed, first be a philosopher; one must – precisely also in light of revelation – have immersed oneself in the mysterious structures of creaturely being...» *Theo-Logic: Theological Logical Theory, Volume I/truth of the World*, trans. By Adrian J. Walker (San Francisco, CA: Ignatius Press, 2000), 8. With specific reference to the interplay of theology and philosophy in ethics, see Klaus Demmer, *Moraltheologische Methodenlehre* (Freiburg: Herder Verlag, 1989), especially 119-178.

Let me elaborate, if only briefly: the notion of «saturated phenomenon,' a central *topos* in Jean Luc Marion's phenomenology, points to a reversal of intentionality with reference to the pretentions of a self-grounding subjectivity, the a priori ground of consciousness, whether self– or transcendental consciousness. Before it comes to itself as self-determining, the subject is already called into reciprocity, already «appealed to.» As Kevin Hart explains, the human being is rather «l'*interloqué* or, better, *l'adonné*, the gifted one, the one who receives phenomena and builds selfhood from them, the devoted one whose being is thoroughly a posteriori and who has no a priori horizon but one that is saturated by givenness.»[23] For the Christian believer, *in the flesh* of Jesus, not beyond or besides it, the presence of the absolute breaks forth. This flesh, with all its vulnerability and frailty, the *humanity* of Jesus, is the medium of God's presence, of Christ's divinity. As Klaus Demmer puts it, the starting point of any theological articulation of the *intellectus fidei* is the a priori synthesis of being and history, now become an a posteriori event in Jesus Christ.[24]

Such statement, the most essential to the Christian tradition, is paradoxical to the Greek mindset, to the paradigm of *transcending* unity.

Relative to this paradigm, Christianity seems rather to be defined by a kind of radical *immanence*, the surrender to the claim that the reality of God lies precisely in the embodied, historically defined, visibility of Jesus. This is why modes of Platonic and neo-platonic thinking will be both a blessing and a curse for the subsequent articulation of a Christian metaphysics of embodiment. A blessing, for the flesh of Jesus can indeed be understood as the icon, the symbol, of his divine presence. A curse, for such imaging can also be interpreted, as in Gnosticism and Docetism, only as an illusory portrait of divinity, one that is imperfect and ephemeral, not fully real.

---

23 Kevin Hart, «Introduction,» in Jean Luc Marion, *The Essential Writings*, ed. By Kevin Hart (New York: Fordham University Press, 2013), 25.

24 «Muss nicht jede theologische Reflexion methodisch bei der Geschichtswerdung Gottes einsetzen? Und was bedeutet ein solcher theologischer Denkansatz für den «intellectus fidei»«? Ist nicht die apriorische Synthese von Sein und Geschichtlichkeit in Jesus Christus zum aposteriorischen Ereignis geworden?» in Klaus Demmer, *Sein und Gebot: Die Bedeutsamkeit des transzendentalphilosophischen Denkansatzes in der Scholastik der Gegenwart für den formalin Aufriss der Fundamentalmoral*, (München: Schöning, 1971), 5.

There is a «realism of embodiment» in Christianity that is the direct result of its incarnational structure. «The glory of God is the human being fully alive» (*gloria Dei homo vivens*), will proclaim Ireneus in the 2nd Century, and this precisely with an emphasis on the embodied self. Furthermore, the care for the vulnerable body, as in the care of the sick, will take such central place in the Christian tradition and in the medical imagery that articulates its meaning (*Christus medicus*) to become the paradigmatic expression of agape, of loving service to the other. Unannounced and unexpected, a divine presence reveals itself in the frailty of the sick body. The body, in the Christian tradition, is ultimately sacramental.[25] Let me further clarify the anthropological implications of such a theology of embodiment. I want to stress two distinct yet complementary points. They both converge upon the recognition of the expressiveness of the body, relative to (1) the identity of personal presence, and (2) the alterity of its gifted origins.

The Identity of Personal Presence

To say that the body expresses the identity of personal presence, not in its individual parts, but in its «unified totality,» is to overcome any form of dualism, either tensional, as we find in the first paradigm, or real, as we find in the second. The body is not the separated substance existing next to, or behind, some mysterious personal presence (the soul), but the very incarnation of that presence in the «communication of properties» (*communicatio idiomatum*) that defines the fleshed embodiment of the self: in the person, the material is the spiritual, the spiritual is the material.[26]

---

[25] See the comprehensive study, both exegetical and theological, of Ulrike Kostka, *Der Mensch in Krankheit: Heilung und Gesundheit im Speigel der modernen Medizin* (Münster: LIT Verlag, 1999).

[26] Consider the following observations by theologian Adrian Walker, who integrates Aquinas's understanding of the soul as the substantial form of the body, into a larger theology of the body: «... The substantial unity of the intellectual soul and the body, grounded in the *actus essendi* that encompasses both but is identifiable with neither, includes a kind of reciprocal though asymmetrical interpenetration of the two components without separation or confusion. In other words, the unity of the human composite includes a circumincessive *communication idiomatum* thanks to which the body and the

It is in the logic defined by this identity that we can understand the statement of Gabriel Marcel: «I *am* my body in so far as I succeed in recognizing that this body of mine *cannot*, in the last analysis, be brought down to the level of being this object, *an* object, a something or other.»[27] Nothing that belongs to the body in the person is secondary to her, as if expressing something like the «animal,» and therefore inferior, part in her. In virtue of its own expressiveness, all of the body is rehabilitated, so to speak, in its ontological meaning and value. Against the dualistic paradigm of modernity, the countercultural model of Christian anthropology points to the body as much more than raw material. In this paradigm rather, the body becomes the intentional field of a possible expressiveness of the subject, of its motivations, intentionality, and emotions. The ethical claims to bodily integrity, privacy, respect, even reverence for the body are grounded in the presupposition that, as a document of the church puts it, «in the body and through the body, one touches the person herself in her concrete reality.»[28]

The Alterity of Gifted Origin

I have almost come to the end, but still need to make sense of the other facet inherent in the experience of embodiment. For what has been highlighted, so far, speaks of the last reserve of the body as mine, of its most intimate incommunicability. There are things I experience in my body – pain, pleasure, for an example, I can never communicate fully. The language of the body is in the end, less

---

      intellectual soul can each enter into the innermost core of the other without destruction or mingling, «"Sown Psychic, Raised Spiritual": The Lived Body as the Organon of Theology,» in *Communio: International Catholic Review* 33 [Summer 2006], 203-215, at 207, footnote 8.

27   Gabriel Marcel, «The Existentialist Fulcrum,» in Richard Zaner and Don Ihde, eds., *Phenomenology and Existentialism* (New York: Putnam Capricorn, 1973), 214. On the symbolic meaning of the body, see also Virgilio Melchiorre, *Corpo e persona* (Genova: Marietti, 1987), 37-91.

28   Congregation for the Doctrine of Faith, *Instruction on Respect for Human Life in Its Origin and on the Dignity of Procreation* (1987), no. 3. For a commentary on the document, see *Gift of Life: Catholic Scholars Respond to the Vatican Instruction*, Edmund D. Pellegrino, John Collins Harvey, and John P. Langan, eds. (Washington, DC: Georgetown University Press, 1990).

articulate than words: laughter or tears, a scream, a sigh, a yawn. And so I celebrate the idiocy of my body, its radical sharing in the deepest recesses of my hidden interiority. What then of its irreducible resistance? What of its otherness?

The question was posed at the beginning of this essay, but it re-emerges now, even more radical and forceful, for it asks to be understood differently, beyond the reductionism of dualistic opposition. What could then mean to experience the alterity of the body, its being *other*, at the heart of its deepest immediacy to myself? If such alterity is a sign, or a pointer, what does it allude to? Let me conclude with a text, which, I suppose, rather than providing a final answer, will trigger a host of new questions. It is a text of Nietzsche, from his *Thus Spoke Zarathustra*, in the paragraph titled «Of the Despisers of the Body» (62):

> What the sense feels, what the spirit perceives is never an end in itself. But sense and spirit would like to persuade you that they are the end of all things: they are as vain as that.

Could the body, in its most radical otherness, be the symbol, the ultimate, in fact the most real, of our being given to be? Could it be that in our embodied condition – in the flesh that nourishes our joy and suffering, pain, and pleasure – there lies the trace of the source that releases us into being, the subtle allusion, most often forgotten, at times denied, of the gift that we are, not from ourselves, for how could we credit to ourselves the price of our own indebtedness, but from another. Some call it life. Some others dare to call it God.

CHAPTER 14
CAN A ROBOT BE A PERSON?
De-*facing* Personhood and Finding it Again with Levinas

*De ce terrible paysage
Tel que jamais mortel n'en vit
Ce matin encore l'image
Vague et lointaine, me ravit*

*Le sommeil est plein de miracles!
Par un caprice singulier
J'avais banni de ces spectacles
Le vegetal irrégulier,*

*Et, peintre fier de mon genie,
Je savourais dans mon tableau
Le 'enivrante monotonie
Du metal, du marbre et de l'eau.
(Baudelaire)*[1]

The question «Can a robot be a person?» has emerged of late in the field of bioethics. It is a fitting provocation, an instigation to think impelled by technological advances, as have been the many ethical issues posed by developments in medicine over the past century.[2] Robots can do almost everything: discharge complex logical

---

1 «This morning I am still entranced – By the image, distant and dim – Of that awe-inspiring landscape – Such as no mortal ever saw. Sleep is full of miracles! Obeying a curious whim, I had banned from that spectacle – Irregular vegetation – And, painter proud of his genius – I savored in my picture – The delightful monotony – Of water, marble, and metal» («Parisian Dream,» *The Flowers of Evil*, fleursdumal.org/poem/228).
2 The emergence of robotics has triggered reflections at different levels, mostly with a concern for the epistemological and ethical dimensions of the impact of robots and artificial intelligence on human life. The anthropological reflection (in the sense of a *philosophical* anthropology) is less developed. A general introduction to the problem is found in Phil Husbands, «Robotics,» in *Cambridge Handbook of Artificial Intelligence*, ed. K. Frankish and W.M. Ramsey (Cambridge: Cambridge University Press, 2014), 269-95; Patrick

operations, resolve algorithmic puzzles impossible to the average human brain, carry out operations on command, and even act *of their own accord*, posing the issue of whether their endowments, either in the cognitive field or in the sphere of autonomous decision-making, might not make them closer to us than we think.³ Thus the question: can a robot be a person?⁴

The question points to a doubt, a puzzlement about the nature of personhood with respect to its attributions. Our world has become the abode of *homo technologicus*, the impersonal space of the Neuter, as Levinas might put it: a world of objects exposed to the totalizing gaze of science, whose *outlook has become a doing*.⁵

Lin, Keith Abney, and George A. Bekey, eds., *Robot Ethics: The Ethical and Social Implications of Robotics* (London: MIT Press, 2012); Margaret A. Bode, *The Philosophy of Artificial Intelligence* (Oxford: Oxford University Press, 1990). In a critical vein, Hubert L. Dreyfus, *What Computers Still Can't Do: A Critique of Artificial Reason* (Cambridge, MA: MIT Press, 1994) and guy Vallancien, *Homo Artificialis* (Paris: Michalon, 2017).

3   Japanese scientist Hiroshi Ishiguro, possibly one of the most renowned in the field, speaks of «human-robot symbiotic society.» See Hiroshi Ishiguro, «Studies on Interactive Humanoids,» in *Robo-Ethics: Humans, Machines, and Health,* ed. Vincenzo Paglia and Renzo Pegoraro (Rome: Pontifical Academy for Life, 2020), 67-102.

4   Although often robots, humanoids, and artificial intelligence are considered alike, this is not necessarily the case. To begin, one ought to distinguish between so called embodied and non-embodied machines. Furthermore, machines, whether embodied or non-embodied, may be provided with some form of artificial intelligence. They are either «stupid» – i.e., programmed to work automatically – or «intelligent» – that is, endowed with increasing cognitive and decisional abilities. Roberto Cingolani sums up the meaning of the distinctions in question thus: «The availability of increasingly powerful calculation machines (provided with sight, touch, and biomechanical abilities), making realistic the assumption that robots are characterized by performances increasingly closer to those of the human» (Roberto Cingolani. «Robots and Intelligent/Autonomous Systems: Technology, Social Impact, and Open Issues,» in Pegoraro and Paglia, *Roboethics*, 34).

5   Though Levinas is critical of Heidegger, the latter has offered a powerful critique of technology (cybernetics) as the inevitable destiny of metaphysics. We dwell in the *Gestell*, in the «framework» provided by the totalizing outlook of technology and the forgetfulness of being. See Martin Heidegger, *The Question Concerning Technology and Other Essays*, trans. William Lovitt (New York: Harper, 1977), 14-17. Another interesting take on the relation of technology to modernity is Monette Vacquin, «The Monstrous as the Paradigm of Modernity? Or Frankenstein, Myth of the Birth of the Contemporary,» *Diogenes* 49/3, no. 195 (2002): 27-33. For a general philosophical

The suggestions advanced by Japanese scientist Hirosi Ishiguro are telling. For years, Ishiguro has worked toward the creation of interactive robots, specifically, robots with a human appearance or *androids*, on the premise that «we empirically know the effect of appearance is as significant as behaviors in communication.»[6] To tackle the problem of appearance and behavior, two approaches are necessary: one from robotics, and the other from cognitive science. The cross-disciplinary framework emerging from such interaction is *android science*, whose goal is the creation of robots with human-like appearance, movement, behavior, and perception. Although further study is needed to address meta-level cognitive functions that more intelligent human-friendly robots might perform (intelligence embodiment, multi-modal integration, intention/desire, consciousness, and social relationships), the goal of scientific research, according to Ishiguro, is to «develop companion robots that can pass the Total Turing Test as a scientific and engineering goal.» The premise of the research in question is that the robot, having passed all the tests that evaluate its «total human-likeness,» will be «accepted as a member of our society.»[7]

The difficult question at stake here is «recognition.» The robot is not the person, but its human-like appearance and behavior allows the *other* to the robot, potentially a conversation partner, to be fooled into thinking it is a person.[8] The relation between human and robot thus rests on a kind of game of pretense: though one knows the robot is not human, he/she can still deal with it as a social partner. This seems to be the hypothesis tested out experimentally by scientists like Ishiguro:

---

analysis of the phenomenon of technology, Don Ihde, *Technology and the Lifeworld* (Bloomington: Indiana University Press, 1990).

6   Hiroshi Ishiguro, «Android Science: Toward a New Cross-Interdisciplinary Framework,» in *Robotics Research*, ed. S. Thrun, R. Brooks, and H. Durrant-Whyte (Berlin: Springer, 2007), 118.

7   Ishiguro, «Studies on Interactive Humanoids,» 72. The Total Turing Test (TTT) allows one to compare a robot manipulated by human operators and an autonomous robot controlled by developed technology.

8   *Geminoid*, a tele-operated android, can transfer the presence of the person to distant places. Tellingly, Ishiguro suggests that «through tele-operation, the operator – myself – could adapt to the Geminoid body and accept it as my own body» («Studies on Interactive Humanoids,» 74).

If a human consciousness recognizes the android as a human, he/she will deal with it as a social partner even if he/she consciously recognizes it as a robot. At that time, the mechanical difference is not significant; and the android can naturally interact and attend to human society. Verification of this hypothesis is not easy and will take a long time. However, it is an important challenge that contributes to developing deeper research approaches in both robotics and cognitive science.[9]

One may wonder what is the vision driving the research in question. The «human-robot symbiotic society,» what awaits us at the end of the experiment, is not a more *human* society, further humanized by the presence of robots, but the turning of humans into *inorganic* intelligent life.[10] The evolution at stake is thus somewhat reversed with respect to a teleological movement geared toward the human being. The trajectory rather entails a progressive liberation of the latter from any «flesh body» and the openness, made possible by technology, to a diversity of bodily forms that «may allow us to evolve further.» Evolution by technology pushes us beyond the limitations of life. «The ultimate aim of human evolution is immortality, achieved by replacing flesh and bones with inorganic material. Organic bodies are not a pre-condition for human existence in today's world....Humans come from, and return to, inorganic material. The human is currently almost a machine. We humans are going back to an inorganic state in the near future...we are trying to be an inorganic intelligent lifeform....We can choose any kind of life forms.... That is, we can be released from the constraint of the human body.»[11]

Can this world still be a world of persons? If so, what does it mean to retrieve a proper understanding of the person's singularity in a spectacle of world-objects? Is not the attempt to expand the notion of personhood to include robots a function of the totalizing tendency in which everything becomes an object? The levelling of the difference between person and machine would then signal the failure to account for the subject's separation from the world of objects, its *exteriority* to any totalizing pretense.

---

9   Ishiguro, «Android Science,» 127.
10 Ishiguro, «Studies on Interactive Humanoids.»
11 Ishiguro, «Studies on Interactive Humanoids,» 94-100.

I take some of the notions elaborated by Levinas, whose echo is already evident in my *incipit*, as relevant to the question I am addressing in this paper. Mine will be less a direct confrontation with him, based on exegetical precision and punctual textual references, and more a personal appropriation of his mode of thinking, with which I find myself attuned.[12]

I begin with something like an archeological reconstruction of personhood in modernity, in order to locate the context out of which the question posed – «can a robot be a person?» – might take on meaning. Descartes, Hume, and Kant are the most important exponents of the story, their position emerging in direct contradiction to the classical metaphysics of the person, such as one finds in Thomas Aquinas. I see Levinas as having a complex relation with modernity, at once defined by a positive retrieval of Descartes and Kant, and critical of the anthropological dualism effected by the Cartesian *cogito*.[13] Levinas rejects the rationalist

---

12 Levinas has been a «companion in thinking» for many years. I have dedicated my STL thesis to Paul Ricoeur («Antropologia ed etica nella *philosophie de la volonté* di Paul Ricoeur,» Pontificia Universita' Gregoriana, 1985) and my doctoral dissertation in theology to another phenomenologist, Dietrich von Hildebrand. See Roberto Dell'Oro, *Esperienza morale e persona: per una reinterpretazione dell'etica fenomenologica di Dietrich von Hildebrand* (Roma: Pontificia Universita' Gregoriana, 1996). Levinas has been an interlocutor for Ricoeur. On this, see Richard Cohen, *Ethics, Exegesis, and Philosophy: Interpretations after Levinas* (Cambridge: Cambridge University Press, 1992), 283-325. As for Dietrich von Hildebrand, the proximity to Levinas is evident at the level of their general sensibility, and this in spite of apparent reciprocal ignorance. Still, in a quasi-biographical narrative, concerning his encounter with phenomenology, and how he came to study with Edmund Husserl, Levinas mentions his friendship with Jean Hérring, who was Husserl's student in Göttingen, together with «an entire circle of young thinkers.» Since Hildebrand was part of it, Levinas must have known, at least, of the name. See Hans Reiner Sepp, ed., *Edmund Husserl und die Phänomenologische Bewegung. Zeugnisse in Text und Bild* (Freiburg/München: Karl Alber, 1998), 27-33. For a comparison between the two thinkers on their philosophy of love, see the introduction of John F. Crosby to Dietrich von Hildebrand's *The Nature of Love* (South Bend: St. Augustine's Press, 2009), xxxi. More broadly, the interesting article of Alexander Montes, «Toward the Name of the Other: A Hildebrandian Approach to Levinasian Alterity,» *Questiones Disputate* 10, no. 1 (2019): 82-109.
13 More than on the various dualisms asserted by Descartes, Levinas focuses on the latter's idea of the infinite, as providing the point of entry into the meaning of the relation between same and the other: «This relation of the same

perspective of a *bodiless mind*, a person reduced to her cognitive capacities, no less than the empirical version of a *mindless body*, a person reduced to the external stimuli of sensations and impressions registered by the mind.

Contemporary bioethics, however, especially in the Anglo-American version of it, is mostly defined by such an understanding of personhood: the person does not «come to mindfulness» out of its bodily conditions, nor does she persist, when no longer conscious, in bodily presence. According to Peter Singer, one of the main voices in bioethics, personhood is transitional: it passes from being to being like a «thing,» as long as certain dimensions of actual empirical consciousness are present. Thus, a dolphin, a chimpanzee, a higher mammal, perhaps a robot can be a person.[14]

On the other hand, as Levinas suggests, to be a person is to be «manifested in the exteriority of the face, which is not the disclosure of an impersonal Neuter, but *expression*, that is, the presence of an infinite idea that always exceeds the idea of the other in me.»[15] If so, a robot cannot be a person. In what follows I try to say why I think this is the case.

Person: A Historical Reconstruction

Thomas Aquinas: the Person as Spirit Incarnate

For Aquinas, who comments on Aristotle's metaphysics, philosophy begins in *wonder*.[16] Wonder is the attitude that throws us back

---

    with the other, where the transcendence of the relation does not cut the bonds a relation implies, yet where these bonds do not unite the same and other into a Whole, is in fact fixed in the situation described by Descartes in which the "I think" maintains with the Infinite it can nowise contain and from which it is separated a relation called "idea of infinity"» (Emmanuel Levinas, *Totality and Infinity: An Essay on Exteriority*, trans. Alphonso Lingis [Pittsburgh: Duquesne University Press, 1969], 48).

14   For a survey of various positions on personhood in bioethics, see the chapter on moral status in Tom L. Beauchamp and James F. Childress, *Principles of Biomedical Ethics*, 7th ed. (New York: Oxford University Press, 2013), 62-100.

15   Levinas, *Totality and Infinity*, 50-51.

16   The statement is originally from Plato, who lets Socrates say: «I see, my dear Theaetetus, that Theodorus had a true insight into your nature when he

onto ourselves, in stunned astonishment at the sheer being there of things. Later on, such astonishment will engender perplexity about the meaning of things. Prior to the activity of questioning perplexity or doubt, wonder entails a kind of trust, a confidence (*fides*) in the natural goodness of reality, a confidence which is a love: being is promising and good (Genesis says: «God saw that it was very good»). For Thomas, who inherits the insights of the entire Christian metaphysical tradition, being is the miracle of gratuitous generosity, a «being there» without explanation, out of a source that gives. Creation is a gift of the Origin whose «coming into being» remains *in excess of* the ontological relations that define being in its «becoming,» such as form and matter, formal and final causality, potentiality and actuality, etc. A being that is becoming already presupposes its «being given into existence,» now charged with the promise of further development.[17]

But what does it mean «to be»? Thomas refers to two aspects of being: the «in-itself aspect» of being (*substance*), and its «towards-others aspect» (*relationality*). «To be» is «to exist,» and to exist is to be *an-integrity-in-relation*. One might say that existence is the actualization of the energy of being, the gift from a source that offers itself, and whose communicative aspect is being participated to each existent.[18] Consider the following quotation from Gerald Phelan:

---

said that you were a philosopher; for wonder is the feeling of a philosopher, and *philosophy begins in wonder*,» Theaetetus 155c-d, in *The Dialogues of Plato*, trans. B. Jowett (New York: Random House, 1937), vol. 1, 157. For Aristotle, «It is owing to their wonder that men both now begin and at first began to philosophize,» *Metaphysics* 982b, in *The Basic Works of Aristotle*, trans. Richard McKeon (New York: Random House, 1941), 692. According to Thomas Aquinas, wonder (*admiratio*) is «a kind of desire (*desiderium*) for knowledge; a desire which comes to man when he sees an effect of which the cause either is unknown to him, or surpasses his knowledge or faculty of understanding» (*Summa theologiae* I-II, q. 32, a. 8).

17   The third proof, so called «on possibility and necessity,» speaks of the contingency of being. See Thomas Aquinas, *Summa theologiae*, I, q. 2, a. 3.

18   «*Being* means that-which-has-existence-in-act....Now any designated form is understood to exist actually only in virtue of the fact that it is held to *be*.... It is evident, therefore, that what I call *esse* is the actuality of all acts,» in *An Introduction to the Metaphysics of St. Thomas Aquinas*, ed. James F. Anderson (South Bend: Regnery/Gateway, 1953), 22.

The act of existence (*esse*) is not a state, it is an act, the act of all acts, and therefore must be understood as act and not as a static definable object of conception. *Esse* is dynamic impulse, energy, act – the first, the most persistent and enduring of all dynamisms, all energies, all acts. In all things on earth, the act of being (*esse*) is the consubstantial urge of nature, a restless, striving force, carrying each being (*ens*) forward, from within the depths of its own reality to its full self-achievement.[19]

Of course, one cannot fully grasp all this without reference to its theological underpinning. Thomas talks about being, but ultimately thinks about God, the Christian God of creation.[20] In this conceptual framework, the person also finds her place. Unlike other beings, the person is not just being-in-itself but being-coming-to-itself in self-presence (Thomas speaks of *reditio*, return unto itself).[21] The person is a mindfulness of being in its totality, a complete openness to its fullness and, thus, most fully being. Indeed, the person is the most perfect of substances, because of her ecstatic openness to the

---

19 Gerald Phelan, «The Existentialism of St. Thomas,» *Selected Papers* (Toronto: Pontifical Institute of Medieval Studies, 1967), 77. Quoted in Norris W. Clarke, *Person and Being* (Milwaukee: Marquette University Press, 1993), 9.
20 To my knowledge, there is no reference to Thomas Aquinas in Levinas. The question of whether the former might not fall into Heidegger's general condemnation for the Western metaphysical tradition as «onto-theological» might have influenced Levinas' own way of reading the tradition. Jean-Luc Marion, a student of Levinas, is keen in rescuing Thomas Aquinas from the accusation. Consider the following: «To think *esse* starting from God, but not in inverse order (in the way of *metaphysica* and of Heidegger as well), allows Thomas Aquinas to free the divine *esse* from its – tangentially univocal – comprehension starting from what philosophy understands by being, entity being of the entity, in a word to mark the distance – an «infinitely infinite distance» – from the creature to God (Pascal)....One could say that such (divine) *esse* keeps within itself the transcendence that opposes the act of being to the *esse commune* of entities' ... Therefore, God without being (at least without *this* being) could become a Thomistic thesis.» The quotations are from the essay titled «Thomas Aquinas and Onto-theo-logy,» in Jean-Luc Marion, *The Essential Writings*, ed. Kevin Hart (New York: Fordham University Press, 2013), 306-307.
21 «*Illa quae sunt perfectissima in entibus, ut substantiae intellectualis, redeunt ad essentiam suam reditione complete,*» Thomas Aquinas, *De Veritate*, q. 1, a. 9.

totality of being, an openness actualized in knowledge and will.²²
Such openness is not just a function of one particular aspect of the
person, say the mind. It is the function of the entire being that is the
person in the unity of its principle – i.e., its soul (for Thomas, like
Aristotle: *anima est quodammodo omnia*) – because the soul is the
form of the body *(anima forma corporis)*.²³

This is an important point: when thinking about the person,
Thomas always points to the unity of body and soul. In encountering the body, one encounters the person in her totality, in her
spiritual presence. The person is, in this sense, «spirit incarnate,»
as Rahner would say, an integrity of being, a substantial unity. In
this integrity of being the person is also infinite openness, infinite
intentionality: *homo capax Dei*.

One might summarize by saying that for Thomas Aquinas the
person entails three dimensions: to be an existent, that is, an integrity of being unto itself (incommunicable substance); actively open to
other being (substance-in-relation); and passively receptive to the
totality of being (receptivity). Incommunicability, relationality, and
receptivity point to a balance between passivity and activity, *passio*
and *conatus*, in the person.²⁴

Here is where also the question of *potentiality* in the person
comes into play. To be a person as substance-in-relation means
that the person (like being itself) is always in a process of becoming: openness and receptivity are the conditions for growth and
dynamic passing from potency to act. To be a person is always
to be in-potency, in a process of progressive actualization toward
a more perfect fulfillment. To be a person is to be an energy of
transcendence.

The Thomistic synthesis will undergo a shift of paradigm as it
moves into modernity, with important changes in the philosophical
and theological disposition of the late Middle Ages, the age of nominalism. First, with respect to the concept of God: the stress is now

---

22 The point is central in the reinterpretation of Rahner and so-called «transcendental Thomism.» See Karl Rahner, *Spirit in the World*, trans. William Dych (New York: Continuum, 1994).
23 For a systematic analysis of Thomas Aquinas' anthropology, see the classic work of Sofia Vanni Rovighi, *L'antropologia filosofica di San Tommaso D'Aquino* (Milano: Vita e Pensiero, 1965).
24 Clarke, *Person and Being*, 25-110.

on God's transcendence, on his distance from the world, as well as on his will, which is inscrutable and absolute – unlike Thomas, who sees the eternal law as the height of rationality. There are also changes in the concept of nature: both the distance of God and the inscrutability of his will leave the world devoid of signs of his presence. The world becomes «de-sacralized» or *secularized*, but, because of this, also available for the exploration of man, his engagement with things, and the general project of natural discovery. Nature can no longer hide – contra Heraclitus, for whom «nature loves to hide.»[25]

God is beyond nature, infinitely powerful, and inaccessible. Now stripped of all the signs of divine communication, being is no longer conceived as communicative (*actus essendi*, act of existence), the energy of relationality, but as a being-unto-itself, distant and cold. A progressive impoverishment in the understanding of being sets in.[26] Consider as an example the loss of an analogical understanding of reality and the univocal reduction to particularity, taking place especially with Duns Scotus.[27]

One aspect of the change is paramount: the emphasis on *essence* rather than *existence*, and the subsequent loss of the notion of being as energy of communication. Being is no longer seen in the fullness of its over-determinacy, in its plenitude (Jacques Maritain speaks of the «generosity of being»), but as a thing-like presence, retracted unto itself, an object facing a subject. The essentialization in question signals the end of a teleological understanding of nature, the denial of its spontaneous finality. Nature knows no intrinsic destina-

---

25 *Physis kruptesthai philei*. The fragment belongs to Heraclitus' *ipsissima verba* in the Diels-Kranz collection (B123). See Jonathan Barnes, *Early Greek Philosophy* (London: Penguin, 1987), 122.

26 A thorough historical analysis can be found in Emerich Coreth, *Metaphysik: Eine Methodisch-Systematische Grundlegung* (Innsbruck: Tyrolia, 1980), especially 15-47. The impoverishment in the conception of being, following the nominalistic «effectual history,» and in critical dialogue with Heidegger, has been argued by Gustav Siewerth, *Das Schicksal der Metaphysik von Thomas zu Heidegger* (Einsiedeln: Johannes, 1959). See also Etienne Gilson, *Being and Some Philosophers* (Toronto: Pontifical Institute of Medieval Studies, 1952).

27 «Being is said in many senses» (*to on legetai pollakos*), says Aristotle in *Metaphysics*, Book IV, 2 (*The Basic Works of Aristotle*, 732). Similarly for Thomas, *ens multipliciter dicitur*.

tion. It is not a dynamic system that gives itself, as in the etymology of *physis*, but a static object, to be grasped by a subject.

The Person as Subject and the Dualism of René Descartes

The story of modernity begins with a skeptical puzzlement about the nature of reality, the place of God in it, and the destination of the self. The modern *Denkform* is new with respect to its historical predecessors. If the classical tradition of metaphysics, from Plato to Aristotle and Aquinas, begins in wonder, Descartes's philosophical discovery, *cogito ergo sum*, emerges in the wake of doubt, and functions as the condition of possibility for the reconstruction of an image of the world now certain, finally resting on secure foundations beyond any skeptical assault.[28] There are losses and acquisitions following such a shift of paradigms.

The losses: Descartes no longer understands being in its character of act, energy, power, communicative to mind. Furthermore, being is no longer endowed with intrinsic value; as such, it was able to speak to the mind. Now a reversal of directionality in the relation between mind and being sets in: the mind dictates to being the conditions of its meaningfulness. «I think,» that is, the epistemological certainty, precedes and grounds «I am,» that is, the ontological constitution.[29]

I said there are also important theoretical acquisitions in this shift. The most important is the turn to subjectivity, an «anthropological turning point» which, later on, Kant will compare to a Copernican revolution: to be a person is to be a subject. In this revolution, the language of substance gives way to the language

---

28 The reference is to the *Discourse on the Method*, Part IV, but the statement recurs in the *Second Meditation* as well. See *The Philosophical Works of Descartes*, trans. Elizabeth S. Haldane and G.R.T Ross, vol. 1 (Cambridge: Cambridge University Press, 1981), 102.

29 Marion has highlighted the ambivalence of Descartes's metaphysics, especially with respect to the idea of God. His focus is on Descartes's doctrine of «eternal truths,» which cannot be ascribed both to the world and to God. Thus, all theology, understood as meaningful talk about God, is no more than a blank space or a white page in Descartes's writings. See Jean-Luc Marion, *Sur la théologie blanche de Descartes. Analogie, creation des vérités éternelles et fondement* (Paris: Presses Universitaires de France, 1981).

of subjectivity, self-reflexivity and interiority. The person is not just a substance, even if the most perfect, in ontological continuity with the others. It is an entirely different being, discontinuous with the rest of creation, because of its ability to think. Pascal will say: «Man is only a reed, the most feeble thing in nature, but a thinking reed.»[30]

To the subjectification of substance follows the objectification of being: being becomes an object for a mind that is a subject. In its neutrality, being is simply a raw reserve of resources available for human exploitation.[31] Without a *telos*, an end (final cause), nature cannot account for its meaningful origin (*arche*). Nature becomes a neutral thereness, stripped of value, without formal or final cause. It is the realm of effective causality, a network of forces linked together by mechanic interaction.

There is no denying that the main conquests of science rest upon such an understanding of reality. In his reconstruction of the entire trajectory of modern thought, especially in terms of its scientific advances, Edmund Husserl offers the following account:

> The exclusiveness with which the total world-view of modern man lets itself be determined by the positive sciences and be blinded by the «prosperity» they produced, meant an indifferent turning away from the questions which are decisive for genuine humanity. Fact-minded science excludes in principle precisely the questions which man finds the most burning: questions of the meaning or meaninglessness of the whole of human existence.[32]

In such a *Weltanschauung*, questions of meaning will, subsequently, be bracketed as unimportant – in fact, be expunged from the epistemic drive toward verifiability. What then remains for the subject, when facing such a mechanized understanding of reality? It sees itself as other to the mechanical world or at least as irreducible to it. The world is purposeless, but the subject is purposeful,

---

30   Blaise Pascal, *Pensées*, trans. A.J. Krailsheimer (London: Penguin, 1995), 347.
31   William Desmond insists on such dialectic of subjectification and objectification in modernity. See his *Ethics and the Between* (Albany: SUNY Press, 2001), 17-47; and *God and the Between* (Oxford: Blackwell, 2008), 17-45.
32   Edmund Husserl, *The Crisis of European Sciences and Transcendental Phenomenology*, trans. David Carr (Evanston: Northwestern University Press, 1970), 5-6.

active with respect to being, not passive. It will provide being with the value being does not possess intrinsically. With Kant, the subject turns into «a self-assertive subjectivity.» The person becomes the source of value, a noumenal being endowed with infinite value, with dignity.[33]

The separation of mind and being at the ontological level effects important changes in the understanding of the person. For the person too will now be defined by an intrinsic split, a separation between body and soul. The person is no longer grasped in the unity of a single substance but dissolved in the dualism of two separate substances. From the idea of the human being as a substantial unity of body and soul, as incarnate being, a separation sets in between soul (mind) and body. The soul and the body will be seen as separated from one another, with the soul losing its original meaning of principle of life and unity. Furthermore, the soul progressively becomes intellectualized as «mind,» and this in opposition to the body.

The separation of mind and body opens up a dualism and a reduction in the understanding of the person: the person is her mind, independent of her body. The separation also comes with a new attribution of value: the mind is higher than the body. Here Descartes inherits the dualistic theological logic of modern devotionalism (*devotio moderna*), with its emphasis on the denial of the body. What emerges is a kind of deracinated person, a personhood without body, a personhood reduced to its cognitive faculties. To be a person no longer means to grow into the biological space defined by a specific life-principle (the soul), but to be able to actualize the faculties that are proper to the traits of a «thinking substance.» The person will be such only because *cogitans*, a capacity to think; and this independently of the body, whose connection with the mind is now viewed as entirely accidental. The human body, now separated from the spiritual principle, becomes like a machine (*res extensa*), which entails the loss of the incarnate self, and the intellectualization of the spirit. As the «other» substance in the human composite, the body is left open to the manipulative intervention of the superior principle. The emergence of anatomy

---

33 Desmond, *Ethics and the Between*, 17-47.

in medicine focuses on the body as inert entity (*Körper*), rather than lived-reality (*Leib*).

We need to understand the profound implications of the dualism in question. We are inheritors of such dualism, and I will say that to ask about the personhood of a robot is to fall *de facto* into the trap set by such pre-comprehension. One can look at the consequences of the dualism in question from two angles: either the angle of a *bodiless mind* or the angle of a *mindless body*. We have here the two developments that follow Cartesian dualism: rationalism and empiricism. Let me begin with the latter.

The Deconstructed Self of David Hume

The empirical line, pursued by Hume and the British empirical tradition, develops all the way to nineteenth century utilitarianism. The body receives sensations, sensible impressions from experience and, progressively, the mind builds a sense of identity out of the congeries of such impressions.

What is the person? It is the product, the net result of external stimuli registered by the mind. One can see that because such impressions are seen in their *transitional* capacity to impress the mind, the identity of the person will only be the result of the *psychological*, rather than substantial, ability of the mind to retain impressions through memory. For Hume, the person is not an integral center of being (substance), but the flow of impressions that come and go, insofar as they are retained by memory. Thus, the paradox: there is «person» only insofar as there is actual consciousness of sensations and impressions or, as we might say today, empirical stimuli.[34]

Consider the thought experiments of contemporary bioethicists: when a person loses her consciousness, does she become a different person? Can we have two persons when a patient loses the ability to

---

34 In the *Treatise* (Book I, Part IV, Section VI), David Hume speaks of the self as «nothing but a bundle or collection of different perceptions, which succeed each other with an inconceivable rapidity, and are in perpetual flux and movement» (*A Treatise of Human Nature*, ed. L.A. Selby-Bigge [Oxford: Clarendon, 1988], 252).

reason or to retain memories, etc.? The case becomes particularly acute when it comes to Alzheimer patients.[35]

For Hume, one of the great problems was the experience of sleep: when I am asleep, do I cease to be the person I was? Am I waking up every time a different person? Hume looked into the mind in order to find a unity to the flow of impressions and found nothing. He concluded that *there is no person* as a substantial principle of integrity-in-communication.

If there is a person only insofar as there is *actual* consciousness of sense impressions, then the body is no longer central in providing the condition for the mind's continuity, nor is the embodied-self understood as dynamically growing into its full potential. Consciousness of sense impressions means that both the preconditions of bodily development and the fading of mental functions in a still operative body will not affect the presence of personhood. The person does not «come to mindfulness» out of its bodily conditions, nor does she persist in its no longer conscious bodily presence. Prenatal life will not be personal. Nor will the life of the senile demented be the life of a person.

Two final points, one relative to freedom of the will, the other to the general conception of morality as based on emotions, rather than reason. For Hume there is no freedom of the will because the person, as he understands it, is entirely determined. Freedom of the will is a mask, a deception, at best the epiphenomenon of something else to which it can be reduced. Humean determinism will cast long shadows, all the way to contemporary philosophies of mind, in which the de-personalization of the human being is complete.

A second point: the deconstruction of personhood and the denial of freedom rest upon a conception of morality based on emotions, rather than reason. The feeling of sympathy towards other human beings will become the basis of morality. But how fleeting such a feeling is! The Humean retrieval of emotions, feelings and, in general, of the affective dimensions of the person, is *de facto* equivocal:

---

35 For an early example of the literature on this, see Allen Buchanan, «Advance Directives and the Personal Identity Problem,» *Philosophy and Public Affairs* 17 (1988): 277-302. See Also Helga Kuhse, «Some Reflections on the Problem of Advance Directives. Personhood and Personal Identity,» *Kennedy Institute of Ethics Journal* 9, no. 4 (1999): 347-64.

it stands in the wake of Cartesian dualism and cannot provide a secure basis for moral judgment. Hume will become the father of non-cognitivism.[36]

The Person as Universal Ego and Autonomous Subjectivity: Immanuel Kant

I come to the rationalist line of development, following Cartesian dualism. Kant saw that the empirical ego is parasitical on a more original, *a priori*, notion of the self. The transcendental ego is the condition of all ordered experience *qua* experience of a unitary I.[37] Kant's philosophical anthropology represents a reaction to the fragmented self of Hume: the universal/transcendental ego is the formal capacity to gather the multiplicity, the manifold into an ordered unity.[38] With Kant, we have a more rigorous understanding of the meaning of personhood which, however, still stands within the «effectual history» of Cartesian dualism. Such dualism will be rendered by Kant in terms of a gulf between the phenomenal and the noumenal sphere of reality. To be a person is to belong to this noumenal realm, a realm of rationality and capacity for absoluteness. It is a realm of freedom, rather than natural necessity, such as the one entailed by Newtonian physics.

The noumenal realm also is a realm of value in a world stripped of it, as if the value sucked out of the objective world is now being

---

36 At the same time, the rehabilitation of the emotional sphere in Hume has not remained without important consequences, for example, in certain strands of contemporary feminist ethics. See Annette Baier, «Hume, the Women's Moral Theorist?» in *Women and Moral Theory*, ed. E. Feder Kittay and D.T.Meyers (Totowa, NJ: Rowman & Littlefield, 1987) and Alisa Carse, «The Voice of Care: Implications for Bioethical Education,» *Journal of Medicine and Philosophy* 16, no. 1 (1991): 5-28.
37 See Kant, *Critique of Pure Reason* on the «transcendental deduction.»
38 For the empiricist, the manifold in question can be brought to unity in a derived synthesis which, however, can easily fall apart without an *intrinsic* principle of unity. Such a principle points to a prior and not just derived synthesis, i.e., the «synthetic a priori»: «The empiricist fails to see that a derived synthesis is possible only on condition of a prior synthesis, which is, in this respect, the condition of its possibility,» William Desmond, *Desire, Dialectic, & Otherness: An Essay on Origins,* 2nd ed. (Eugene: Cascade, 2014), 69.

channeled back into the subject, who becomes a being of infinite value. Question: where does the value of the subject come from if the world is valueless? How can something like a being *endowed with value* emerge from such a valueless world?

With Kant, the notion of personhood is clearly recognized in its moral significance. To be a person is to be a moral being able to exercise moral agency. In this resides the dignity of the person, its infinite value. To be a person is to be *autonomous* – i.e., to grant meaning, not to receive it (from God, religion, nature, etc.).

After Kant, the story of autonomy is the story of its progressive radicalization. There is a logical trajectory that runs from Kant to Nietzsche, from the person as power of autonomous decision making to the person as will to power.[39] In a world devoid of intrinsic value, and with an assertive (autonomous) notion of the person, the human also becomes an object. The objectification of the world turns into the objectification of the human. Having mechanized the world, we then end up mechanizing ourselves, looking at ourselves as mechanical systems, machines. The machine is the perfect expression of the modern outlook on reality: it is entirely constructed and available, that is, disposable. It is the perfect «object.»

We too become like constructed mechanisms. We cry freedom of choice and autonomy but fall victim to all sorts of constructionisms: social (Marxism), psychological (Freud), neurological. Think of the attempt of cognitive science to reduce the human mind to the physiological circuitry of the brain, and explain away consciousness in terms of deterministic happening, entirely reducible to neurological functionality.

The essence of the machine: to be entirely passive to our own construction. In the case of the robot, its apparent activity, whether cognitive or practical, is entirely determined by us. Even when robots think for themselves, they do not really think. They only carry out pre-ordained programs based on algorithmic laws of our own devise.

---

39   For this interpretation already, see Henri De Lubac, *Le drame de l'humanisme athée* (Paris: Cerf, 1999); and Romano Guardini, *Das Ende der Neuzeit. Ein Versuch zur Orientierung* (Würzburg: Echter, 1951).

To Be a Person: A Philosophical Account

In light of this historical reconstruction, I now want to offer a more systematic understanding of the person. I begin with a provisional definition, something like a heuristic statement, aware of the fact that Levinas would contest the possibility to «de-fine» a person. As a manifestation of the Infinite, the person escapes definition, at least in the sense that definitions are able to circumscribe objects. No objective definition of the person is possible without passing through a «lived subjectivity,» the *event in being* Levinas calls «psychism»: this is a *way of being* resistant to totality, in this case the one pursued by the search for definitional boundaries. Perhaps a better way to pose the question is not to enquire «what is a person?» but rather «who is a person?»

I think Levinas would push us to think of the person in terms of *an incarnate singularity, coming to itself, in openness to the Other*. I will parse out these three elements, incarnate singularity, selfhood, and openness to the Other, in order to arrive at an understanding of the person that owes greatly to Levinas, even if in the end it will differ from his in significant ways.

The Person as Incarnate Singularity

To be a person is to be individuated, to be an individual, to be one with oneself (already Boethius and Thomas Aquinas speak of *individua substantia*). Such individuality is expressive of a certain incommunicability (Roman law defines the person as *sui iuris et alteri incommunicabilis*). We are singular beings, though similar to others (brothers, monozygotic twins, or even clones). Levinas stresses this sense of incommunicability, independence, and separation, in terms of a break from any notion of participation:

> One can call atheism this separation so complete that the separated being maintains itself in existence all by itself, without participating in the Being from which it is separated – eventually capable of adhering to it by belief – One lives outside of God, at home with oneself; one is an I, an egoism. The soul, the dimension of the psychic, being an accomplishment of separation, is naturally atheist.[40]

---

40 Levinas, *Totality and Infinity*, 58.

The singularity in question is not the result of a statement of singularity; it is not a vindication of singularity (as «constituted,» it would already belong to «the order of thought»). It is not constructed, but *given in the flesh*, in the body that we have, the «body that we are,» to echo Gabriel Marcel. This singularity is incarnate. Only a human being can be a person – i.e., someone who belongs to the human species – because the singularity in question is embodied in the singularity of a human flesh.[41] Consider the following quotations from *Totality and Infinity*:

> «The sensibility we are describing starting with enjoyment of the element does not belong to the order of thought but to that of sentiment, that is, the affectivity wherein the egoism of the I pulsates.»[42]
> «Sensibility established a relation with a pure quality without support, with the element. Sensibility is enjoyment. The sensitive being, the body, concretizes this way of being, which consists in finding a condition in what, in other respects, can appear as an object of thought, as simply constituted.»[43]

This is relevant in addressing the question of double personality raised in the wake of the empiricist understanding. To say «I» (as we will see later, when talking about the self) is to actualize an energy of being that might still be dormant in the body, but that in its elementality reminds us of our being given to be *in the body that we are*.

---

41 The question of the attribution of the notion of person to spiritual creatures, such as angels, or to God, would take us too far afield and cannot be addressed here. The focus remains on the human person. In critical distance from Heidegger, for whom sensibility means the reversal of praxis over theory, Levinas stresses a more elemental notion of sensibility, which he calls «enjoyment» (*jouissance*): «Sensibility, Levinas discovers, does not first emerge as a praxis caught up in the larger network of «in-order-to» (*das Um-zu*) – the «referential totality» (*Verweisungsganzheit*) – which ultimately implies Dasein. Rather, sensibility is first the sheer enjoyment of sensations, a «carefree» contentment with sensing itself. Embodiment, sensibility, flesh, is, first a self-satisfaction and an enjoyment of elemental sensations, the sun on one's arms, the breeze in the air, indifferent to the higher-level significations of instrumentality and theory,» Richard Cohen, *Ethics Exegesis and Philosophy*, 154.
42 Levinas, *Totality and Infinity*, 135.
43 Levinas, *Totality and Infinity*, 136.

To be a person is to be in a potential, constant state of growth. The person is always the promise of something more. We could say that such elementality cannot yet be objectified; it is pre-objective, in the sense of being felt, rather than determined in itself, as an object. We feel ourselves, first of all, in bodily immediacy. If the dimension of self-insistence is prevalent at this point, it is a self-insistence that is also already a community with others. Levinas brings attention to the phenomenon of the *face*. In my face I say, «I am»; this always also is a «here I am,» that is, «I am with.» One must stress the flow-like, rather than fixed, character of this incarnate singularity.[44] The singularity in question is also a site of flow and passage, undergoing the world. The incarnate singularity is a passion of being.[45]

The Selfhood of the Human Person

The person is a being that comes to itself in self-awareness and reflectivity, in action and the development of moral agency. The self is not just given to itself. It becomes a self, it becomes a subject, indeed, a «thinking self» (Descartes). This coming to self, as Hegel points out, can only be possible in the intermediation with what is other. The third dimension in the definition provided above, that of «openness to the Other,» is the last only in a temporal, rather than logical sense: it already subtends the other two components. The relation to the other is «older» than the relation to the self: «One may legitimately ask oneself whether the internal discourse of the *cogito* is not already a derivative mode of the conversation with the other; whether the linguistic symbolism that the soul uses in «conversing with itself» does not suppose a dialogue with an interlocutor other than itself; whether the very interruption of the spontaneous impulse of thought reflecting upon itself, all the way down to the dialectical alterations of reasoning

---

44 Here is where a metaphysics of substance falters.
45 One will note the difference between this elemental, suffering being (*passio*) and the self-positing ego of transcendental philosophy (*conatus*). Leibniz had a premonition of the meaning of this suffering being when he distinguishes between «perception» and «apperception»: the self as flesh is perception not yet conscious in the distinct sense of apperception.

where my thought separates from and rejoins itself as if it were other than itself – whether this interruption does not bear witness to an *original and foregoing* dialogue.»[46]

I said before that the incarnate singularity already undergoes the world. The body is always a medium of exchange, it is never only «mine»: «The subjectivity of the subject, its very psyche, is a possibility of inspiration. It is the possibility of being the author of what has been breathed in unbeknownst to me, of having received, one knows not from where, that of which I am the author. In the responsibility for the other we are at the heart of the ambiguity of inspiration.»[47]

First Ethical Selving: Intentionality

We come to ourselves in knowledge and action. The first component speaks to the exercise of mindfulness as an intentional act, a communion with what is other to us, an «object» toward which we move (or «to which we attend,» *ad-tendere*) and yet could not do so, if not because of a mysterious participation already given to us in the onto-logical intimacy of mind and being.

This is why intelligence can never be «artificial»: artificial intelligence is a preordained function wired to carry out certain operations. If we speak of intelligence, and can do so only analogically, we should always distinguish it from the intelligence of a person, a human being. Whereas a machine possesses, at best, «syntactical» capacity, the person is capable of «semantic» appropriation – i.e., is able to understand what she is doing:

> In the sense in which people «process information» when they reflect, say, on problems in arithmetic or when they read and answer questions about stories, the programmed computer does not do «information processing.» Rather, what it does is manipulate formal

---

46 Emmanuel Levinas, *Of God Who Comes to Mind*, trans. Bettina Bergo (Stanford, CA: Stanford University Press, 1998), 146.
47 Emmanuel Levinas, *Otherwise Than Being or Beyond Essence*, trans. Alphonso Lingis (Pittsburgh: Duquesne University Press, 1998), 148-49. «Inspiration» is existence «through the other and for the other, but without being alienation,» (114-15).

symbols. The fact that the programmer and the interpreter of the computer output use the symbols to stand for objects in the world is totally beyond the scope of the computer. *The computer, to repeat, has a syntax but no semantics.* Thus, if you type into the computer «2 plus 2 equals?» it will type out «4.» But it has no idea that «4» means 4 or that it means anything else. And the point is not that it lacks some second-order information about the interpretation of its first-order symbols, but rather that its first-orders don't have any interpretations as far as the computer is concerned. All the computer has is more symbols.[48]

Like Searle, Hubert Dreyfus has been especially critical of the artificial intelligence model of human thinking and cognition understood as disembodied processes. Overall, his is a critique of disembodied artificial intelligence. Dreyfus uncovers a number of false pre-suppositions entailed by dis-embodied artificial intelligence. First, a biological assumption: the brain processes information in discrete operations by way of some biological equivalent of on/off switches. Second, a psychological assumption: the mind can be viewed as a device operating on bits of information according to formal rules. Third, an epistemological assumption: all knowledge can be formalized; what can be understood can be expressed in terms of logical relations. Fourth, an ontological assumption: since all information fed into digital computers must be in bits, the computer model of the mind pre-supposes that all relevant information about the world, everything essential to the production of intelligent behavior, must in principle be analyzable as a set of situation free determinate elements. His conclusion:

> Thus the view that the brain as a general purpose symbol manipulating device operates like a digital computer is an empirical hypothesis which has had its day. No arguments as to the possibility of artificial intelligence can be drawn from current empirical evidence concerning the brain. In fact, the difference between the «strongly interactive» na-

---

48   John R. Searle, «Minds, Brains, and Programs,» in Boden, *The Philosophy of Artificial Intelligence*, 85 (emphasis mine). Searle's conclusion: «The point is that the brain's causal capacity to produce intentionality cannot consist in its instantiating a computer program, since for any program you like it is possible for something to instantiate that program and still not have any mental state. Whatever it is that the brain does to produce intentionality, it cannot consist in instantiating a program since no program, by itself, is sufficient for intentionality» («Minds, Brains, and Programs,» 87).

ture of brain organization and the non-interactive character of machine organization suggests that insofar as arguments from biology are relevant, the evidence is against the possibility of using digital computers to produce intelligence.[49]

Intelligence can only be an embodied act (think of the notion of «emotional intelligence»!), the actualization of knowledge and thinking through which we complete the world, we make the world come to itself, while receiving from the world the gift of a deeper sense of ourselves.[50]

The distinction between human and artificial intelligence becomes even clearer when we analyze basic dimensions of intentionality. Consider desire, imagination, and memory. In the eruption of *desire*, something possible only to an incarnate singularity, there emerges for the self the possibility to be other to itself. Desire reveals the energy of transcendence at work within the self and testifies to the power of self-differentiation in the self. There is more to the self than the identity of the same.[51] When the latter feels itself as lacking, it not only expresses something negative but, rather, more positively brings forth the energy of being in all its richness, exploding in the self in the form of desire. As Levinas says: «Desire is an aspiration that the Desirable animates; it originates from its

---

49 Dreyfus, *What Computers Still Can't Do*, 156.
50 On this, consult the beautiful essay by Herman Krings, *Meditation des Denkens* (München: Kösel, 1956).
51 For Paul Ricoeur to be a self is not the same, *idem* has to become an *ipse*! The issue of the relation between Ricoeur and Levinas is complicated; the two positions ought to be carefully distinguished. To put it briefly, and with reference to a pithy quotation from Ricoeur, the point of disagreement between the two consists in the fact that, for Ricoeur, «Awakening a responsible response to the other's call cannot work except by presupposing a capacity for reception, of discrimination, and of recognition» (*Oneself as Another*, trans. Kathleen Blamey [Chicago: The University of Chicago Press, 1992], 339). See Cohen's comment: «Following an existentialized version of philosophy's transcendental route, for Ricoeur, in contrast, there must always first be self-reflexivity, a capacity in the sense of a base, ground, zero-point, from which and out of which and into which otherness is *correlated*. For Levinas, in contrast, such an insistence on recognition, or on recognizing the priority of recognition, misses accounting for the prior impact which is at once the impact of alterity as such and moral obligation» (Cohen, *Ethics, Exegesis and Philosophy*, 304-05).

"object"; it is revelation – whereas need is a void of the Soul; it proceeds from the Subject.»[52]

In this process, desire and *imagination* are allied. Imagination brings to further clarity the process of othering in the self, for desire acquires specific contours only in imagination. One ought to remember that the process of othering takes place in the intermediation with the other. The self is not fixed, it is a *metaphor*, a carrying across which differentiates itself in the images of itself it images for itself. The self is the metaphor of a carnal mindfulness.

Careful vigilance is required, though, since the process of imagining itself as other to itself is equivocal. It can be seen as an end in itself, a process in which the self goes from one self-formation to the next, never encountering the other, and in so doing, never coming to itself. The self is then in flight from itself and not toward the other.

We end up dissipating the original energy of being, given to us in who we are, when we cannot face the selves we are in promise. *We need to remember* who we are in our self-transcendence. Memory is needed to balance desire and imagination, as well as the quest for self-transcendence: «Memory recaptures and reverses and suspends what is already accomplished in birth – in nature…By memory I ground myself after the event, retroactively: I assume today what in the absolute past of the origin had no subject to receive it and had therefore the weight of a fatality.»[53]

Memory is the persistence of elemental self-awareness in the passage of transcending or becoming: «Memory as an inversion of historical time is the essence of interiority.»[54] It is the return of the self to itself in the passage of becoming. I am talking here about memory in a non-objective way: not so much in terms of the process of remembering things, but as a kind of non-objective function, grounding our sense of interiority. Self-transcendence is not only externally directed: memory opens up the self to its inner otherness.[55] How is one to speak of artificial intelligence in terms of desire, imagination, and memory?

---

52 Levinas, *Totality and Infinity*, 62.
53 Levinas, *Totality and Infinity*, 56.
54 Levinas, *Totality and Infinity*, 56.
55 I am thinking of the discovery of the unconscious or Dostoevsky's «underground man» in *Notes from the Underground*: the ground of autonomy turns out to be a groundless abyss!

Second Ethical Selving: Action

In action we come to ourselves as moral agents. This is first of all an openness, a response (*Wertantwort*) to the world and call of values which, for Levinas, is being revealed in the face of the Other, its vulnerability and indigence. To act morally is to transcend oneself; better, to embark in a movement of *transascendence*: «The metaphysical movement is transcendent, and transcendence, like desire and inadequation, is necessarily a transascendence.»[56]

This in two ways: we transcend ourselves in what we become, when acting morally. This is why the «doing» involved in acting (*praxis, agere*) is different from the doing of production (*poiesis, facere*). In the latter, we do something that brings forth an external being, an external object. In the former, the effect is not outside the agent, but on the agent itself. The effect of moral action is the achieved integrity of agency.[57] The robot produces effects, but does not act in the sense above, thus it cannot be a moral agent.

The transcendence in question, however, is ultimately toward the other as other. For this reason, there can be an ambiguity at play here, when the process of moral performance turns into a journey of self-achievement, whereby the other is sought only as a function of one's fulfillment. We can desire the other out of a lack, now seen not so much as impelled by the energy of the source that originated us, but driven by the sense of lack that, negatively, determines the other for-self.

One might call this kind of transcendence «self-oriented.» Hegel understands the subject in the process of its becoming thus, as self-determining negativity: the other is for-self, like in the master-slave dialectic of the Phenomenology, a story that repeats itself in Sartre's dialectic of masochism and sadism.[58] Deformation takes place with the stifling of the plenitude of excess, which is the origin prior to the

---

[56] Levinas, *Totality and Infinity*, 35. The term «transascendence,» as Levinas clarifies in the footnote, is from Jean Wahl.

[57] The point, with the distinction between *poiesis* and *praxis*, is obviously Aristotelean.

[58] See G.W.F. Hegel, «Independence and Dependence of Self-Consciousness. Lordship and Bondage,» in *The Phenomenology of Mind*, trans. J.B. Baillie (New York: Harper & Row, 1967), 228-40; Jean Paul Sartre, *Being and Nothingness: An Essay in Phenomenological Ontology*, trans. Sarah Richmond (London: Routledge, 2018), chap. 3: «Concrete Relations with the Other.»

lack of the self-oriented self, culminating with Nietzsche in the will to power willing itself for the purposes of its own self-glorification. Perhaps, with Levinas, one can envisage another way. For him «metaphysics does not coincide with negativity.»[59]

The Person as Openness to the Other

We need to understand the fulfillment of the self differently. While the self for sure is *penia* (poverty), it is also excess, because *porous* to the sourcing power.[60] The openness of transcendence can be *agapeic* (other-oriented), rather than self-oriented: it can be an openness for the sake of other-being, rather than self-being. The «agapeic self» breaks the circle of mediation and returns to the origin (God?) as a source of infinite energy. Impelled by the generosity of the origin, the self opens itself up to what is other to itself, breaching the circle of self-mediation. In self-oriented transcending, the self is more than the transcendence. It is *transascendence*: in other-oriented transcending, transcendence is more than the self.

A dialectical mediation is at play, one that becomes an inter-mediation, rather than a sublation of the other to the self (à la Hegel). The space between self and other rests intact as the middle space between infinitudes: to the inward infinitude of the self, there corresponds the infinitude of the other. The mediation between such a plurality of infinitudes cannot be a self-mediation, in which one subordinates the other to itself. As the singularization of communicative being, the self *as self* cannot be reduced to will to power but will have to be understood as the *willingness to give itself* up for the other. Herein lies the paradox: loss of self is finding oneself. The «agapeic self» is dis-interested, in the sense of transcending self-interest into the middle (*inter-esse*). It is also *hetero-archic*: subject to the other, and because of this, a subject. In free obedience to the other, the self finally finds itself. For Levinas this openness to the other is the ultimate meaning of fecundity: «The I springs forth without returning, finds itself the self of another: its pleasure, its pain is pleasure over the pleasure of the other or over his pain – though not

---

59   Levinas, *Totality and Infinity*, 41.
60   Think of the story of love in Plato's *Symposium*.

through sympathy or compassion. Its future does not fall back upon the past it ought to renew; it remains an absolute future by virtue of this subjectivity which consists not in bearing representations or powers but in transcending absolutely in fecundity.»[61]

The openness to the world, chiefly the world of the other, is an act of love, a fulfillment of reciprocity. To be a person is to love, because this is what «the incarnate singularity of a self, open to the Other» ultimately does: it actualizes itself beyond itself, in the responsibility for the other that is both a response (responsibility comes from *respondere*) and a release. A responsibility: I become myself, I become a moral agent, because I see myself commissioned by a call. «I am summoned as someone irreplaceable. I exist through the other and for the other, but without this being alienation.»[62]

I can be many things, and yet fail myself, when failing to heed the particular call to which life calls me (this forms the nucleus of truth in situation ethics). Such call is singular: it is not a general call for «the humanity in me,» as Kant would have it; nor is it a response reducible to the production of a good, not even «the greatest good for the greatest number» of utilitarianism: «Freedom is borne by the responsibility it could not shoulder, an elevation and inspiration without complacency. The for-the-other characteristic of the subject can be interpreted neither as a guilt complex (which presupposes an *initial* freedom), nor as a natural benevolence or divine "instinct, nor as some love or some tendency to sacrifice."»[63]

Whence the moral call then? Why should one heed it? Why be moral, in the end? Kant put the question in terms of a difference between a «hypothetical» and a «categorical» imperative. He had a premonition of the issue at stake here but was ambivalent about recognizing an alterity that summons the autonomy of the self, lest falling into heteronomy again. A communion of autonomous beings will be possible, for Kant, only in the «kingdom of ends.» But this is only a postulate, an exigence of our thinking, not a reality we can know.[64]

---

61   Levinas, *Totality and Infinity*, 271.
62   Levinas, *Otherwise than Being or Beyond Essence*, 114.
63   Levinas, *Otherwise than Being or Beyond Essence*, 124.
64   Immanuel Kant, «Transition from Popular Moral Philosophy to a Metaphysics of Morals,» in *Grounding for the Metaphysics of Morals*, trans. James W. Ellington (Indianapolis: Hackett, 1981), 43.

The question here is the question of God as the ground of morality. An important question, and not only for Kant.⁶⁵ At stake is not only the question of God as the law-giver, who grants the moral imperative its absoluteness. It is rather a question of *release*: the release of our own freedom into the reciprocity of love, out of the love that generates us into being. We love because we are being loved. Generated into love, in the space of goodness predisposed for our enjoyment, we are capable of a freedom-for-the-other beyond autonomy, in the generosity of service. This is to be a moral being. This is, ultimately, what it means to be fully human, to be a person.

If I see correctly, Levinas has a tendency to moralize the relation to God and, subsequently, the notion of creation. He sees God as the infinite *in* the face of the other, but *not as the ground* of love for the same, which opens the same to the other. Because the other is the Master, «The interlocutor is not a Thou, he is a You: he reveals himself in his lordship. Thus, exteriority coincides with a mastery. My freedom is thus challenged by a Master who can invest it.»⁶⁶

There is moral earnestness in this God, but this is hardly a God of love! A different notion of creation is also needed. Levinas sees, correctly, that creation is a freeing of the person into her autonomy but does not quite understand such freeing against an ontology, a metaphysics of goodness. If this is the conclusion, then I am already *beyond* Levinas.⁶⁷

---

65 Recall Ivan's argument in Dostoevsky's *The Brothers Karamazov*: «If you were to destroy the belief in immortality, not only love but every living force that maintain the life of the world would at once be dried up....For every individual...who does not believe in God or immortality, the moral law of nature must immediately be changed into the exact contrary of the former religious law.» Fyodor Dostoevsky, *The Brothers Karamazov*, trans. Constance Garnett (New York: Barnes and Noble Classics, 2004), 71. The quotation is in Book 2, chapter 6 [«Why Is Such a Man Alive?»].
66 Levinas, *Totality and Infinity*, 101.
67 As William Desmond puts it, «Levinas seems to reiterate again and again, not only the horror of the *il y a*, but the *evil of being* in the relentless self-insistence of the *conatus essendi*. ...His *version* of Plato's Good beyond being dictates a saving trauma and reversal from myself to the other. But *is there not an evil in that ethical good that sees being as evil?*... We have not quite been released to the agapeic "It is good"» (*Art, Origins, Otherness: Between Philosophy and Art* [Albany: SUNY Press, 2003], 162). I have articulated the implications of such a metaphysics of the good for the field of bioethics in

Concluding Unscientific *Postscript*

At the end of this long detour on Levinas and the philosophy of the person, I return to the initial provocation and the question «can a robot be a person?» What to make, now, of the dream of an immaterial universe, the mad project instigated by a notion of robotics in which the human being becomes, in the end, «the subjugated subject»?[68]

In his 1920 novel *R.U.R.*, Czech writer Karel Capek describes the world imagined by scientist Rossum, a world of artificial workers (the word «robot» derives from the Slavic root for «work») intelligent and indefatigable, but incapable of feelings: «Rossum's Universal Robots,» hence the acronym for the novel's title, «R.U.R.» Although the intention of Rossum is to liberate human beings from the slavery of work and make them the masters of creation, the dream eventually fails: once all their needs are satisfied, human beings no longer have to work, and thus cease to reproduce. The robots, however, revolt and kill the humans, and their leader proclaims, «The time of man is over. A new world begins. The reign of robots.»[69]

We are not there yet of course. And there is no need to think that robotics as a scientific enterprise has to necessarily end in transhumanist madness. Still, the retrieval of a personalist philosophy capable of highlighting the essential difference between person and machine provides an important buffer to any scientific totalizing pretense. As long as this is the case, «the time of man is *not* over yet.»

Levinas's phenomenological account provides important insights that remain closed to any scientific gaze, including the one articulated by Ishiguro and his vision of a «human-robot symbiotic society.»[70] For Levinas, as for Merleau-Ponty, phenomenology «is from the start a forswearing of science.»[71] Thus the «unscientific»

    Roberto Dell'Oro, «On the Ultimate That is the First: Thinking Beyond (Bio)ethics,» *Gregorianum* 100, no. 3 (2019): 621-47.

68 Thus Rémi Brague, in his reconstruction of modernity. See his *The Kingdom of Man: Genesis and Failure of the Modern Project*, trans. Paul Seaton (Notre Dame: University of Notre Dame Press, 2018), 160-68.

69 See reference to Capek's novel in Brague, *The Kingdom of Man*, 167.

70 Ishiguro, «The Studies on Interactive Humanoids.»

71 Maurice Merleau-Ponty, *Phenomenology of Perception*, trans. Colin Smith (Oxford: Routledge, 1962), ix.

nature of this conclusion, which only calls for the suspension of those premises that make it impossible to see the person's singularity in a spectacle of world-objects. In this paper, I have attempted to provide something like an archeological reconstruction of such premises, only to show that the dualism they subtend falls short of accounting for what the person is: «an incarnate singularity, coming to itself, in openness to the Other.»

CHAPTER 15
PASSIO ESSENDI: ON SUFFERING

In this chapter, I discuss the section on suffering in a recent book by the Pontifical Academy for Life, *La gioia della vita*.[1] I will place my comments within the context of the text at large, retrieving some of its central conceptual premises. It is important to recognize the necessity of reading the phenomenological dimensions of suffering not simply in their descriptive character, but in their existential and ultimately ontological significance. This is a central insight, which responds to the deeper intentionality of the text. Indeed, in the text, the paragraphs about suffering stand within the backdrop of the reflections developed in the preceding part, under the heading of «Being Born and the Joy of Life Received». What is developed there in terms of the condition of «being born» represents something like an *archeological* moment, whose meaning re-emerges at the end, in the part about «Eschatology and the Drama of Life,» when the revelatory nature of the *eschaton* brings into full evidence the truth of the origins.

In light of this polarity, between the beginning and the end, between *arche'* and *eschaton*, suffering finds its proper meaning. In all its various forms – from the sting of physical pain (dis-ease) to the pangs of meaninglessness (the «sickness unto death» of Kierkegaard), suffering breaks through the ontological texture of life with its enigmatic power of disruption. Yet, such disruption need not be a contradiction of life and its *gaudium*, if seen as the symbolic reminder of the fact that we *undergo* life in a condition of «being given to be.»

---

[1] Carlo Casalone, Maurizio Chiodi, Roberto Dell'Oro, Pierdavide Guenzi, Anne Marie Pelletier, Pierangelo Sequeri, Marie-Jo Thiel, Alain Thomasset, La gioia della vita. *Un percorso di etica teologica: Scrittura, tradizione, sfide pratiche* (Roma: Libreria Editrice Vaticana, 2024). The part in question bears the title «Il soffrire e la vita "messa alla prova"», 130-137. A general introduction to the text in Roberto Dell'Oro and Therese Lysaught, «Theological Ethics of Life: A New Volume by the Pontifical Academy for Life,» *Journal of Moral Theology* 2 (2022): 65-77.

Implicitly, this realization presupposes something like a «reversal of intentionality» in the way we think of our relation to life. Spontaneously, we assume that «to live» is to be in a position of *active* engagement toward life, either in terms of our mindful relation to it, for an example, in knowledge, or in the practical effort to shape our destiny, as in the moral life. And yet, no intentional or practical relation to life would be possible, if not on the premise of a prior openness to it. Such openness, as a transcendental condition of possibility, is not self-determining, the result of an autonomous act of «construction.» Rather, it is given in the intimate communion with being that defines our ontological constitution. As such, it is received in the gift of life itself.

One might say that, in the *pathos* of life, in what Michel Henry refers to as «Life's primitive suffering of itself,» we are instructed about life's radical *givenness*. In this sense, the particular *pathos* of suffering becomes a dimension of the broader *pathos* of life, and the sign that our «effort to be,» with all its connotations of activity, is foregrounded by a more original receptivity.»[2]

The point finds a number of iterations in contemporary philosophy, especially in phenomenology. However, insofar as it bespeaks a *constitutive* dimension of the human condition, it possesses a metaphysical quality, for it conveys something deeply rooted in our understanding of what it means «to be» at all.

Consider, again, Michel Henry: «The singular Self that I am experiences itself only within the movement by which Life is cast into itself and enjoys itself in the eternal process of absolute self-affecting... *The Self self-affects itself only inasmuch as absolute Life is self-affected in this Self.* It is Life, in its self-giving, which gives the self to itself. It is Life, in its self-revelation, that reveals the self to itself. It is Life, in its *pathetik* embrace, that gives to the Self the possibility of pathetically embracing itself and of being Self.»[3]

---

[2] The notion of life self-affectivity as originary, and thus grounding, with respect to the secondary, and thus derivative, nature of intentionality, represents a central insight of Henry's *corpus*. More explicitly in *Incarnation: A Philosophy of Flesh*, trans. Karl Hefty (Evanston, IL: Northwestern University Press, 2015), and *I Am the Truth: toward a Philosophy of Christianity*, trans. Susan Emanuel (Stanford, CA: Stanford University Press, 2003).

[3] Michel Henry, *I Am the Truth*, 107.

Paul Ricoeur, in *Oneself as Another*, makes his own the assertion of Spinoza, in the Book III of *Ethics*: «The power or conatus by which each thing perseveres in its own being, is nothing but the given, or actual, essence of the thing itself.»[4]

Commenting further on Spinoza's statement, Ricoeur adds: «In this sense, the power to act can be said to be increased by the retreat of *passivity* tied to inadequate ideas... This conquest of activity under the aegis of adequate ideas makes the work as a whole an *ethics*. Thus there is a close connection between the *internal dynamism worthy of the name of life* and the power of the intelligence, which governs the passage from inadequate to adequate ideas. In this sense, we are powerful when we understand adequately our, as it were, horizontal and external dependence with respect to all things, and our vertical and immanent dependence with respect to the primordial power that Spinoza continues to name God.»[5]

In his metaphysical trilogy, William Desmond has rephrased the issue in terms of a dialectical movement between the «effort to be» (*conatus*) and the more original porosity to being that signals our undergoing life.[6] He refers to it in terms of the «passion of being» (*passio*). To retrieve such porosity toward being is to overcome the strictures of self-determining thought and a subjectivity become, especially in modernity, self-assertive. It also entails unclogging the resistances toward a prior, generative dimension of life, from

---

4 Baruch Spinoza, *Ethics*, trans. Samuel Shirley (Indianapolis: Hackett, 1982), 109.

5 Paul Ricoeur, *Oneself as Another*, trans. Kathleen Blamey (Chicago: The University of Chicago Press). 316. The emphases in the text are mine. The question of how the *passivity* evoked by Spinoza, and to which Ricoeur refers, grounds the *conatus* by which each thing perserveres in its own being, is beyond the scope of this paper. Also, the plausibility of Ricoeur's own interpretation of Spinoza is not the issue. On the other hand, the emphases on the dialectical relation between *conatus* and *passio* in Spinoza remains interesting to note for the sake of the argument I am developing. With respect to Spinoza, Ricoeur is concerned with «showing the connection between the phenomenology of acting and suffering self and the actual and potential ground against which selfhood stands out. For me, this connection is Spinoza's *conatus*,» Paul Ricoeur, *Oneself as Another*, 315.

6 The dialectic of *passio* and *conatus* appears early on in Desmond's work, but finds systematic articulation especially in *God and the Between* (Blackwell, 2008) and *The Intimate Universal: The Hidden Porosity Among Religion, Art, Philosophy, and Religion* (New York: Columbia University Press, 2016).

which spring all dimensions of activity in the self, whether theoretical (intentionality) or practical (action). The *conatus* is derivative of a more primordial *passio*.[7]

The attestation of the ontological primacy of *passio* over *conatus* opens to the mysterious *recognition* of the suffering of others, and, furthermore, of the sharing in their suffering. Here the *passio* becomes *com-passio*. The point can be fully made only if the recognition in question is seen within a phenomenology of intersubjectivity, and more deeply, a «relational anthropology,», in which the other belongs to the very constitution of the self.

This entails overcoming any «egological» or «atomistic» hermeneutics of personhood, including the «expressive individualism» prevalent in bioethics today.[8] Here the «self» is posited *a priori*, that is, independently of the encounter with the other. In this perspective, the relation of self and other is resolved in terms of an externality of the other to the self.[9]

The issue, once more, stresses the close relation between the ethical and the anthropological/ontological dimension. The ethical gesture of solicitude toward the other finds its ultimate condition of possibility in the ontological dimension of receptivity. One might even say that the *condition of possibility* of the recognition of another's suffering is the ontological intimacy of self to the other at the heart of self itself. The possibility of empathy rests on the ontological openness, in fact, the *love of being* (*passio*) that becomes a *love of the other* (*com-passio*).

---

[7] This is what perhaps Thomas tries to express with the dialectic of *irascibile* and *concupiscible*, though the interpretation of the former in terms of desire, and so activity, might not give due recognition to the ontological dialectic here at stake.

[8] Charles Taylor speaks of «atomism» and points to the general character of the modern self in terms of a «buffered self» in *A Secular Age* (Cambridge, MA: Harvard University Press, 2007). «Expressive Individualism» is a term coined by Robert Bellah et al, in *Habits of the Hearts: Individualism and Commitment in American Life* (Berkeley: University of California Press, 1996)

[9] The externality in question differs from Levinas' insistence on «separation.» On the interplay of «internal» and «external» relations as defining the ontology of «things,» which includes persons, also William Desmond, *Being and the Between* (Albany: SUNY, 1995), 318-322: «Exteriority has to be rethought in terms of community in irreducible otherness that gives things the space in which they become themselves,» at 320.

The point just made is relevant in order to properly articulate the meaning of the Christian economy of suffering, according to which one is asked to partake in Christ's suffering. One risks understanding such partaking, once more, in too *activist* a sense, whereas what is at stake in this profound attitude of «dissolution of oneself» into the suffering of Christ is a «letting be» that makes the undergoing of suffering *porous* to the more original undergoing of Christ himself.

Our suffering *can be redeemed* only because assumed in the suffering of Christ. He does the action. We only partake in it. Only under this condition, that is, as a partaking in the original action of Christ, *in his gift for us*, we can also come to see *suffering as a grace* rather than a curse.[10]

Finally, keeping in mind the dialectic of *passio* and *conatus* might also help understand better the tension between sickness and health. The text refers to the W.H.O. statement about health as a «state of complete physical, mental and social well-being...»

The definition is not without problems. For sure, too «expansive» a definition of health risks opening the doors to a complete medicalization of the human experience. One thinks of the potential implications of such understanding of health with respect to certain requests for assisted suicide and euthanasia, now justified by an appeal to a notion of «unbearable suffering» that extends to loss of meaning, and loses sight of the proper goals of medicine.[11]

---

10 One must be alert to the linguistic ambiguity of language and suffering, which I feel is lost in translation, when the original Italian *dolore* is rendered with the English *pain*. There follows a kind of binarism of suffering and pain, whereas, in Italian, the term *dolore* means both in a more embracing fashion. This is especially important when reading the meaning of the Christian revelation. The statement «Christ saves us through pain» (177) is clearly ambiguous and implicitly legitimizes the well-known accusation by Nietzsche, for whom Christianity is a religion of masochistic self-annihilation, a slave religion. The beautiful book of Dorotee Sölle, *Suffering*, trans. Everett R. Kalin (Philadelphia: Fortress Press, 1975) is a significant theological attempt to rescue Christianity from any ideological mystification of suffering.

11 The point is stressed, recently, by Carlo Casalone in his piece for *La Civilta' Cattolica*. See «La discussion parlamentare sul suicidio assistito,» *La Civilta' Cattolica* vol. 1 (2022): 143-156 (retrieved at https://www.laciviltacattolica.it/articolo/la-discussione-parlamentare-sul-suicidio-assistito/). The article references the situation of Belgium concerning what Casalone calls «polipathology,» i.e., «a situation in which suffering does not derive from a specific

I would defend the deeper intention of text by saying that it does indeed introduce the W.H.O. definition, but only *in obliquo*, so to speak, with a rhetorical, rather than assertive, function. *In recto* the intention of the document is to offer a more holistic notion of illness, throwing into relief the existential conditions that account for illness as *lived experience*, irreducible to the reductionism of disease. Such an understanding of illness clearly presupposes a notion of *embodiment* in which the body is not only a «body-object» (*Körper*), but a «lived-body» (*Leib*), a flesh.

In so doing, the text responds to the need, already highlighted at the beginning of these considerations, to throw into relief the phenomenological dimensions of suffering in their existential and, ultimately, ontological significance.

---

illness, as the law requires, but from a combination of various and nuanced dysfunctions.»

PART III
THEOLOGICAL EXPANSIONS

CHAPTER 16
THEOLOGICAL DISCOURSE
AND THE POSTMODERN CONDITION
The Case of Bioethics

The scope of this chapter is admittedly ambitious. Indeed, to confront the function of theological discourse in bioethics is to engage in a risky exercise in cultural interpretation. The underlying presumption here is that any theological discourse – in bioethics or in any other field – constitutes de facto a hermeneutic exercise. As such, it requires from the theologian the personal commitment to one's religious tradition no less than the keen insight of an open philosophical mind. This is nothing new for anyone committed to the scientific articulation of theological perspectives. The one defining this paper is that of moral theology. Its systematic effort has historically reflected a double concern, one for the internal coherence of the system, and one for the new intellectual challenges of the times.[1]

Like many other disciplines, bioethics partakes in the cultural fragmentation and the complexity of what has come to be known as the «postmodern condition.» My intention is to provide, first, an analysis of postmodernism as a cultural condition in its relation to the field of bioethics. Second, to interpret and assess the plausibility of that particular solution to the challenge posed by the postmodern condition consisting in the reduction of ethical rationality to a function of political regulation. Third, to underline the public function of theological discourse, and to plead for the necessity of ethical dialogue across different moral traditions

---

1 This has led Josef Ziegler to define moral theology in view of its history as a «restless science.» See Josef G. Ziegler, «Moraltheologie, IV. Geschichte der Moraltheologie,» in *Lexikon für Theologie und Kirche* (Freiburg: Herder Verlag, 1962), vol. VII, 618-623.

Facing the Condition: Postmodernism Defined

The cultural situation of our time seems to be characterized by a condition of fundamental dispersion. The fragmentation of knowledge and the incredible amount of information now available in each discipline make it impossible for anyone to master but a limited material. Ethics, of course, is no better off than other sciences traditionally counted among the *Geisteswissenschaften*. For moral philosophers or theologians confronting any issue in their own field of expertise, the subjective feeling is many times one of dismay. Perhaps this is why we have come to simply accept the limited space of recognized notions and partial perspectives which amount to our definition of reality and to feel comfortable within the familiar walls of specialized yet windowless monads.[2] To put it in the words of Karl Popper «we are prisoners caught in the frameworks of our theories, our expectations, our past experiences, our language.»[3] Such an atmosphere of epistemological «incommensurability»[4] is certainly due to the growing complexity internal to each science and appears to be more frightening in a field like bioethics.[5]

Yet, what can account for the very fact that sciences are in flux? For those committed to the traditional theory of truth, the actual state of the sciences «corresponds» to the increasing complexity of reality. Such an explanation relies upon the presupposition that, since knowledge faithfully mirrors the nature of things, the *ordo*

---

2   On the condition on knowledge in highly developed societies see Jean-Francois Lyotard, *The Postmodern Condition: A Report on Knowledge* (Minneapolis: University of Minnesota Press, 1984). The book which appeared originally in 1979 can be considered the *manifesto* of postmodern thought.
3   Karl Popper, «Normal Science and Its Dangers,» in Imre Lakatos and Alan Musgrave, eds., *Criticism and the Growth of Knowledge* (Cambridge: Cambridge University Press, 1970), 56.
4   As Richard Bernstein suggests, «"incommensurability" was thrust into the center of Anglo-American philosophic debate because of Thomas Kuhn's provocative book, *The Structure of Scientific Revolutions*.» See Richard J. Bernstein, *The New Constellation: The Ethical-Political Horizons of Modernity/Postmodernity* (Cambridge, Massachusetts: The MIT Press, 1992), 59.
5   For a general overview: Ronald A. Carson and Chester R. Burns, eds., *Philosophy of Medicine and Bioethics: A Twenty Years Retrospective and Critical Appraisal* (Dordrecht: Kluwer Academic Publishers, 1997).

disciplinae has to reflect the *ordo rerum*. Even if difficulties of an ontological nature have historically been raised against a realistic theory of truth, I find it impossible to grapple with epistemological questions without *any* ontological reference of some sort.[6] For this reason, the hypothesis that relates the increasing complexity of ethics as a science to the growing complexity of reality as such can be assumed to be, at least, a promising heuristic devise.

Postmodernity is the category currently in use for labelling this complexity. In its broad and, at times, equivocal meaning such a category points in the direction of a general mood or *Stimmung*, to borrow the suggestive interpretation of Richard J. Bernstein. Its content can be defined, negatively, as the definitive overcoming of the Modem philosophical and scientific agenda characterized by the optimism of critical argumentation; positively, by the recognition of a structural fragmentation or a nihilistic contextualism which defies any illusion of totality.[7]

The theoretical indeterminacy of postmodernism as a philosophical label contrasts with the clear and very real dimensions of the problems it creates at a practical level. Two are particularly important and worthy of deep reflection. First, the problem of bringing together the plurality of moralities – so called moral pluralism – under the common denominator of a shared ethos. Secondly, the difficulty of finding a level of discourse which can engage the differences among moral traditions in a real confrontation of substance.[8]

---

6   On truth as correspondence (*Korrespondenztheorie*) see the presentation of Lorenz Bruno Puntel, *Wahrheitstheorien in der neueren Philosophie*, (Darmstadt: Wissenschaftliche Buchgesellschaft, 1993), 26-40.
7   For an interpretation of postmodernism Zygmunt Bauman, *Postmodern Ethics* (Oxford: Blackwell, 1993). The author, a «critical sociologist», defines the novelty of the postmodern approach to ethics as consisting «not in the abandoning of characteristically modern moral concerns, but in the rejection of the typically modern ways of going about its moral problems...The great issues of ethics...have lost nothing of their topicality. They only need to be seen, and dealt with, in a novel way,» *Ibidem*, 4. Also David C. Thomasma, «The Post-Modern Challenge to Religious Sources of Moral Thinking,» in Earl E. Shelp, ed., *Secular Bioethics in Theological Perspective* (Dordrecht: Kluwer Academic Publishers, 1996), 51-74.
8   See the historic and systematic interpretation of the problem given by Alasdair MacIntyre, *After Virtue*, 2nd edition (Notre Dame: University of Notre Dame Press, 1984).

The first difficulty pertains to the definition of the mores we live by and the moral climate that structures the realm of praxis. The second refers more precisely to the theoretical reconstruction of morality, both at the level of ethical discourse and of public policy. Relying upon an analysis of different typologies of moral argumentation, MacIntyre observes:

> ...debate between fundamentally opposed stand points does occur; but it is inevitably inconclusive. Each warring position characteristically appears irrefutable to its own adherents; indeed in its own terms and by its own standards of argument it *is* in practice irrefutable. But each warring position equally seems to its opponents to be insufficiently warranted by rational arguments. It is ironic that the wholly secular humanistic disciplines of the late twentieth century should thus reproduce that same condition which led their nineteenth-century secularizing predecessors to dismiss the claim of theology to be worthy of the status of an academic discipline.[9]

One way of solving the problem in postmodern moral debate is the attempt to bridge the gap of cultural fragmentation by surreptitiously reducing the language of morality to the language of law. The consequence is the progressive obfuscation of the distinction between the legal and the moral. The field of bioethics, in particular, offers ample testimony to the fact that, once laws have been established, concerns for the specific ethical dimension of any issue at stake seem to lose momentum altogether.

The tendency to sublate ethics under the law rests upon the assumption that dialogue on moral convictions will only separate us, whereas the law can bring us back together under the banner of unifying social rules. This assumption not only discourages moral discourse at a practical level; it actually changes its nature by neutralizing the content of conversation across different moral traditions. Moral discourse will take place on presuppositions of political correctness only if it abstains from reference to specific values, virtues, or goals for action. In the end, moral discourse becomes formalistic and empty. The universe of morality – so goes

---

[9] Alasdair MacIntyre, *Three Rival Versions of a Moral Enquiry: Encyclopedia, Genealogy, and Tradition* (Notre Dame: University of Notre Dame Press, 1990), 7.

the argument – is too varied and complex for being univocally traceable. An ethical discourse capable of laying out a territory of discussion where differences can meet and confront each other is a priori excluded from the theoretical agenda of ethics. At best, ethics can provide a geography of procedural conditions upon which the differences among moral traditions co-exist without necessarily coming into contact with one another. Rather than on values and contents – inevitably different and irreconcilable – the moral discussion can focus on rules of reciprocal engagement which allow each moral discourse to take place in a safely protected, yet totally separated, moral universe.

Searching for a Solution: Postmodernism Confronted

The solution adopted by Tristram Engelhardt in his *The Foundations of Bioethics* can be assumed as an exemplary one.[10] It strikes for internal coherence and boldness in drawing all the consequences of its premises. Taking as his starting point the conflictive and irreconcilable condition of postmodern moral pluralism, Engelhardt pleads for a neo-Kantian version of formalism in ethics. His view is based upon the conviction that, since no particular moral vision can provide the ground for a moral consensus, a pure conditional concern for peaceableness cannot meet the crisis we live with. Rational arguments in favor of an ethics of the good, the restoration of a particular moral principle, or the recommendation of a defined set of values, would inevitably exacerbate the conflict of opposite interpretations rather than establish the conditions for a peaceable community.[11] Therefore, the task of bioethics should be understood

---

10 H. Tristram Engelhardt, *The Foundations of Bioethics*, 2nd edition (New York: Oxford University Press, 1996). The recently published *The Foundations of Christian Bioethics* (Lisse: Swets and Zeitlinger, 2000) does not fundamentally alter Engelhardt's view; in fact it can be considered the logical articulation of his system with specific reference to the question of the place of religious traditions *vis à vis* secular bioethics.
11 One may think of the difficulty in rehabilitating the notion of beneficence. See Edmund D. Pellegrino and David C. Thomasma, *For the Patient's Good: The Restoration of Beneficence in Health Care* (New York: Oxford University Press, 1988).

«as a disclosure, to borrow a Kantian metaphor, of a transcendental condition, a necessary condition for the possibility of a general domain of human life and of the life of persons generally.»[12]

Given its regulative function, bioethics cannot logically pre-determine the language of common morality by surreptitiously favoring a particular content over another. The language of bioethics has to be neutral, the «lingua franca» of a society concerned with health care, but incapable of converging around a common viewpoint. Recurring to a different metaphor – this time borrowed from Wittgenstein – Engelhardt defines the task of bioethics as «a disclosure of the minimum grammar involved in speaking of blame and praise with moral strangers, and for establishing a particular set of moral commitments with an authority other than through force.»[13]

The neo-rationalism of this position is only apparently at odds with the diffused epistemological skepticism characterizing the postmodern *manifesto*. In my opinion, Engelhardt's reference to Kant is altogether a loose one when one considers that the «purity», and therefore formality, of practical reason in Kant functions as a transcendental condition for the possibility of morality itself, not only as a condition for an agreement on its possible content.[14] Unlike the former, the latter does not possess the character of a transcendental necessity, or a categorical imperative, but only that of a legally binding factuality, a purely hypothetical imperative. As Engelhardt puts it: «I use *transcendental* here to identify an argument that lays out the conditions for the possibility of a major domain of human experience of action. As defining conditions, they are a priori.»[15] Yet, the a priori character does not explain the whole nature of the Kantian transcendental. As a condition of possibility, it establishes the rational ground for what (*was*) is meaningful expe-

---

12   H. Tristram Engelhardt, *Foundations of Bioethics*, 70. See also idem, *Bioethics and Secular Humanism: The search for a Common Morality* (London: SCM Press, 1991).
13   H. Tristram Engelhardt, *Foundations of Bioethics*, 70.
14   As for the reference to Wittgenstein, I will not discuss here the appropriateness of reading Wittgenstein through the lenses of Kant. Engelhardt seems to rely upon Wittgenstein's use of transcendental arguments. Yet, the analogy between «transcendental condition» and «minimum grammar» makes sense only against the backdrop of an identical ontology. Perhaps, this pushes the general analogy between Kant and Wittgenstein too far.
15   H Tristram Engelhardt, *Foundations of Bioethics*, 94, note 82.

rience, not simply the fact (*dass*) that such a ground can be agreed upon as meaningful. Engelhardt – I may suggest – is reducing the function of transcendental rationality, with respect to the realm of morality, to the function of an abstract rational agreement as a condition of possibility for the existence of social rules. After all, there is nothing in the nature of this a priori condition that tells us why these rules should be considered specifically moral.[16] Although formal, the Kantian transcendental is, on the other hand, not purely procedural. In this sense, the solution of Engelhardt remains within the framework of postmodern ideology.

Warren Reich has suggested reading Engelhardt's position in light of its «mythological» premises. Indeed, the original story underlying this kind of ethics could be easily seen as fitting the Hobbesian model.[17] The nature of the agreement that constitutes the basis for ethics is understood as an alternative to force in resolving moral controversies. This, of course, presupposes that the relation between different moral perspectives can be only a conflictive one. Moreover, such a difference is taken to immediately signify potential for controversy rather than for reciprocal integration and growth.

If the hopes for societal openness to dialogue and reciprocal understanding among human beings can be rendered, at best, with the metaphor of ethics as a shield to the use of force, then our understanding of the human condition remains essentially defined by the notion of *homo homini lupus*. It is certainly clear why such a pessimistic anthropology offers too unstable a ground for ethical conversation. The formalism of a minimalist agreement is a necessary

---

16 Phenomenological ethics has successfully shown that moral norms are essentially different from, say, traffic signs. Unlike the latter, the former are not only the artificial product of a social construction, but the mediation of a transpersonal value claim which binds the conscience of a moral agent. The difference, however, cannot be sufficiently appreciated by a purely pragmatic approach to the meaning of moral experience. See Max Scheler, *Formalism in Ethics and Non-Formal Ethics of Values* (Evanston: Northwestern University Press, 1973), 163-328.

17 Warren T. Reich, «Alle origini dell'etica medica: mito del contratto o mito di cura?», in Paolo Cattorini and Roberto Mordacci, ed.,. *Modelli di medicina: Crisi e attualita' dell'idea di professione* (Milano: Europa Scienze Umane Editrice, 1993), 35-59. The reference is to Thomas Hobbes, *Leviathan* (London: Penguin Books, 1985, 182-205).

condition for the tolerant co-existence of unrelated moral ideals, but certainly not a sufficient one in making the positive interaction on their content a process which is meaningful and worthy at the same time.

As for Engelhardt's thesis, such formalism does not exhaust the whole story of ethics. It defines only its «secular» profile, the logical, *i.e.*, rational, constraint based on respect for the freedom of the moral agents involved in moral discourse without establishing the correctness of any particular moral sense. The regulative character of a general secular bioethics is to function as the logic of pluralism, as the means that allows the peaceable negotiation of moral intuitions to occur. Yet, to keep the Kantian metaphor going, if intuitions without concepts are blind, concepts without intuitions remain, for their part, inevitably empty. Translated into Engelhardt's nomenclature: without some moral sense of a concrete nature, the moral life of persons could have overcome the conceptual blindness of unrestrained devotion to a particular idea of the good, but only at the cost of *any* sense of the good itself. The retrieval of such a sense capable of shaping the moral vision and the conduct of individuals requires the reference to a concrete context, an historical moral community different from the formal and, ultimately, pragmatic realm of society as such. Whereas *community* designates the free association of individuals through a common and concrete view of the good, *society*, on the other hand, denotes the association of individuals who may pursue a number of important goals together without sharing that common vision.[18] The nature of society is essentially pragmatic: it sets forth the conditions for building a limited yet necessary consensus. Society brings together various communities around common goals and tasks thereby originating a community

---

18 As for its etiology, the distinction between *Gesellschaft* and *Gemeinde* goes back to the XIX century German sociologist and philosopher Ferdinand Tönnies who, in turns, influenced Max Weber's popularization of the terms. Whether this distinction can be borrowed from Weber independent of its underlying methodological presuppositions – one of them being the so called postulate of «avalutativity» or *Wertfreiheit*, namely, the impossibility of a scientific analysis concerning value judgments in the realm of praxis – can be left unquestioned for now. See Max Weber, «Science as a Vocation,» in *From Max Weber*, ed. H.H. Gerth and C Wright Mills (New York: Oxford University Press, 1946), 155.

of a higher order. In a certain way, societies are also communities; only, the term society underlines the perspective of a consensus that does not exist, and yet, needs to be fashioned in order for society to function as a texture of different communal traditions.[19]

Investigating the Presuppositions: Postmodernism Interpreted

The distinction between society and community just referred to translates *de facto* into the separation between the reasonable claim of a general secular morality, as such open to rational control and verification, and the theoretical indeterminacy of moralities upheld by particular moral communities, by definition impervious to objective criteria. In the end, the paradoxical situation of moral discourse becomes one of structural fluctuation between theoretical rationalism on the one hand, and dogmatic fideism on the other. Such a fluctuation entails not only a logical contradiction, but also an obvious methodological schizophrenia. It favors a cognitivist presumption for metaethical rationality at the *transcendental* level of society, while reducing the identification with the *categoreal* ethos of a particular community to a pure matter of taste. Indeed, «since there are no generally sustainable, secular arguments to establish as canonical a particular view of the good life and content-full moral obligations, and as long as the ways of life under consideration respect the freedom of the innocent, then, from outside, the choice among any particular moral community will appear to be similar to an aesthetic choice.»[20]

The impartial, unprejudiced, and non culturally biased reasoners of society whose only interest are the logical strength and force of rational arguments will suspend in a radical and quite impersonal *epoche'* any residue of intelligence at the threshold of their moral communities. Within specific communities of this geography of

---

19 H.T. Engelhardt, *Foundations of Bioethics*, 74-84.
20 *Ibidem*, 75. This holds true also for the theological vision underlying particular moral traditions. Indeed, the very meaning of theological reflection seems to be aesthetical in essence. See H. Tristram Engelhardt, «Looking for Good and Finding the Abyss: Bioethics and Natural Theology,» in Earl E. Shelp ed., *Theology and Bioethics: Exploring the Foundations and Frontiers* (Dordrecht: Kluwer Academic Publishers, 1985), 79-91.

moral discourse one must abandon the hope of ever reconciling the loyalty to a moral vision with the consistency of arguments, the fascination with values and virtues with a truly personal and responsible commitment to them.[21] Perhaps, like the Kantian moral agent pulled between the pure aspiration of noumenal freedom and the opposite *Neigung* to phenomenal nature, the moral subject of this post modern situation has remained a divided self, citizen of two completely different worlds.[22]

In order to overcome the problems posed by our postmodern condition it seems imperative to rethink the meaning and the purpose of ethical dialogue across different traditions and within the realm of secular society as such. This renewed agenda of ethics moves between the Scylla and Charybdis of a twofold dead end: the reduction of ethical rationality to a purely procedural function of political regulation and the intellectual impotence toward an incommensurable pluralism which implicitly legitimizes the relativity of different ethical points of view.[23]

It is not my purpose here to directly take up the burden of a philosophical reconstruction. I will plead, however, for the need of ethical conversation across traditions on a theological basis. As it seems, the post modern solution to the question of moral pluralism does not imply the end of religious discourse in ethics. It actually creates the conditions for its articulation while astutely marginalizing its meaning.[24]

---

21   The entire phenomenology of *value response* (*Wertantwort*) should be retrieved at this point in its profound meaning. See Dietrich von Hildebrand, *Ethics* (Steubenville, OH: Hildebrand Press, 2020). For a thorough analysis, Josef Seifert, «Dietrich von Hildebrands philosophische Entdeckung der "Wertantwort" und die Grundlegung der Ethik,» in *Truth and Value: The Philosophy of Dietrich von Hildebrand* (Bern: Peter Lang, 1992), 34-58.
22   The task of a reconstruction of this postmodern «shattered *cogito*» in the direction of personal identity is attempted by Paul Ricoeur, *Oneself as Another* (Chicago: The University of Chicago Press, 1992). For a penetrating historical analysis see Charles Taylor, *Sources of the Self: The Making of Modern Identity* (Cambridge, Mass: Harvard University Press, 1989).
23   See Antonio Da Re, *L'etica tra felicita' e dovere: L'attuale dibattio sulla filosofia practica* (Bologna: Dehoniane, 1986).
24   See Stephen E. Lammers, «The Marginalization of Religious Voices in Bioethics,» in Allen Verhey, ed., *Religion and Medical Ethics: Looking Back, Looking Forward* (Grand Rapids, MI: William B. Eerdmans, 1996), 19-43.

However, the real problem is not so much the place of a theological ethics in a bioethical discourse become entirely secular in nature. After all, theology has been addressing the question of secularization for a long time.[25] Rather, the challenge for theology is that of making sense of its own object by engaging in universal ethical dialogue. This aspect is particularly important to moral theology. Relying upon the theoretical possibilities of variable natural law theories, moral theology has always presumed to argue on the basis of universally binding notions of reason and rationality.[26] Let me mention but one example: at the beginning of his *Medical Ethics*, Edwin Healy states quite clearly the conviction that the natural law provides the basis for moral principles virtually open to the recognition of every reasonable person:

> All men ... are called upon to obey the natural law. Hence it matters not whether one be a Roman Catholic, a Protestant, a Jew, a pagan, or a person who has no religious affiliation whatsoever; he is nevertheless obliged to become acquainted with, and to observe the teachings of, the law of nature. In the present volume all the obligations which are mentioned flow from the natural law, unless the contrary is evident from the context.[27]

Of course, one should recognize that the neo scholastic version of natural law was heavily shaped by the epistemological presumptions of rationalism and the scientific ideal of an ethics *more geometrico demonstrata*.[28] This particular paradigm of natural law

---

Also Lisa Cahill, «Theology and Bioethics: Should Religious Traditions Have a Public Voice,» *The Journal of Medicine and Philosophy* 17 (1992): 263-272.

25 The names of Friedrich Gogarten, Dietrich Bonhöffer, and Harvey Cox among Protestant theologians, Karl Rahner, Yves Congar, Marie-Dominique Chenu, and Johannes Baptist Metz among the Catholics should be mentioned. See the entry «Secularization» by Albert Keller in *Sacramentum Mundi* (New York: Herder and Herder, 1968), vol. VI, 64-70.

26 See Russell Hittinger, «Natural Law,» in Warren T. Reich, ed., *Encyclopedia of Bioethics*, 2nd edition (New York: Simon and Schuster Macmillan, 1995), vol. 4, 1805-1812.

27 Edwin F. Healy, *Medical Ethics* (Chicago: Loyola University Press, 1956), 7.

28 «What seems clear is that Roman Catholic bioethics in the first half of this century had become *reductionistic*, making the biological aspect of human life absolutely determinative; *static*, viewing nature as independent of histo-

shows the *philosophical* limits of a moral theology for too long fixated on a particular understanding of rationality, one whose absoluteness was claimed at the expense of historical awareness and hermeneutical criticism; nevertheless, it bespeaks the *theological* need to articulate the universal meaning of revelation in a language accessible to everyone. What grounds such a need is not so much the canonization of an essentialist model of rationality to be sanctioned as the only one, once and for all, but rather the dialectic between the historical particularity of the Christological event with its ethical implications and the universality of the moral experience.[29] Conversation across different traditions is grounded upon the awareness that the universality of the Christological event can be mediated in an anthropological system of coordinates accessible to everyone *if* the Christological event has to represent the eschatological truth of what does it means to be human. This is lucidly expressed by Klaus Demmer:

> What must be presented is a coherent concept of revelation in which the mediation of theological and anthropological categories constitute the central concern ... Faith elicits reflection from moral reason; it inspires it. It does so in that it produces in the believer, through his understanding and interpretation of revelation, an understanding of himself and of the world. It is that understanding of one's self and of the world that is at the root of all ethical insights and expressions. The moral reason of the believer is bound to this self-understanding and understanding of the world.[30]

The believer's self-understanding unfolds along the line provided by a new anthropological basis. Notions like the intrinsic dignity of the human person, the fundamental equality and fraternity among all people, the quality of history as a recommendable space

---

ry; and *authoritarian*, making reason subservient to the ecclesial authority,» Edward Collins Vacek, «Catholic "Natural Law" and Reproductive Ethics,» *The Journal of Medicine and Philosophy* 17 (1992): 329-346, at 329.

29 See on this question Klaus Demmer, *Moraltheologische Methodenlhre* (Freiburg: Herder, 1989) 26-29, 178-193.

30 Klaus Demmer, «Theological Arguments and Hermeneutics in Bioethics,» in Edmund D. Pellegrino, John P. Langan, amd John Collins Harvey, eds. *Catholic Perspectives on Medical Morals: Foundational Issues* (Dordrecht: Kluwer Academic Publishers, 1989), 103-122, at 107.

of moral creativity and meaning, the reality of illness, suffering and death as the ultimate challenge to the extension of human hope: all these represent the essential articulations of the anthropological system generated by faith. Such a system neither secures the agreement on specific issues among different moral traditions, nor does it establish *a priori* a univocal context for their solution. The implications of faith do not prevent, but rather stimulate practical reason in the difficult task of interpreting and assessing the meaning of circumstances and contexts.[31] Agreement on moral issues is more the result of conversation among participants of historically determined communities of discourse, than the presupposition for that conversation to occur. The success of such a conversation rests upon the continuous hermeneutics of the anthropological implications of faith.[32]

If bioethics is understood as a purely procedural enterprise ultimately geared toward the creation of public rules, an ethics which attempted to convey a theological meaning would be excluded from the outset as relevant to the moral discourse *per se*. Like any other language dealing with particular contents, the theological one is ultimately bound to live in a separate limbo, and make sense for those who already understand it. Yet, how could theology exhibit any truth claim without already presupposing the universal conditions of possibility for the recognition of its meaning? If the autonomy of moral reasoning is deemed to shatter in a Babel of moral languages and inescapable confusion, then theology can only give up its epistemological program of universal communication. An apology of practical reason appears to be quite unfashionable a goal in a time when the critical task has grown weaker and its paradigms

---

31 The importance of casuistry is vouched for by Thomas A. Shannon and James F. Keenan, *The Context of Casuistry* (Washington, D.C.: Georgetown University Press, 1995).

32 What is at work here is, obviously, an explicit option for the plausibility and the importance of a transcendental turn in theology. From a programmatic point of view see Karl Rahner, *Foundations of Christian Faith* (New York: Seabury, 1978), especially 1-90. For the articulation of such a transcendental turn in moral theology the most systematic work remains Klaus Demmer's *Sein und Gebot: Die Bedeutsamkeit des transzendentalphilosophischen Denkansatzes in der Scholatisk der Gegenwart für den formalin Aufriss der Fundamentalmoral* (München: Schöning, 1971). Also Franz Böckle, *Fundamental Moral Theology* (New York: Pueblo Publishing, 1980), 15-63.

fragmented. Yet, I believe, the real task of a postmodern theology is also philosophical in nature. It pleads for the rehabilitation of moral reasoning, the creation of an ecumenical dialogue across traditions, the confidence in the true experience of otherness in general.

Theology and the Analogy of Religious Discourse

The need for a clear articulation of theological discourse in bioethics meets with the problem of a univocal definition of theology.[33] Books and articles appearing under the general rubric «theology and bioethics» present very often a methodological instability of the language used.[34] Terms like «theology», «religion», «spirituality», etc., function many times as interchangeable variables. It is no wonder that such linguistic indeterminacy triggered the sarcastic comment of a philosopher otherwise sympathetic to the meaning of theological traditions: «Theologians still owe it to the rest of us to explain why we should not treat their discipline as we do astrology or phrenology.»[35] Even if the terms in question point in the direction of a univocal meaning, they undoubtedly belong to different semantic areas. Therefore, a general attitude of circumspection in the use of the terms must be recommended.

---

33   See Hubert Doucet, «Un théologien dans le débat en bioéthique,» in *Le Supplément: Revue d'éthique et de théologie morale* 202 (1997): 17-37. The entire issues of this journal, which reflects the point of view of French speaking ethicists is entirely dedicated to the question of the relationship between moral theology and bioethics.

34   See, for example, Earl E. Shelp, *Theology and Bioethics*. The book constitutes a very valid attempt to confront the issue of the general meaning of theological discourse in bioethics. Yet, in my opinion, it presents the disadvantage of dealing with completely different pre-understandings of theology.

35   Alasdair MacIntyre, «Theology, Ethics, and the Ethics of Medicine and Health Care,» *The Journal of Medicine and Philosophy* 4 (1979): 435-443. As for the substance of the statement, the impression of a methodological vacuum in theology would have been hardly shared by anyone familiar with literature. Among the many publications already present at the time on the question of theological epistemology see, for instance, David Tracy, *Blessed Rage for Order: The New Pluralism in Theology* (New York: Crossroad, 1975), or Wolfhart Pannenberg, *Theology and the Philosophy of Science* (Philadelphia: Westminster, 1976). More recently, and with a particular focus on moral theology also Klaus Demmer, *Moraltheologische Methodenlehre*.

My contention is that «religion», «theology», «spirituality» etc., are *only* analogous terms. It should be noted that I am not concerned here with religious discourse in general, but more precisely, with the theological articulation of that religious discourse. Since I take theology to be a scientific enterprise, its epistemological status provides the rules for engaging in an interdisciplinary field of reflection like bioethics.[36] Of course, all the sciences are historical constructions. They undergo a growing complexity which forces them to make room for «shifts in paradigm» and to redefine the internal coherence of their system. This holds true for theology no less than for all the other sciences.[37] For the Christian traditions, the task of theological reasoning remains rooted in the historical revelation culminating in the Christological event. Theological reasoning presupposes the identification with a horizon of transcendence whose meaning is not suspended in order to be objectively thematized, but rather accompanies and embraces the historical activity of theological reasoning itself. The task of a theological rationality consists in showing how faith's claim to universality objectively corresponds to the anthropological a priori intrinsic to faith. Insofar as it searches for its own intelligibility faith sets up a critical investigation which adds nothing from the outside, but discloses the conditions for its intelligent recognition from within.[38]

---

36 According to Gilbert Hottois «Définir la bioéthique est une enterprise périlleuse. Son apparition récente, sa localisation interstitielle plus ou moins accentuée et les enjeux idéologiques qu'elle véhicule lui confèrent une identité instable et controversée,» idem, «Bioéthique,» in Gilbert Hottois and Marie-Hélène Parizeau, eds., *Les mots de la bioéthique* (Bruxelles: De Boeck Université, 1993), 49. Indeed, of the two terms, bioethics is the one characterized by a certain degree of epistemological indeterminancy, and not theology. What accounts for the definition of bioethics as a defined discipline remains open to question. See Massimo Reichlin, «Observations on the Epistemological Status of Bioethics,» *The Journal of Medicine and Philosophy* 19 (1994): 79-102.
37 See Thomas R. Kopfensteiner, «Historical Epistemology and Moral Progress,» *Heythrop Journal* 33 (1992): 45-60. Also Klaus Demmer, «Das theologische Argument und der Paradigmenwechsel,» *Freiburger Zeitschrift für Philosophie und Theologie* 34 (1987): 65-89.
38 This understanding of theology relies upon the classical definition of theology as «*fides quaerens intellectum*,» faith seeking understanding. However, the Anselmian formula leaves open the question *how* faith seeks understanding. For an example, the theologies of Karl Barth and Karl Rahner have

This general definition of theology, being purely formal, does not exhaust the materiality of different theologies, or the pluralism of theologies. It opens up the conditions for understanding the historical character of theology, the specific nature of its own rationality, the structural reference to the event of God's revelation and to its effectual history.[39] The universality of this event shields the task of theology from any self-centered particularism and sectarianism. The theologian is a believer, and like the believers of the community he belongs to, he serves as an ecclesial intellectual. Yet, the function of theology cannot be reduced to the internal hermeneutics of a particular group, a closed symbol system with reference only to the shared but limited universe of a tradition. Precisely because of the truth claims of religion, theology must develop *public* criteria for such affirmations. The search for public criteria of truth claims identifies, in my opinion, the specific function of theology.[40]

Searching for a Theological Model

I am very well aware of the fact that, by emphasizing the «publicness» of theology, I am also taking stock – rather explicitly – of

---

both taken their starting point from the Anselmian *motto* with two opposite results. «Faith seeking understanding» takes Barth in the direction of theological positivism, whereas in Rahner it leads to a systematic reconstruction of theology along the line of an anthropological turning point. The most penetrating comments on the issue remain those of Hans Urs von Balthasar, *The Theology of Karl Barth* (New York: Holt, Rinehart, and Winston, 1971).

39   On the problem of theological pluralism see David Tracy, *The Analogical Imagination: Christian Theology and the Culture of Pluralism* (New York: Crossroad, 1981). With particular reference to the issue of pluralism in moral theology Bruno Schüller, *Pluralismus in der Ethik: Zum Stil wissenschaftlicher Kontroversen* (Münster: 1988), 27-44.

40   Klaus Demmer writes: «So ist der Christ als Christ, und der theologische Ethiker als Theologe in die denkerische Mitverantwortung genommen. Es ist dies eine Anfrage an das demokratische Wahrheitsethos, das keine Ausflucht duldet, es sei den, man operiere stillschweigend mit einer doppelten Wahrheit. Wer von dieser Herausforderung versagt, ist als Demokrat nicht existent,» idem, «Ethische Wahrheit in der Demokratie: Eine Herausforderung an Moraltheologie und kirchliches Lehramt,» in *An-Denken Festgabe für Eugen Biser* (Innsbruck: Verlag Styria, 1998), 321-328, at 324. Also idem, «Ethique et Genetique Humaine,» in *Actes II Symposium du Conseil de l'Europe* (Strasbourg: Les Editions du Conseil de l'Europe, 1994), 83-90.

a relevant methodological debate in theology today. I do not intend to indulge in the contrasting solutions to the issue at stake, but only to point out that the theological scenario mirrors, quite literally, the *status questionis* referred to early in the bioethical section. Indeed, even if the discourse takes place at different levels, and to some extent, in different fields of research, the central question remains exactly the same. It is the question of pluralism and its challenge to our ability to converse across different traditions and different cultures without becoming strangers to one another.

The need to form a new and inevitably complex theological strategy that will avoid privatism by articulating the genuine claim of religion to truth seems, therefore, the central methodological issue. The alternative is very much the same observed above: either let each theological tradition dissolve into some lowest common denominator, analogous to Engelhardt's «minimum grammar», or accept a marginal existence as one interesting but purely private option: «neither alternative is acceptable to anyone seriously committed to the truth of any major religious tradition.»[41]

Yet, if no serious theologian can accept the confinement to pure privatism, the interpretation of religion as possessing an «internal» rather than «external,» and thus public, justification, represents to many one of the plausible answers to the question raised by postmodern pluralism. In his provocative *The Nature of Doctrine*, for example, George Lindbeck argues that particular religious traditions are like particular languages that shape and determine our experience of the world.[42] In a similar fashion Stanley Hauerwas emphasizes the importance of a communal narrative as a primary element of identification with the meaningfulness of values and virtues.[43]

---

41 David Tracy, *Analogical Imagination*, XI. The central concern of David Tracy's reflection has been to reframe the proper task of Christian theology in the face of cultural pluralism. For a useful overview see T. Howland Sanks, «David Tracy's Theological Project: An Overview and Some Implications,» *Theological Studies* 54 (1993): 698-727.
42 George A. Lindbeck, *The Nature of Doctrine: Religion and Theology in a Postliberal Age* (Philadelphia: The Westminster Press, 1984).
43 Stanley Hauerwas, *A Community of Character: Toward a Constructive Christian Social Ethics* (Notre Dame, IN: University of Notre Dame Press, 1981).

Insofar as it takes a critical stand toward the rationalism of a theology too easily influenced by the Enlightenment paradigm of reason as trans-historical, trans-cultural, and therefore, abstract medium of communication, the so-called «cultural-linguistic» model provides a welcome correction. The paradigm of reason inherited from Modernity is, indeed, unsustainable, both philosophically and theologically. As in the classical tradition of manualistic apologetics within the Catholic tradition, and in the various editions of neo-scholastic theologies, the reasonableness of faith is claimed at the cost of revelation's essential historical nature.

If, on the other hand, it implies a fundamental suspicion toward the publicness of theology and its necessary engagement in universal communication, the «cultural-linguistic» model must be faced with a necessary *caveat*. Indeed, by giving up the idea of any public claim to truth in the name of a «soft» version of theological rationality, the «cultural-linguistic model» favors indirectly the marginalization of the religious traditions for which it intends to speak. Such a process of marginalization, however, represents the capitulation of theological thinking to ideological relativism, and the subjectification of religious experience to the meter of individual taste. In reality, the publicness of theology is an expression of its peculiar responsibility in serving the intelligence of faith and the vision of the good entailed by it. Such a responsibility is the more imperative the bigger the issues confronting humankind. Are religious traditions to be engaged at all in the bioethical discussion on cloning, stem cell research or the applicability of ever perfected reproductive technologies? Are the intuitions of the good which define the moral treasure of every great religious tradition to be dismissed from the public debate as fundamentally irrational, or will they be looked at as reservoirs of meaning and wisdom on what it means to be human? If one of the tasks of theology is to produce public discourse, then religious claims must be made shareable and remain open to interpretation.

Following David Tracy, one could suggest that the trend toward the privatization of religion in postliberal societies presents an obvious analogy with the marginalization of art.[44] Religion, like art,

---

44 *The Analogical Imagination*, especially 12-13, 109-115.

has become merely a matter of taste without any claim to validity.[45] Indeed, the inability to grasp the meaning of art as a manifestation of truth couples with the reductivist notion that religion can spare any public claim to truth, thereby releasing the postmodern believer of any intellectual responsibility toward the object of his faith. One can shop around looking for a religious identity in the same way in which he does for other commodities. The only criterion: «*de gustibus non disputandum est.*»[46]

In this perspective, theology can be totally dispensed with the elaboration of public criteria for its affirmation. At best, it may be called upon to convey the purely internal coherence and aesthetic appeal of a religious identity whose meaning, in turn, remains inaccessible to anyone still foreign to the language of the group. Yet, how could one ever come to speak a particular language without partaking of the horizon of understanding which that language entails? And how will this very horizon become accessible other than through language itself?[47]

Reclaiming the Meaning of Theological Argument

I have been arguing all along against a «loophole solution» to the problem of postmodernism, one which marginalizes moral traditions and their meaning by canonizing a particular version of rationality. Oddly enough, rationality is invoked as the only viable means of agreement on the one hand, and on the other, it is deemed incapable of conveying the ethical content that shapes the life and language of particular traditions. The theory of a two-tier morality translates into a two-tier version of practical rationality. A «strong»,

---

45 For an historical and systematic account of the question of truth as it emerges in the experience of art, see the masterful analysis of Hans Georg Gadamer, *Truth and Method* (New York: Continuum, 1994), 1-169.
46 Yet, as Gadamer points out, «The concept of taste implies *a mode of knowing*. The mark of good taste is being able to stand back from ourselves and our private preferences. Thus, taste in its essential nature, is not private but a social phenomenon of the first order», *Truth and Method*, 36.
47 On language as horizon and medium of understanding see, again, *Truth and Method*, 383-481. Also David Tracy, *Plurality and Ambiguity: Hermeneutics, Religion, Hope* (Chicago: The University of Chicago Press, 1987), especially 47-65.

but purely procedural level of moral reasoning, free of content and essentially neutral, regulates the public sphere of society. A «soft» level of moral discourse, one taking place within each moral tradition, establishes the coherence of particular moral languages within specific moral communities. The separation between the two levels makes sure that moral contentions and frictions within society are solved by peaceable agreement rather than by force, and more importantly, that concrete moral agents defined by a particular history and moral tradition never speak to one another.

The appeal of this solution to the undeniable puzzle of our postmodern condition lies in its simplicity. It protects the only publicly recognized faith's article of liberal societies, the autonomy and freedom of individuals, while creating the conditions for private identification with the values and virtues each individual feels worthy of ethical pursuit. For anyone interested in ethical dialogue – that is, in what precedes and follows the agreement on specific public policies and universally binding laws – the solution virtually reduces the whole task of ethics to a strategy of political co-existence. Yet, how does such an agreement come into place? How can it be fashioned if not by means of ethical dialogue?

What is at work, here, is nothing but the old positivistic prejudice toward an ethics of meaning and the philosophical articulation of a language which transcends the realm of empirically verifiable factuality.[48] The prejudicial dimension of this paradigm lies in its concept of science and the consequent dilemma it forces upon ethics: if the language of ethics is scientific, then it must be reduced to neutral conditions of verifiability virtually accessible to anyone. Or it is purely subjective and value laden, then it escapes the criteria of control exhibited by any scientific claim.

A theological methodology borrowing such a solution – whether implicitly or explicitly – must be aware of the underlying version of rationality it surreptitiously introduces into theological discourse. By making room for religious taste at the expenses of theological meaning, such a paradigm of rationality simply sanctions the end of theology altogether. Two basic consequences will inevitably fol-

---

48 For an analysis of how the development of bioethics has hindered a positive attention to meaning see chapter 2 of this book, «Method and Meaning: In Search of a Richer Bioethics.»

low: theological ethics will have to abstain from publicly articulating its truth claims and from fashioning their particular content in universally accessible arguments. Second, theological traditions will be forced into intellectual isolation, thereby losing the ability to converse with one another across confessional boundaries. A rampant new wave of non-ecumenical discourse, particularly in theological ethics, is bound to take possession of the scene on the presupposition that, by speaking for itself, each tradition can at least make clear what it stands for.[49] Yet, making clear does not necessarily imply making sense. A theological claim is, by definition, one that states a position while taking into account the interlocutor's ability to understand its meaning. The need for a universal articulation of theological discourse is not decided by a conditional situation of cultural homogeneity, but by a structural necessity internal to the object of theology itself, *i.e.*, the universal meaning of God's revelation.

Conclusion

After defining the so-called postmodern condition, I have concentrated my attention on one of its possible solutions, one which reduces the meaning of moral argumentation to a formal agreement on procedure, and marginalizes ethical contents to a matter of private taste. I have, then, turned my attention to theology, and highlighted the appeal of a methodological interpretation that emphasizes the «cultural-linguistic» thrust of theological discourse. Theological ethics finds itself at the crossroad of two fields and two analogous solutions, one emerging from the field of ethics, the other from theology. I have argued against a model which homologizes theological ethics to a particular linguistic system, and pleaded for ethical dialogue across different moral traditions. The possible form of this dialogue, the conditions of possibility under which it might take place, and, ultimately, its content, constitute, in my opinion, the main challenge faced by each major theological tradition today.

---

49 See H. Tristram Engelhardt «Toward a Christian Bioethics,» *Christian Bioethics* 1 (1995): 1-10.

CHAPTER 17
RECASTING FUNDAMENTAL MORAL THEOLOGY
Notes on
Klaus Demmer's Christological Anthropology

A group of Italian scholars have recently dedicated a *Festschrift* to Klaus Demmer, on the occasion of his 80th birthday.[1] With the book, they celebrate the accomplishments of their *Doktorvater* by way of a careful reconstruction of his thought and articulate his relevance for today's moral theology. One can predict Demmer's importance will continue to grow, quietly and unassuming, like the person he is. The mustard seed of his long *magisterium* has formed generations of students, now teachers all over the world. Still, with the exception of two book translations, Demmer remains a rather unknown thinker to an English speaking audience.[2] The purpose of this paper is to provide a general introduction to his fundamental moral theology, with a particular focus on the mediating function of moral anthropology. There is no pretension to offer a comprehensive analysis, which would have to take into account an impressive scholarly output of twenty five books and almost two hundred scholarly papers.[3] Rather, only some aspects of Demmer's thought

---

1 A. Fumagalli – V. Viva, ed., *Pensare l'agire morale. Omaggio italiano a un maestro internazionale: Klaus Demmer*, Cinisello Balsamo 2011.
2 James Keenan edited Roberto Dell'Oro's translation of *Einfahrüng in die Moraltheologie* an unpublished manuscript, in 2000. See *Shaping the Moral Life. An Approach to Moral Theology*, Washington (D.C.) 2000. Brian McNeil has recently translated a second book in English: *Living the Truth. A Theory of Action*, Washington (D.C.) 2010.
3 A complete bibliography in A. Fumagalli – V. Viva, ed., *Pensare l'agire morale* (cf. nt. I), 233-264. The most comprehensive study of the thought of Klaus Demmer is M. Wolfers, *Theologische Ethik als hadlungsleitende Sinnwissenschaft. Der fundamentalethische Entwurf von Klaus Demmer*, Freiburg 2003. Extensive treatments of Demmer's thought can be found also in the following works: J. Römelt, *Personales Gottesverständnis in heutiger Moraltheologie auf den Hintergrund der Theologie von K. Rahner und H.U. von Balthasar*, Innsbruck – Wien 1988; P. Bubmann, *Fundamentalethik als Theorie der Freiheit. Eine Auseinander-setzung mit römisch-katholischen*

will be highlighted. Having briefly situated Demmer's intellectual biography in the context of post-Vatican II theological development, this paper will look at Demmer's moral theology as the attempt to unpack the ethical implications and the «effectual history» of the Christological event.

Perhaps the final *Gestalt*, to use a term dear to Hans Urs von Balthasar, of Demmer's moral theology, seen in all its systematic coherence, as well as its nuanced facets, might be understood as a «transcendental hermeneutics» of the Christian moral tradition. The starting point of any theological articulation of the *intellectus fidei* is the a priori synthesis of being and history, now become an a posteriori event in Jesus Christ.[4] The ontological singularity of God's incarnation determines theology's *Denlifonn*, defined, in Demmer's case, by the creative retrieval of the Thomistic tradition in dialogue with modem and contemporary philosophy – from Kant to Hegel, to XX century thinkers such as Heidegger, Gadamer, Ricoeur, and Habermas.[5] Moreover, the designation «transcendental hermeneutics» conveys a sense of the twofold commitment driving the retrieval: to the unrelenting interpretation of the historical *particularity* of the Christian faith tradition; and to the reconstruction of the conditions for *universal* moral communication beyond the limits of one's specific theological affilia-

---

Entwurfen, Güterlsloh 1995, 133-240; and G. Mattai, «Etica ed esperienza di fede. Un'introduzione alla teologia morale di K. Demmer», *Asprenas* 15 (1993) 587-596.

4   The statement in question stands, somewhat programmatically, albeit in an interrogative fashion, at the very beginning of Demmer's *Habilitationsschrift*. «Muss nicht jede theologische Reflexion methodisch bei der Geschichtswerdung Gottes einsetzen? Und was bedeutet ein solcher theologischer Denkansatz für den «intellectus fidei»«? Ist nicht die apriorische Synthese von Sein und Geschichtlichkeit in Jesus Christus zum aposteriorischen Ereignis geworden?». *Sein und Gebot: Die Bedeutsamkeit des transzendentalphilosophischen Denkansatzes in der Scholastik der Gegewart für den formalen Aufriss der Fundamentalmoral*, München 1971, 5.

5   Interestingly, in a footnote of *Fundamentale Theologie des Ethischen*, a book Eberhard Schockenhoff has termed his «magistrales Alterwerk», Demmer pleads for a Thomism that is open and capable of constant integration. See *Fundamentale Theologie des Ethischen*, Freiburg 1999, 34, nt. 76. Schockenhoff's statement can be found in his «Moraltheologie im Zeichen der Schwachen Vernunft? Neuere Entwicklungen der Fundamentalmoral und der theologischen Lebensethik», *Theologische Revue* 97 (2001) 447.

tion.⁶ In the works of our author, hermeneutical and transcendental modes of thinking intermediate with one another beyond any methodological separation, reflecting a concern for the internal coherence of the moral theological system, on the one hand, and for the constant integration of new philosophical perspectives, on the other, in critical dialogue with culture and society.

Klaus Demmer: A Brief Intellectual Biography

Klaus Demmer, a German by origin, is now a *professor emeritus*, residing in Münster, Germany, after more than thirty years of teaching at the Pontifical Gregorian University in Rome. At the Gregorian, Demmer studied with Josef Fuchs, and in 1961 completed a doctoral dissertation on the Christological foundations of natural law in St. Augustine.⁷ Ten years later, Demmer received his university habilitation in Innsbruck, Austria, working with Emerich Coreth and Otto Muck on a systematic articulation for moral theology of the transcendental philosophical method appropriated by the neoScholastic tradition in the wake of Joseph Marechal's groundbreaking *Le point de départ de la métaphysique*.⁸

In his second book, *Sein und Gebot*, and already at a relatively young age, Demmer established himself not only as an impressive interpreter of a school of thought that had already found develop-

---

6   «Das partikuläre Heilsereignis in Jesus Christus und die Universalität der sittlichen Vernunft wollen denkerisch miteinander vermittelt sein». K. Demmer, *Deuten und Handeln. Grundlagen und Grundfragen der Fundamentalmoral*, Freiburg 1985, 86.
7   The dissertation was published in *Analecta Gregoriana*. See K. Demmer, *Jus caritatis. Zur christologischen Grundlegung der augustinischen Naturrechtslehre*, Analecta Gregoriana 118, Roma 1961. The dissertation on Augustine represents an original intuition for Demmer, which sets the tone for his later theological work. There is a beautiful statement in Melanie Wolfers' interview with Demmer: the Augustinian *noli foras ire* as the driving force of his mature *Subjektphilosophie*! See M. Wolfers, *Theologische Ethik* (cf. nt. 3), 196.
8   J. Maréchal, *Le point de depart de la métaphysique. Lecons sur le development historique et théorique du problème de la connaissance*, Cahier 1-5, Bruxelles 1944-1949. For a general overview of Maréchal's enterprise, see O. Muck, *The Transcendental Method*, trans. by D. Seidenticker, New York 1968, 27-161.

ments in the areas of metaphysics, philosophy of religion, and dogmatic theology – especially with the works of Johannes B. Lotz, Emerich Coreth, Bernhard Lonergan, and Karl Rahner, but also as a creative thinker who was among the very first to see the importance of transcendental and hermeneutical approaches for the systematic articulation of a renewed moral theology.[9] A year after the publication of his *Habilitationsschrift*, in 1972, Demmer was called to the chair of moral theology at the Pontifical Gregorian University, where his career as a teacher and scholar will flourish until his retirement, in 2003.

In his many books, Demmer has taken up not only questions of method, virtually touching upon all the main issues of post-Vatican II moral theology, but also showed how a fundamental moral theology nourished by a concern for both the *metaphysical* foundations and the *historicity* of moral truth can find fruitful applications in the areas of bioethics, social ethics, spirituality, and pastoral theology.

Perhaps one way to introduce the thought of Klaus Demmer is to contextualize it by reference to the general shift in post-Vatican II theology, known in the language of Peter Eicher, as the «anthropological turning point».[10] Karl Rahner, one of the main exponents of such turn, explains its meaning thus:

> Theology wants to tell man what he is, and what he still remains even if he rejects the message of Christianity in disbelief. Hence theology itself implies a philosophical anthropology, which enables

---

9   Demmer's *Habilitationsschrift* can be considered the first systematic application of Rahner's transcendental method to the field of moral theology. Bruno Schüller acknowledged, if somewhat critically, the importance of the book and the creativity of Demmer's interpretation in his review for the journal *Theologische Revue* 69 (1973) 136-138. Schüller's critique speaks to the problematic transition, in *Sein und Gebot*, from the transcendental to the categorial level of concrete moral judgments for the articulation of norms for actions. However, rather than recognizing the critical realism of Demmer's epistemology, one that is congruent with his (neo) Scholastic background, Schüller reads Demmer's transcendentalism through the lenses of Kant's categorical imperative. For a study of Demmer's appropriation of the transcendental method see R. Dell'Oro, «La filosofia di K. Demmer. Sull'infrastruttura transcendentale del suo pensiero», in A. Fumagalli – V. Viva, ed., *Pensare l'agire morale* (cf. nt. 1), 55-82.

10 See P. Eicher, *Die anthropologische Wende: Karl Rahners philosophischer Weg von Wesen des Menschen zur personalen Existenz*, Freiburg 1970.

this message of grace to be accepted in a really philosophical and reasonable way, and which gives an account of it in a humanly responsible way.[11]

As for Rahner, so also for Demmer, anthropology is the hermeneutical *locus* and the point of entry to the theological question.[12] Thus the basic concern of moral theology becomes, under what conditions can faith speak to the universality of moral experience? How can the disclosure of moral truth in the Christological event represent a meaningful alternative to the reality of moral conflict experienced by every human being?

The Meaning and Function of Moral Normativity

According to Demmer, truths about the human being are implications of truths about God's free and forgiving self-communication, reaching its eschatological measure in the Christ-event. Christian morality will thus be understood by Demmer as the practical mediation of a renewed anthropological understanding flowing from the encounter with the Christ-event and its «effectual history» in the life of the Christian tradition.

To posit a structural correlation between theological anthropology and ethics entails a shift in moral discourse: first and foremost,

---

11 K. Rahner, *Foundations of Christian Faith. An Introduction to the Idea of Christianity*, New York 1978, 25. For a recent assessment of Karl Rahner's contribution from the specific perspective of moral theology see K. Demmer, «Zur Bedeutung Karl Rahners für die Moraltheologie», *Theologie und Glaube* 94 (2004) 537-550. In 1984, Demmer participates with a long study to the moral theological *Festschrift* for Karl Rahner, on the occasion of his 80th birthday. See K. Demmer, «Sittliche Anspruch und Geschichtlichkeit des Verstehens», in H. Rotter, ed., *Heilsgeschichte und ethische Normen*, Freiburg 1984, 64-98. For an exhaustive assessment of Karl Rahner's relevance to fundamental moral theology see E. Guggenberger, *Karl Rahners Christologie und heutige Fundamentalmoral*, Innsbruck – Wien 1990. This book originated as a dissertation at the Pontifical Gregorian University under the direction of Demmer.

12 Such anthropological turning point must be understood in Christological terms: «Anthropologie ist jener privilegierte hermeneutische Ort, an dem Theologie umgeschlagen, mithin greifbar wird. Sie ist [...] christologisch, wobei Menschwerdung und Kreuz zusammen gehören». K Demmer, «Zur Bedeutung» (cf. Nt. 11), 546.

one is «invited» to share in an anthropological ideal. Of course, the latter articulates itself in models of moral conduct; yet, the «good news» of Christian moral discourse consists in the fullness of meaning disclosed by a renewed human teleology, from which any prescriptive or normative functions depend. Norms are always relative to an anthropological self-understanding articulated in ideologies of human fulfillment:

> The question, «what should I do?» is thus secondary to the question, «what can I be?». For from constraining the person, thwarting her expectations for happiness and fulfillment, moral truth represents the promise of existential possibilities whose function is to open up, rather than close off, even higher possibilities of freedom.[13]

The meaning of the shift just mentioned can be grasped when reflecting on how the communication of moral claims – be they values, ideals, or virtues, takes place through the mediation of normative language. Indeed, there is a certain *phenomenological* primacy to the normative language: this is how we come to face moral claims and learn to structure our response to them. Yet, from an *ontological* point of view, the moral language of rules and norms is secondary to the reality of human *action*, namely, the communication and concrete living of ethical insights and notions of human flourishing that substantiate the moral *experience* of a given community.

Experience ought to be understood here not just in its empirical meaning; rather, as the interpreted sedimentation of lived intuitions of the good, which have matured historically within a moral tradition through the progressive confrontation and solutions to situations of conflict. To say that moral norms are *grounded* in experience means recognizing that they do not come down directly from an «ethical heaven», as various forms of neo-Platonic value moralities, from Scheler to von Hildebrand might suggest.[14] Rather, norms express

---

13 K. Demmer, *Die Wahrheit Leben. Theorie des Handelns*, Freiburg 1991, 11. All translations of the German texts quoted are mine.

14 In an ethics of value, the preoccupation for the «transcendent» dimension of the moral imperative tends to reduce moral norms to their eidetic ideal types. This tendency can be seen, for an example, in the phenomenological ethics of Dietrich von Hildebrand. For an articulation of the contrast between realistic-phenomenological and transcendental-hermeneutical approaches to moral experience see R. Dell'Oro. *Espeirenza morale e persona. Per una*

the concrete existential potential of a moral community articulated through trials and errors over time. One could say that moral norms convey a standard of «moral performance», a consensually defined moral threshold behind which the community stands as a definition of its moral identity. Because they are an interpretation of experience, norms are also being kept in constant movement by the very experience that grounds them. Looking at ultimate goals and at notions of fulfillment in this perspective entails recognizing that the morally *right* is also an articulation of the morally *possible*.

The discussion on moral norms, however, needs preliminary clarifications. This is so because moral normativity is *analogical*, rather than univocal: it entails different levels of meaning, which, in turn, must be analyzed in their own specific nature. Moral principles, such as the first principle of practical reason, the Categorical Imperative, or the Golden Rule, cannot, by themselves, provide standards for action. They can do so when the substance of their moral normativity is unpacked, or «broken down», in concrete moral norms.[15] The path to concreteness in the language of morality signals a move from the *formal* orientation of moral principles, aimed at the goodness of the moral agent, to a progressively *content-filled* determination of moral norms, directed at the rightness of the action.

As Demmer points out, «formal» or «abstract» are not synonymous with «lacking in content». On the contrary, the most formal level of normativity is «overdeterminate», rather than

---

reinterpretazione dell'etica fenomenologica di Dietrich von Hildebrand, Roma 1996, especially 157-204. «la fenomenologia hilebrandiana sarà completamente rivolta all'analisi dell'intuizione cognitiva della verità morale (*Sosein*), ma non potrà dire nulla circa la forma storica della sua attuazione pratica (*Dasein*). L'articolazione del momento teoretico e di quello pratico della verità morale resta dal punto di vista fenomenologico abbastanza oscuro» *Ibid.*, 200. For the critique of value phenomenology, see already H.G, Gadamer, «Das ontologische Problem des Wertes», in *Human Sciences and the Problem of Values*, ed. K. Kuypers, The Hague 1972, 17-31.

15 We need images to articulate the meaning of this process. See Th. R. Kopfensteiner, «The Metaphorical Structure of Normativity», *Theological Studies* 58 (1997) 331-346. Demmer previleges the notion of «sedimentation», especially with reference to a community of moral discourse: «So ist es nicht missverständlich zu sagen, Handlungsnormen seien *geronnener* Konsens innerhalb einer sittlichen Kommunikationsgemeinschaft». K. Demmer, «Sittlicher Anspruch und Geschichtlichkeit des Verstehens», in H. Rotter, ed., *Heilsgeschichte und ethische Normen* (cf. nt 11), 75 (the italic is mine).

«indeterminate». For this reason, transcendental and categorial dimensions already intermediate with one another, beyond any rationalistic separation.[16]

Consider the relevance of the last observation when applied to the distinction between «transcendental» and «categorial» norms. The former refer to norms whose meaning is to appeal to moral values and, consequently, directly speak to the moral intentionality of the agent. In categorial norms, on the other hand, the language of values is enriched with reference to particular goods, which now become the intended object of a specific action. In the passage from intention to action, from goodness to rightness, values intermediate with goods in a creative synthesis that ultimately aims not only at the pursuit of partial goals, but at the complete fulfillment, the *ordinate dilectio*, of the moral agent.[17]

Moral norms, now clearly understood as action norms, will be the result of a kind of «probing» of experience, in which both standards of practicability and consequences of action converge into the definition of the morally right. Far from being understood as fixed determinations, moral norms will be seen as «signposts», to echo the language of Heidegger, in the historical between. Their epistemic status remains that of *prima facie* rules[18], namely, rules that have a presumption of validity, as long as their claim can be supported by experience. This might happen in two ways: first in relation to a change of the circumstances envisioned by a norm as the concrete conditions of its own practicability; and, second, in relation to a possible shift in the standards of freedom presupposed by the norm.

---

16   This can be understood only in light of transcendental philosophical premises and a notion of being as *perfectio omnium perfectionum*. «In [den obersten sittlichen Prinzipien] ist ein Höchstmass an erreichbarer Abstraktion erreicht. Das bedeutet nicht [...] vollendete Inhaltslosigkeit [...] weil [der Mensch] inhaltlich – transzendental und kategorial vermittelt – verfasst ist. Denkt er über Inhalte». K. Demmer, *Fundamentale Theologie* (cf. nt. 5), 147.

17   The reference is to Augustine, *De doctrina christiana*, 1, 28, in CCL 32, 22: «Sittliche Werte gehen mit humanen Güter eine Schöpferische Synthese ein, und dies mit dem Anspruch die umfassende Volledung des Handelnden zu befördern». K. Demmer, *Fundamentale Theologie* (cf. nt. 5), 156.

18   «Es scheint es nicht abwegig Handlungsnormen als «Trampelpfade» der sittlichen Vernunft zu bezeichnen: Spekulative Reflexion und wahrnehmende Erfahrung fliessen zusammen». K. Demmer, *Fundamentale Theologie* (cf. nt. 5), 157.

The practicability of a moral norm traditionally considered true might be falsified by the evidence of changed historical circumstances or the existential condition of the community to whom the norm is directed.[19] When this happens, a shift in paradigms sets in:

> Each science attempts to remember its own past. The guiding question is whether or not fundamental paradigms have been exhausted in their ability to provide adequate explanations. Only when this has been determined is it possible to responsibly speak of a shift of paradigm. One can discover new paradigms; indeed this happens when the chasm between empirical verification and theoretical explanation becomes disjointed.[20]

Tom Kopfensteiner has shown the importance of such development in the meaning of moral norms by reference to the change in the interpretation of the principle of totality.[21] Issues pertaining to the ethics of transplantation provide a context for the application of the principle. Traditionally, the principle rests upon a physicalistic and individualistic understanding, seeing the removal of an organ in the body as «mutilation» justified by the axiom *pars propter totum*. A body part can be surgically mutilated only for the good of the whole. Thus the logic of mutilation justifies removing a cancerous organ, but not a healthy kidney. In the first case, the beneficiary of the mutilation is the sick person; in the second, the receiver of the donor's gift. However, in 1944, Bert Cunningham made possible

---

19 John T. Noonan has shown the logic of moral development in relation to the cases of usury, marriage, slavery, and religious freedom. See «Developments in Moral Doctrine,» in J.F. Keenan – Th. A. Shannon, *The Context of Casuistry*, Washington (D.C.) 1995, 188-204. More broadly A.R. Jonsen – S. Toulmin, *The Abuse of Casuistry. A History of Moral Reasoning*, Berkeley 1988.
20 K. Demmer, «Theological Argument and Hermeneutics in Bioethics» in E.D. Pellegrino – J.P. Langan – John Collins Harvey, ed., *Catholic Perspectives in Medical Morals. Foundational Issues*, Dordrecht 1989, 103. Demmer's moral epistemological analysis of «paradigms» is inspired by Th. S. Kuhn, *The Structure of Scientific Revolutions*, Chicago 1962. For an intensive confrontation with Kuhn's thesis see K. Demmer, *Moraltheologische Methodenlehre*, Freiburg 1989, especially 34-52.
21 See Th. R. Kopfensteiner, *Paradigms and Hermeneutics*, Dissertation at the Pontifical Gregorian University, Rome 1988. Also «Historical Epistemology and Moral Progress», *The Heythrop Journal* 33 (1992) 45-60.

a further expansion of the principle.[22] His examples concerned the attempt to transplant an ovary from one woman to another and the more common transplantation of a cornea from a seeing person to a blind one. To the question how could such operations be justified on the basis of the principle of totality, he invoked the theological notion that all humans are members of the Mystical Body of Christ: «There exists an ordination of men to one another and as a consequence, an order of their members to one another [...] Thus we contend that men are ordinated to society as parts to the whole and, as such, are in some way ordinated to one another».[23] One can see that in a different context, defined by the discovery of new medical possibilities, an expanded understanding of the notion of therapy, and the development of the field of transplantation, also the principle of totality and the norms supported by it undergo a «shift of paradigm». In the end, moral norms safeguard the integrity of freedom by protecting *historically* defined standards of moral performance from possible erosion. Moreover, they articulate goals for action, making progressively more transparent their relation to the ideologies of human fulfillment from which they depend.[24]

The articulation of the historical conditions for both the understanding and the concrete «practicability» of moral norms constitute the specific task of hermeneutics.

Hermeneutics maintains the tradition alive: it safeguards reached standards of knowledge and freedom whose erosion would bring about the danger of individual and societal self-destruction.[25] At the

---

22 B. Cunningham, «The Morality of Organic Transplantation», *Studies in Sacred Theology* no. 86 (1944). For an analysis of Cunningham's thesis, D.F. Kelly, *The Emergence of Roman Catholic Medical Ethics in North America. An Historical – Methodological – Bibliographical Study*, New York – Toronto 1979, 332-341. See also the intriguing study of Barbara Newman, «Exchanging Heart: A Medievalist Looks at Transplant Surgery», *Spiritus. A Journal of Christian Spirituality* 12 (2012) 1-20.

23 B. Cunnningham, «The Morality of Organic Transplantation» (cf. nt. 22), 63. Of course, Cunningham put restrictions on his thesis, such as that the donor cannot be the cause of his own death. For example, a father cannot give his heart for the health of his child.

24 Th. R. Kopfensteiner, «Science, Metaphor, and Moral Casuistry» in J.F. Keenan – Th. A. Shannon, *The Context of Casuistry* (cf. nt. 19), 207-220.

25 «Hermeneutics concerns itself with the intelligibility, and as a result of that, with the practicability of moral norms, insofar as it tries to penetrate the historical ground that has originated them». K. Demmer, *Sittlich Handeln aus*

same time, hermeneutics possesses an emancipatory and progressive function, insofar as it pushes the tradition forward toward the articulation of new paradigms for action.[26]

Moral Truth as Sinnwahrheit

In light of the previous reflections on the meaning and function of moral normativity, it becomes clear that the moral truth conveyed by concrete norms is always the mediation of two dimensions: one orienting the moral agent to a value attunement; the other measuring such orientation with the concrete reality of moral goods to be protected for the sake of the agent's integrity and dignity:

> Human beings learn by reflecting on their own reactions in the face of challenging demands. In this way, the concept of moral value can be concretized within one's own *experienced context*. This concretization does not invalidate the universality of values. The morally true (*das sittliche Wahre*) is always also the possible (*das Mögliche*. But what is possible is grasped through the experience of one's own freedom.[27]

On various occasions in his writings, Demmer takes a special care in highlighting the distinctive phenomenological character of moral truth as *Sinnwahrheit*, i.e., truth pertaining to meaning.[28] He

---

Verstehen. Strukturen hermeneutisch orientierter Fundamentalmoral, Düsseldorf 1980, 11.

26 Demmer is aware of the debate, in hermeneutical philosophy, between Gadamer and Habermas precisely on this point. See J. Habermas, «Der Universalitätsanspruch der Hermeneutik», in *Hermeneutik und Ideologiekritik*, Frankfurt 1971, 120-159. In the *querelle* between Gadamer and Habermas, Demmer positions himself very carefully, in an attitude of receptivity to the merits of both. Thus, he will retrieve some aspects of the Habermasian critique, without losing control of his own epistemological premises. See M. Wolfers, *Theologische Ethik* (cf. nt. 3), 168-190. Also K. Demmer, «Wahrheitsanspruch und Hermeneutik christlicher Praxis», in W. Lesch – A. Bondolfi, ed., *Theologische Ethik im Diskurs*, Tübingen 1995, 144-162.
27 K. Demmer, «Sittlich Handeln aus Erfahrug», *Gregorianum* 59 (1978) 677. More broadly on experience and truth see R. Dell'Oro, «Esperienza e verità morale», *Rivista di teologia morale* 109 (1996) 63-82.
28 See especially *Die Wahrheit leben* (cf. nt. 13), 67-97. Also in a more systematic fashion, «Wahrheit und Bedeutung: Objektive Geltung im moraltheologischen Diskurs», *Gregorianum* 81 (2000) 59-99.

often laments the fact that moral theology hardly seems to have any interest in the theoretical discussion on truth; as a result, it rarely engages in serious confrontation with contemporary theories of truth, feeling rather at home in the Scholastic definition it inherits from the tradition. Unfortunately, in so doing, it precludes itself from becoming aware of several difficulties affecting a naïve use of the notion of truth, especially one that fails to distinguish between the truth of empirical states of affairs, which presupposes an objective distance between subject and object, and historical, anthropological, and ethical truths, which entail a reference to a meaning recognized and interpreted by a subjectivity.

Normally, moral theology relies upon Thomas' definition of truth as the *adeguatio intellectus et rem*, the conformity of the knowing intellect to its object (*De Veritate* q.I, a. 1). This definition conveys an understanding of truth in terms of a formal notion expressing a relation; as such, it presupposes as its condition of possibility the ontological truth of things, manifesting itself to a *receptive* mind, open to the manifestation of the object. Although certainly not wrong, the above definition must be further nuanced when applied to moral truth. This can be done by looking at the analogical relation of knowledge and truth.

The knowledge of an object entails an act of receptivity on the part of the mind toward the object, a receptivity which, however, is grounded in the active anticipation (*Ausgriff*) of the totality of being. Following Karl Rahner's reinterpretation of Thomas Aquinas' metaphysics of knowledge, Demmer sees the relation between *intellectus possibilis* and *intellectus agens* as one in which the former always presupposes the activity of the latter.[29] In an analogous way, the truth of the object is already grounded in the active anticipation of its meaning by the mind, against whose horizon the truth of this object can be recognized as such, i.e., as true.

What is meant with this analogy? As *Sinnwahrheit*, as truth pertaining to meaning, moral truth is, by its very nature, the truth «intended» by an agent, not by a detached spectator. Moral truth is the truth that measures one's lifeproject, the happiness and moral fulfillment pursued by a moral subject in a delicate balance of personal expectations and disposition to sacrifice. It is the particular function

---

29 K. Rahner, *Spirit in the World*, New York 1994.

of practical reason to envision such *telos*, not once and for all, as if this were something like an eidetic form, or the object of an intellectual intuition (*schauen*); rather the *telos* is envisioned (*sehen*) historically, through a dynamic process in which reason continually probes and discovers new possibilities for action. The distinction between *schauen* and *sehen*, as Johannes B. Lotz points out, highlights the difference in the intentionality of the object in question. Whereas the former refers to a direct intuition of the object in a Platonic (and perhaps Husserlian) fashion, the latter conveys the sense of a grasping that presupposes the historicity of the mediation.[30] Moral reason is *practical* because its knowledge of the truth is always energized and sustained by freedom's desire of the good.[31] In Thomistic terms, the anticipation of the mind (*excessus mentis*) toward being as true always corresponds to an anticipation of the will (*excessus voluntatis*) toward being as good.[32]

---

30   J.B. Lotz, *Transcendentale Erfahrung*, Freiburg 1978. The rejection of a metaphysical intuition articulated by Lotz mirrors Rahner's position in *Spirit in the World* (cf. nt. 29): «Absolute being [...] is not intuited, that is, it does not come before the gaze of an intellectual intuition in its own self, but is affirmed simultaneously as the condition of the possibility of the objective of knowledge of *that* existent which alone is represented to the intuition [...] as the proper object of the one human knowledge» (398). In his Introduction to the English translation of *Spirit in the World*, F.P. Fiorenza explains: «since (Rahner) is aware that all human knowledge is related to sense intuitions, he rejects those philosophical positions which maintain that a metaphysics of transcendence is possible because of a special and immediate intuition of a metaphysical object, be it an eternal truth or an objectively conceived absolute being». «Karl Rahner and the Kantian Problematic» xliii.
31   The practical nature of truth, as well as its beauty is already presupposed by the unity of the transcendentals. On this aspect insists especially von Balthasar: «because it is a formal relation of correspondence, (the true) displays a certain correctness, but this is not yet the same as saying why anyone should care about this adequation. Thus, even the light of truth could seem cold and joyless if it did not also have the warmth of the good». H.U. von Balthasar, *Theo-Logic. I. Truth of the World*, trans. A.J. Walker, San Francisco 2000, 221.
32   «Dicendum quod verum et bonum in se invicem coincident, quia et verum est quoddam bonum, et bonum est quoddam verum; unde et bonum potest considerari cognitione speculative, prout consideratur veritas eius tantum» (*De Veritate* III, q. 3, ad 9). Also «Vis cognitiva non movet, nisimediante appetitive» (*Summa Theologiae* 1, q.20, a.1, ad 1). On this, see the beautiful study of K. Riesenhuber, *Die Transzendenz der Freiheit zum Guten*, München 1971.

Finally, the singular character of moral truth can be recognized when seen as the result of an inter-subjective or communal engagement. Although truth cannot be reduced to consensual agreement or social constructionism (*consensus nonfacit veritatem*), it is nevertheless important to recognize that moral experience intermediates the historical meaning of values and ideals relative to a specific life-world and its interpretive framework. Engaging the «communicative» and «discourse ethics» of Habermas and K.O. Apel respectively, Demmer recognizes that standards of moral performance can only be gained through a «process of communication» within a community of moral discourse in which all the participants bring to the table their own contribution, in a spirit of freedom and fairness. In such *ideal* community, echoing the Kantian «kingdom of ends», theoretical presuppositions must be disclosed in order to exclude positions that ground their claim on the basis of privilege or other uncontrollable sources, and to insure the best results in terms of intellectual honesty and transparency.[33]

How to understand, in the end, the relativity of moral truth to freedom? More specifically, how can the «transcendental reduction»[34] of moral truth to freedom avoid the pitfall of a possible *reductionism* of moral truth to a pragmatic function of self-preservation and self-interest?

It is important to keep in mind that with the term «freedom» Demmer does not denote freedom of choice in its empirical facticity, but rather a fundamental disposition toward the good that constitutes the ontological condition for the possibility of each concrete and free choice. Following Rahner, Demmer terms *transcendental* freedom the radical disclosure or self affirmation of this original source that grounds and sustains empirical freedom. In its dynamic actualization (*Vollzug*), however, freedom is not pitted against

---

[33] The hermeneutic rendition of the transcendental framework can be found in K. Demmer, *Sittlich handeln aus Verstehen* (cf. nt. 25), with special reference to the articulations of «meaning as the original ground of ethical normativity» (51-65), «nature as established (*festgehaltener*) meaning» (66-79), «history as creative mediation (*Vermittlung*) of meaning and nature» (80-102), and «moral decision as the positing (*Setzung*) of history» (103-130).

[34] The term «reduction» is, of course, technical, in that it refers to the transcendental movement of grounding. See K. Demmer, *Sein und Gebot* (cf. nt. 4), 15.

its own metaphysical and, ultimately, theological ground, as in the Kantian version of autonomy; rather, it is released to its own identity as love. Each moral decision can be considered good when it mediates the transcendental ground of love upon which it rests; only thus can each human decision articulate, in a historical way, the radical openness (*Vorgriff*) of human freedom to its own fulfillment (*finis ultimus*).[35]

Every moral claim is an expression of the moral agent's autonomy for, to borrow from the language of Hegel, like the spirit, so the person is *in itself* by nature, but becomes *for itself* through the exercise of freedom. Moral experience can be understood, ultimately, as the journey (*ex-perior*) of becoming oneself fully, a journey whose destination cannot be reached without trials and tribulation (*peiros*). In a profound way, we are our freedom, in the sense that our own identity (*actus primus* in Scholastic terms) can be understood only in the act of a radical openness to the source that gives us to be, a source giving out of love and whose call is the call to love.

To see the essence of freedom in its transcendentality, however, entails recognizing, rather than «suspending», the «historical» and «embodied» determinations that always define human freedom. More specifically, freedom finds in the interpretation of «nature» the material conditions that point in the direction of its own *telos*. Demmer reinterprets the classical notion of *natural law* in «personalistic» terms. The nature presupposed by natural law can only be the «nature of the person»: here, the constancies of the good, to which the natural law alludes, become morally relevant only be-

---

35 «Die aristotelisch-thomanische Idee des Letztziels als die wesensmässige Vollendug des Menschen ist zentraler Bezugsspunkt der Ethik Demmers. Den *finis ultimus*, der als transzendentaler Reflexionsbegriff die Möglichkeitsbedingung partikulärer Handlungsziele angibt und in jedem Willensakt als das «Erstgewollte» wirksam ist, personalisiert Demmer und bindet ihn in die Geschichtlichkeit der Existenz ein». M. Wolfers, *Theologische Ethik* (cf. nt. 3), 231, note 13. On the Thomistic understanding of *finis ultimus* see R. Pannier, «Aquinas on the Ultimate End of Human Existence», *Logos: A Journal of Catholic Thought and Culture* 3/4 (2000) 169-194. More generally, J.F. Keenan, *Goodness and Rightness in Thomas Aquinas's Summa Theologiae*, Washington (D.C.) 1992. For a reinterpretation of the notion of *finis ultimus* in contemporary moral theology, J. Fuchs, *Moral Demands and Personal Obligations*, Washington (D.C.) 1993.

cause they mediate the person's practical destination. The essentialistic contraction of natural law, Demmer notes, is a possible, but certainly not a necessary tendency of the tradition. It finds its clear appearance in the late scholastic thinking of the 16th century, but cannot be imputed to Thomas Aquinas:

> An historical overview (of natural law theory) cannot pass over the names of Gabriel Vasquez and Francisco Suarez.; both are responsible – albeit with different accents – for increasingly turning natural law into a rigid metaphysics [...] They share a concern [...] for the certitude of knowledge and the security of action to cut the ground away from any voluntaristic and subjectivistic tendency; the absoluteness and objectivity of moral claims are now metaphysically anchored. Manuals of moral theology were deeply influenced by this mindset [...] Ultimately we could read the contemporary effort to retrieve the genuine Thomas as a reaction against the restrictions of the tradition, in particular its objectivistic and essentialistic notion of metaphysics.[36]

Against this essentialistic tendency, Demmer maintains that *material*, i.e., non-formal universally binding principles of natural law are grounded in an inductive process of discovery, articulating the meaning of fundamental human goods which are not posited in the abstract, once and for all; rather, they will be progressively discovered and appropriated as essential to the realization of personal identity.[37] From a transcendental perspective, the objectivity of natural law is thus «reduced» to transcenden-

---

36 K. Demmer, *Shaping the Moral Life* (cf. nt. 2), 39-40. Also, «The notion of an absolute metaphysical human nature that was the immediate and universal rule of morality (*natura metaphysica et absoluta hominis tamquam regula proxima et homogenea moralitatis*) [...] served as an immutable criterion and thus became the pillar of neo-Scholastic natural law doctrine». *Ibid.*, 13. The essentialism of modern developments of natural law is also highlighted by the 2009 document of the International Theological Commission, À la recherché d'une éthique universelle: nouveau regard sur la loi naturelle, especially nn. 28-33. A general assessment of the document is provided by A. Vicini, «The Search for a Universal Ethics: The International Theological Commission's 2009 Document on Natural Law», *Concilium* (2010/3) 111-17. The entire issue in question is dedicated to a discussion of natural law from multiple perspectives.

37 On the epistemological dimensions of this process of discovery and appropriation, see the beautiful study of Th. R. Kopfensteiner, «The Metaphorical Structure of Normativity» (cf. nt. 15).

tal *subjectivity*. Moral norms of natural law owe their normative force to freedom because they serve the *true* fulfillment of freedom, a fulfillment which is only possible as an experience of transcendence, relative to the epiphany of an absolute good in the midst of history. The experience of the good remains bound to the dimension of *historicity* because as humans we are given over to the condition of temporality, within which we find ourselves, and which we try to mold into concrete historical projects, though we never become completely masters of our temporal condition (*Befindlichkeit*). To be ethical is to be in the milieu of the good, between the conditional goods we find and create in the web of relativities, and the unconditional good that is shown or intimates itself in the happening of temporality, and to whose promise we respond in the lives we live.[38]

The original mediation of the Christ-event by the primitive community, which finds its first interpretive articulation – both in terms of narrative and speculative Christology – in the New Testament writings, speaks to the faith in the disclosure of God's final revelation as the disclosure of an unconditional good for humankind.[39] Such a disclosure is both reassuring and unsettling on the one hand, it speaks to the depth of human freedom and to its ability to recognize the promise of the good intimated in the Christ-event. At the same time, it speaks of a judgment on the tension experienced by freedom to make the experience of the good into a self-serving project, rather than an experience of personal and social liberation.

In the end, the Christian *form* of freedom might be reformulated in terms of a de-mystification of autonomy: freedom released to its own true self in love, freedom beyond autonomy. An ethics grounded in the Christological event is an ethics of the other in which being-for-the-other becomes a free service for the good of the other, beyond the calculative self-interest of utilitarian pru-

---

38 On the transcendental-hermeneutical reinterpretation of natural law as the «law of the person» see especially K. Demmer, «Natur und Person. Brennpunkte gegenwärtiger moraltheologischerAuseinandersetzung», in B. Fraling, ed., *Natur im ethischen Argument,* Freiburg 1990, 55-86.
39 On the contribution of narrative and speculative Christologies see K. Demmer, *Deuten und Handeln* (cf. nt. 6), 78-130 (Christologie und Anthropologie), and *Fundamentale Theologie* (cf. nt. 5), 43-76 (Von der Christozentrik zur christologischen Anthropologie – Handeln in der Nachfolge).

dence, beyond the determination of autonomous will. One might call this *agapeic* service, a service that serves the good of the other out of a release of freedom towards the other, a release that is an overflow of generosity.[40]

The Anthropological Coordinates of Fundamental Moral Theology

So far, I have analyzed Demmer's understanding of fundamental moral theology in its *formal* dimensions. A contribution to the universality of moral experience, however, presupposes faith's ability to publicly mediate its vision of the good. The contribution of moral theology to the public articulation of moral discourse directly flows from the content of faith, understood here as *fides quae*. Such content should not be immediately identified with a set of moral norms. Rather, the «effectual history» of Christian revelation aims, first, at the anthropological conditions of moral normativity. One could say that the moral reason of the believer operates, now, on the basis of renewed *transcendental* conditions.[41] Faith opens up new dimensions of moral creativity and competence, which affect both the selfunderstanding and the freedom of the moral agent.

Demmer speaks, in this context, of the *maieutic* function of faith in relation to moral reason. More specifically, the *intellectus fidei* can ground the anthropological implications of faith because the latter remains, as the First Vatican Council puts it, an *obsequium rationi consentaneum*.[42] Although one might encounter within moral theology different interpretive models of the relation between faith and moral reason, it remains undisputable that faith generates a particular anthropological pre-understanding. One might think here, most obviously, of the different assessments on

---

40 For a demystification of autonomy and the articulation of a metaxological ethics of generosity and service in a philosophical perspective see the work of the Leuven philosopher William Desmond. In particular, *Ethics and the Between*, New York 2001.
41 «Die Sittliche Vernunft des Glaubenden arbeitet in Begründung wie Einsicht unter gewandelten transzendetalen Bedingungen». K. Demmer, *Deuten und Handeln* (cf. nt. 6), 82.
42 See Denzinger-Schönmetzer, 3009.

the analogy of faith and reason between Catholic and Reformed theologies, but also of the polarity within the Catholic moral tradition itself between an «autonomous morality in a context of faith» and «an ethics of faith».[43] Faith elicits reflection from moral reason and inspires moral reason. In doing so, it produces in the believer a *specific* horizon of meaning, something like a «system of anthropological coordinates» (Edward Schillebeeckx), from which one sees himself or herself and the world. Taking the lead from Heidegger's analytic of *Dasein*, hermeneutic philosophy has articulated the notion of «horizon» to define, as Gadamer puts it, «the range of vision that includes everything that can be seen from a particular vantage point».[44]

The *Christological* dimension of such horizon must be emphasized for the moral implications of faith are not simply the result of a doctrinal «deduction», but spring from the very self-understanding of Jesus, to which, through the church, the believer partakes. In Christ, the believer is led to an encounter with a radically new and utterly intense consciousness of God, ultimately defined by Jesus' relation to «his heavenly father».[45] Jesus' self-understanding represents like the focal point of a continuous hermeneutical exercise of memory, understanding, and interpretation in the ecclesial history of the Christian tradition. Jesus is the *mens* from which the various dimensions of the Christian horizon of meaning unfold.

If so, the system of anthropological coordinates cannot simply be identified with a set of «doctrinal themes», such as creation,

---

[43] For an overview of the debate between «autonomous morality in a context of faith» and «ethics of faith» see V. MacNamara, *Faith and Ethics: Recent Roman Catholicism*, Dublin 1985; also R. Simon, *Fonder la morale: Dialectique de la foi et de la raison pratique*, Paris 197) and E. Gillen, *Wie Christen ethisch handeln und denken. Zur Debatte um die Autonomie der Sittlichkeit im Kontext katholischer Theologie*, Würzburg 1989. For Demmer's position on the debate see «Die autonome Moral – eine Anfrage an die Denkform», in A. Holderegger, ed., *Fundamente der theologischen Ethik. Bilanz und Neueransätze*, Freiburg 1996, 261-276 and *Fundamentale Theologie* (cf. nt. 5), 116-127.

[44] H.G. Gadamer, *Truth and Method*, New York 1994, 302.

[45] «Die neutestamentliche Heilsökonomie nimmts ihren Anfang in der Begegnung mit jener Neuheit, Reinheit und Intensität des Gottesbewusstsins, wie es im Selbstverstandnis Jesu und in der einmaligen Intimität seines Verhältnisses zum himmlischen Vater zu geschichtlicher Greifbarkeit und Anschaulichkeit aufsteigt». K. Demmer, *Deuten und Handeln* (cf. nt. 6), 90.

fall, incarnation, redemption, eschatology, etc.[46] Unlike «doctrinal themes», which seem to presuppose a model of revelation in terms of a *theoretical* disclosure of truths, the notion of «anthropological coordinates» points to the importance of the anthropological mediation in the Christian revelation. Thus, the pre-understanding in question is not so much a doctrinal theme, defined once and for all, but a dimension of the believer's self-understanding, whose theological meaning remains to be articulated *historically*.[47]

The first element in the Christian horizon of meaning is the intrinsic *dignity* of the human person, which is at the root of the principle of autonomy. It directly flows from faith in a personal God as creator, who posits the human being as the event of a free, unmerited and forgiving, and absolute self-communication of God. The Christian tradition has constantly expressed such an understanding of the human person as the image of God, who participates in the new creation of Jesus Christ. Closely connected to this is the awareness of the singularity of one's existence as well as the unrepeatable nature of one's personal history.

This first anthropological implication of faith opens up a special understanding of the uniqueness of each person, of her absolute singularity and incommunicable mode of existing. Though many human beings have existed in the course of history, and are now existing, every person is, as if were, the only one. Every person is a *universale concretum*, a concrete whole, in which there is certainly included the nature of the species with its general characteristics, but also in such a way that this nature is appropriated by the subject

---

[46] For an example of such approach Ch. E. Curran, *New Perspectives in Moral Theology*, Note Dame (IN) 1974, 47-86. On the other hand: «Die Wirklichkeit Gottes wird in Jesus Christus zur geschichtlichen Erfahrungswirklichkeit des Menschen. Und dies nicht nur in isoliert verstandenen wie abstract behandelten Schlüsserereignissen von soteriologischen Relevanz wie Inkarnation und Ostergeschehen. Die gesamte Lebensgeschichte Jesu von Nazareth, wiewohl immer schon in christologischer Deutung vorgestellt, wird umgriffen». K. Demmer, *Deuten und Handeln* (cf. nt. 6), 78.

[47] «The mediating function of the anthropological implications of faith gives rise to a creative and critical potential for knowledge and insight which, in turn, is explained in an historical process. The anthropological implications provide guideposts; they serve as directional signals». K. Demmer, «Theological Argument» (cf. nt. 20), 110).

in an absolutely singular way, so as to transcend that nature.⁴⁸ Thus Romano Guardini defines the singularity of the person as «the fact that she exists in the form of «belonging to herself» (*in der Form der Selbstgehörigkeit*)».⁴⁹

Creation represents only the beginning of God's self-communication. The Christian God, unlike the Aristotelian first principle, does not limit itself to a position of pre-eminence in the causal order of beings. God's relation to human history is, in the Christian faith, one of self-communication and complete solidarity. The Christ-event represents the radical and definitive symbol of God's willingness to enter in a communion of love with human kind. This second theological insight opens up new possibilities of meaning in the moral realm. First, the historical event of the incarnation further articulates the notion of human dignity at an *interpersonal* level. The fact that God becomes man in the person of Jesus Christ lays the foundation for a radical equality among all human beings and for the notion of social solidarity. To be in service of others, especially the most vulnerable and poor, is not just one way of understanding life in society, but one faith indicates as the very truth of our being-in-the-world with others. Secondly, the quality of human history becomes a consideration: «The temporal form of our earthly life includes both a promise and a threat, an indissoluble mixture of hope and fear. Yet we are given the choice to attend so fully to the present that it increasingly becomes for us the beginning, indeed, the very presence, of an eternal future».⁵⁰ Because God shared in the very depth of history, no individual event or dimension of the human ex-

---

48 Von Balthasar has beautifully articulated the problem of the relation between universality and particularity in the essence of the person: «this essence can appear only in the unrepeatably occurring uniqueness of the singular. The person alone is the field of its expression. It remains essentially dependent upon this field in order to reveal itself and to bring its truth to the light of day. The individual alone can show what is man's essence, the scope of his powers, his depth and breadth. Thus, in every instance the individual contains the whole (for he lacks no component of human nature), even though the whole infinitely transcends him (since the whole manifests itself in an infinite number of other appearances)». H.U. von Balthasar, *Truth of the World* (cf. nt. 31), 155.
49 R. Guardini, *Welt und Person. Versuche zur christlichen Lehre von Menschen*, Würzburg 1955, 128.
50 H.U. von Balthasar, *Truth of the World* (cf. nt. 31), 199.

perience can be considered, in principle, meaningless. Particularly the experience of suffering remains open to the dimension of hope in light of Christ's death and resurrection, rescued from a final judgment of absurdity and non sense. Furthermore, the Paschal mystery provides a key to interpreting one's death with all its historical anticipations: if death is not to be the definitive human and moral catastrophe, but rather a passage into final communion with God, then there can be no historical situation that stands outside of this promise and its power to transform.

These are only some of the elements defining «the system of anthropological coordinates». Its relevance to specific ethical issues cannot be determined a priori: it requires further steps accompanying the unrelenting work of practical reason. This is so because

the anthropological implications are neither formal nor lacking in content. However, they have not been fully established with regard to their content. Rather they occupy a middle position. Seen from this vantage point, they furnish a standard criterion. It is the task of autonomous moral reason to achieve the further determination of the content.[51]

To exemplify the main components of this process, which is both deductive and inductive, consider the belief in the goodness of all creation informing the Christian attitude towards life. Historically, it has translated, since the dawn of the Christian tradition, into a new attitude of respect for the poor, the sick, the children; more specifically, in the recognition of their personal status.

This anthropological presupposition, however, needs to be further unpacked in its specific ethical meaning. Thus, it is the function of practical reasoning and moral experience to interpret and articulate the ethical implications of this pre-understanding. General moral principles later developed by the Christian tradition might be seen as plausible inferences of such faith-hermeneutics. In their general meaning, however, these principles need to be «specified» in action norms through a risky process of ethical justification.[52]

---

51 K. Demmer, «Theological Argument» (cf. nt. 20), 109.
52 Because of its historical nature, faith like moral reason itself is «risky»: «Der Glaube nimmt die geschichtliche *Riskiertheit* der sittlichen Vernungt mit je wachsender Konkretisierung nicht ab. Heilsgewissheit darf nicht kurzschlüssig mit Handlungsgewissheit verrechnet werden». K. Demmer, *Deuten und*

One can see that the norm prohibiting the «direct killing of the innocent» stands, for sure, in the effective history of the anthropological a priori mentioned above; it represents, nevertheless, the result of a formal process of argumentation not immediately deducible from that a priori. Moral norms that govern the «taking of human life» will emerge from the intermediation of anthropological coordinates with the various moral goods involved, i.e., from the process of moral reasoning itself.

To conclude, the system of anthropological coordinates stands at the edge of Christian theology, mediating between theological affirmations and moral norms; also, it stands as a positive contribution to public discourse, on the presupposition that the public realm is not just the neutral space to be conquered or won over, and that the participants in the community of discourse of an «open society» are not to be faced as enemies but as partners.[53] To acknowledge such an a priori of communication of both epistemic and moral relevance has nothing to do with relativism: in fact, dialogue between moral agents, whether «strangers» or «friends», to use the distinction *in vogue*, can only function on the presumption that any claim to meaning and truth be, at the same time, an attestation of freedom and respect for the other. The public realm, as we know all too well, is not an ideal community of discourse, but one that is historically determined; thus, it can become subjected to mechanisms of reduction and alienation. In a situation where technology and the market forces have so great a role in molding and transforming our intuitions, feelings and visions of the good for human beings, even reasons and arguments can become merely technical, reflecting strategies for the achievement of goals whose value is measured by an instrumental rather than a specifically moral criterion.

To look at the moral arguments flowing from a Christological anthropology is to realize that they do not stand in isolation, but as articulations of an ultimate vision of the good for human beings. According to Demmer, it is the synthetic meaning of this vision as it

---

Handeln (cf. nt. 6), 129, note 123 (the Italic is mine). The possibility of ethical pluralism, even in the context of faith, finds here its justification.

53 See K. Demmer, *Angewandte Theologie des Ethischen*, Freiburg 2003, especially 243-289 (Staat, Recht und Gewissen – christliche Zeugenschaft in demokratischen Kultur).

unfolds in the plurality of theological resources embedding specific moral arguments that is worth interpreting in the public realm. The contribution of moral theology will be like a thorn in the flesh of public moral convictions, the source of an unrelenting hermeneutic of suspicion inspired by prophetic courage more than indulgence in post-Cartesian doubt. The former calls public discussion on moral questions to the suspension of preconceived judgments and dogmatisms of any kind, making possible the «fusion of horizons» that opens our eyes to a deeper vision of who we are and to what is good for us as humans.

CHAPTER 18
CONSCIENCE AFTER VATICAN II
Theological Premises for a Discussion
of Catholic Health Care

In this chapter, I reflect on the Catholic understanding of the notion of conscience, looking specifically at post-Vatican II theological developments.[1] Thus, the essay will be a kind of background framework for the articulation of contemporary issues in faith-based health care. The latter articulation will remain somewhat latent in my observations. Rather, the focus will be on unpacking different facets of a «pre-comprehension,» so to speak, that is, a set of theological premises about moral conscience that nourish, explicitly or implicitly, normative criteria, attitudes, and value judgments. A hermeneutic reconstruction remains inevitably embedded in specific systematic preoccupations: We revisit the past in order to find better solutions for the present. We do so impelled by the present's historical location and questions, all the time mindful of the fact that new developments, especially when called for by dramatic changes, are never possible without a retrospective outlook on the tradition within which we stand: The consistency of such tradition and its dynamic integrity are at stake.

Recent public debates in health care see characterized by an almost inflationary reference to the notion of conscience. Indeed,

---

1   The literature on conscience is quite vast. For a recent assessment, see James F. Keenan, «Redeeming Conscience,» in *Theological Studies* 76, no. 1 (2015): 129-47. An interesting take on the historical journey of the topic is provided by Richard Sorabji, *Moral Conscience through the Ages: Fifth Century BCE to the Present* (Chicago: University of Chicago Press, 2014). From the international literature, in addition to the works indicated in the individual notes, I have benefited from Eberhard Schockenhoff, *Das umstrittene Gewissen: Eine theoloische Grundlegung* (Mainz: Grünewald, 1990), and Aristide Fumagalli, *L'eco dello Spirito: Teologia della coscienza moral* (Brescia: Queriniana, 2012). Most recently, and with reference to Catholic health care, David DeCosse and Kristin Heyer, eds., *Conscience and Catholicism: Rights, Responsibilities, and Institutional Responses* (Maryknoll, NY: Orbis Books, 2015).

appeal to conscience has become the argument privileged by faith-based organizations, as they defend the right to stand by their own normative principles. For an example, religious conservatives invoke constitutional protection for their conscientious moral judgments. Grounding their claim on the Free Exercise Clause of the First Amendment and the Religious Freedom Restoration Act, they have refused to comply with the contraception mandate enacted by the Department of Health and Human Services in the regulations implementing the Affordable Care Act. The objection of the Little Sisters of the Poor, who are involved in a lawsuit against the Obama administration, is a case in point: For the Sisters, even the most remote connection to the governmental mandate to provide insurance coverage for the contraception constitutes a burden on their consciences.[2]

The Catholic tradition can certainly claim for itself a rich heritage of reflection on the doctrine of conscience. The rhetoric of conscience, however, gives pause, for here a discrepancy seems to emerge between the strong appeal to *institutional* conscience and the relatively lukewarm deference paid to conscience as a dimension of *personal* decision-making. A category that, especially in the magisterial teaching, tends to be looked at with suspicion when predicated of the individual's power to self-determination[3] receives now an all too facile endorsement when invoked as a condition for institutions to operate freely according to their own cherished moral principles. Thus, in what appears like an ambiguous lack of consistency, Catholic rhetoric promptly calls for re-

---

[2] The case is discussed by Cathleen Kaveny in David DeCosse and Thomas Nairn, *Conscience and Catholic Health Care: From Clinical Contexts to Government Mandate* (Orbis Books, 2017). In broad terms, Kaveny points out that the debate about religion, law, and morality in the American public square has shifted from one on the *enactment* of generally applicable laws to one on *exemptions* from such laws.

[3] See *Veritatis Splendor*'s considerations about «creative» conscience. «There seems to be an emphasis within *Veritatis Splendor* on the suppression of conscience and a move of power toward the Magisterium,» writes Jayne Hoose in «Conscience in *Veritatis Splendor* and the *Catechism*,» in *Conscience: Readings in Moral Theology No. 14*, ed. Charles E. Curran (New York: Paulist Press, 2004), 89. For a critical reading of moral theological developments under the pontificate of John Paul II, including *Veritatis Splendor*, see Paul Valadier, Éloge *de la conscience* (Paris: Seuil, 1994).

spect of institutional conscience while undermining the space of freedom that protects *conscientious* decision-making by individuals, whether health care professionals or patients, within its very institutional space. I am tempted to call this state of affairs an «inconsistent ethic of life.»

To be sure, the judgment in question warrants careful scrutiny. It points to the need, on the one hand, to distinguish between the personal and the institutional spheres of conscience, and to respect the relative autonomy of each; on the other, it calls for avoidance of conceptual ambiguity and dangerous double standards. The celebration of conscience constitutes a welcome retrieval of an important, if too often neglected, aspect of the Catholic tradition. The plausibility of such retrieval, however, rests on a coherent rehabilitation of conscience that speaks to the moral agency of mature individuals, and, by *analogical* extension, of institutions alike. Thus, the retrieval will be convincing, provided the rights of the latter are defended as central to an ethos that equally cherishes and cultivates those of the former.

This chapter proceeds in three stages. First, it lays out some central aspects of the doctrine of conscience developed by the Second Vatican Council, especially in the statements of *Gaudium et Spes* and *Dignitatis Humanae*, and shows how the council's «turning point» integrates and corrects important aspects of the traditional understanding of conscience. Second, it looks at the theological conceptual framework that supports a renewed understanding of conscience, with reference, in particular, to the meaning and function of moral normativity, the particular character of moral truth, and the centrality of freedom. Third, I allude, in the conclusions, to some implications of the above reconstruction for Catholic health care today.

The Doctrine of Conscience at a Turning Point: The Second Vatican Council

The council's statements on Conscience, specifically in *Gaudium et Spes* and *Dignitatis Humanae*, speak to «the dignity of conscience» (*GS*, 16), «the excellence of freedom» (*GS*, 17), and the right of the individual «not to be forced to act against conscience, nor be prevented from acting according to conscience» (*DH*, 3). The positive recognition of the centrality of conscience goes hand in hand, perhaps not without contradiction, as Linda Hogan has observed,[4] with a more sober recognition that «as it often happens… conscience goes astray through ignorance, which it is unable to avoid, without thereby losing its dignity» (*GS*, 16). To prevent conscience from embarking along erroneous paths in the perilous journey of life, there stands, unequivocal and clear, the «holy and certain teachings of the church,» to which «the faithful must pay careful attention in forming their conscience» (*DH*, 14).

Despite possible conflicts of interpretations, I stand with those authoritative readers of the council's statements who see them in a clear development, perhaps even a shift in paradigms both in tone and content, with respect to the previous tradition of neo-Scholastic theology.[5] Indeed, one finds in the council's texts an understanding of conscience that interprets its function beyond the limitations of normative concerns, that is, in terms of the relation of conscience and law. In this latter perspective, conscience concerns the practical judgment on the objectivity of the norm; as the «ultimate practical judgment on an act to be made, or an act that has been made» (*judicium ultimo-practicum de actu ponendo vel de actu positio*),[6] conscience tends to be reduced to a post-factum mechanism of praise, when the action is done in conformity to the law, or regret when contradicting it.

Three aspects are central to this traditional understanding of conscience: first, the restriction to an act-centered morality; second, the dualism of subjectivity and objectivity; third, the overall syllogistic character of moral reasoning, understood as a deductive conclusion from clearly defined premises. The focus, to begin with, is on the on-

---

4 Linda Hogan, «Conscience in the Documents of Vatican II,» in *Conscience: Readings in Moral Theology No. 14*, ed. Charles E. Curran (New York: Paulist Press, 2004), 82-88: «Two strands of conscience that had been successfully integrated by Aquinas are present in the documents, not together but as competing accounts.»

5 See especially two classic studies by Domenico Capone on the history of redaction of *Gaudium et Spes* 16: «Antropologia, coscienza e personalita,» *Studia Morale* 4 (1966): 73-113, and «La teologia della coscienza morale nel Concilio e dopo il Concilio,» *Studia Moralia* 24 (1986): 221-49.

6 As quoted in Bruno Schüller, *Die Bergründung sittlicher Urteile*, 2nd ed. (Düsseldorf: Patmos Verlag, 1980), 45.

tic facticity of the act, which will be more clearly defined in its ethical meaning the more it is isolated from the idiosyncratic features of the acting person. Thus, while the norm predetermines the rightness of the act (*finis operis*), conscience speaks only to the goodness, that is, the motivational and intentional structure, of the agent (*finis operantis*). Because the relation between the two is extrinsic rather than intrinsic, conscience functions only as a subjective, rather than objective, determinant of the action's morality. Furthermore, the classic neo-Scholastic tradition distinguishes between «original conscience» (*synderesis*) and «conscience in situation,» each one with their different degree of certainty: the former absolute, the latter fallible. For Thomas Aquinas, *synderesis* is the habit of practical reason, by which one knows the first principles of natural law, that is, do good and avoid evil.[7] Conscience in situation, however, is the act of applying the first principles known in *synderesis* to the conduct at stake. The application in question is virtually equivalent to the conclusion of a syllogism: The major premise expresses the moral law; the minor refers to the act to be done in the particular situation, and for which the law is supposed to obtain; the conclusion states whether the action comes under the moral law. One can see that, for the neo-Scholastic manuals of moral theology, the reasoning of conscience is the form of deductive syllogism.[8]

---

[7] When confronted with the Augustinian tradition, the contribution of Thomas Aquinas is certainly precious in terms of its systematic power. However, it also contributes to what has been labelled an «intellectualization» of the notion of conscience: For Thomas, the measure of action is not so much conscience, but practical reason. To act according to conscience means to act according to *recta ratio*. Furthermore, with respect to the Franciscan school, Thomas stands by the pure *cognitive* character of the judgment of conscience, as he claims that «iudicium conscientiae consistit in pura cognition» (*De Veritate* 17, 1, ad 4).

[8] See Charles E. Curran, «Conscience in the Light of the Catholic Moral Tradition,» in idem, *Conscience*, 3-24. The difference between Thomas and the neo-Scholastic tradition should not go unnoticed. See, for an example, the classic manuals of Merkelbach, Aertnys/Damen, and Zalba. Although the latter develop into the «effectual history» of the former, the two differ in important points, especially with regard to the character of the judgment of conscience in question. Unlike the neo-Scholastic tradition, which emerges from the confrontation with a modern conception of ethics *more geometrico demonstrate*, Thomas distinguishes *determinatio* from *conclusio* in *S. Th.*, I-IIae, q. 95, a.2. See Schockenhoff, *Das umstrittene Gewissen*, 77-82.

When confronted with the previous tradition, the council's statements on conscience cannot fail to impress, not only for their deeper theological quality, but also for the, at least initial, overcoming of the methodological flaws of traditional doctrine. In its various statements, the council looks at conscience not only as a function of practical judgment on individual actions, but as the very definition of personal selfhood. The retrieval of the biblical notion of «heart» constitutes an attempt to root the notion of conscience in a more clearly biblical understanding, which sees in the heart the ultimate space of one's communication with God.[9] Furthermore, the theological grounding supports a more holistic anthropology. By using the heart metaphor, the council reconciles in a higher synthesis the more intellectualistic strain of the Thomistic tradition with the emphasis on the will by the Franciscan school. According to Karl Golser, the council's text even echoes an understanding of conscience developed in the German mystical tradition of Meister Eckhart with the notion of *Seelengrund*.[10] The heart alludes to a complex set of dimensions that synergistically bring together knowledge, decision, and power of ratification by a moral agent.[11] Also, there is a reference to a more inductive, rather than deductive, way of proceeding, alluded to in *Gaudium et Spes* 16, by a call for collaboration with others within society, in the search of common values and criteria for action.

Insofar as it express the whole moral history of the person, with its successes and failures, conscience is more than a purely syllogistic deduction from established premises, one that could function without reference to the moral character of the agent. Rather, conscience is always «in act,» already actualized, so to speak, by a concrete history of freedom, in which moral disposition and knowl-

---

Thomas underscores the autonomy of *practical* reason, and the singularity of practical judgment vis-à-vis theoretical reason. On this, among others, Martin Rhonheimer, *Natural Law and Practical Reason: A Thomist View of Moral Autonomy* (New York: Fordham University Press, 2000).

9   On the Biblical notion of conscience in light of biblical anthropology see Schockenhoff, *Das umstrittene Gewissen*, 48-55.

10 Karl Golser, «Das Gewissen als verborgenste Mitte im Menschem,» in *Grundlagen und Probleme der heutigen Moraltheologie*, ed. Wilhelm Ernst (Würzburg: Echter Verlgag, 1989), 116.

11 For a philosophical anthropology inspired by the «heart tradition» see Andrew Tallon, *Head and Heart: Affection, Cognition, Volition as Triune Consciousness* (New York: Fordham University Press, 1997).

edge condition each other.¹² This is why conscience cannot be fully accounted for in terms of a function of ratification or application: It stands for the moral identity of the subject, for the agency that commits itself to a particular principle of action with a free insight supported by an existential openness to the good, or, conversely by a progressive blindness to it.¹³ Indeed, conscience exceeds mechanical compliance with the norm; it exceeds the concern for correspondence between subjective and objective dimensions of morality: For conscience underlines, and in doing so, grounds, both the objective and subjective presuppositions, and the conditions of possibility of such correspondence.¹⁴ Conscience speaks of the incommunicable «idiocy» of the self, a knowing-of-oneself irreducible to abstract universality.¹⁵ To say that «in her [or his] conscience» a person is given-over-to-herself [or himself] most radically is to

---

12 This is, of course, not a new idea, for Thomas Aquinas already underscores the relation between knowledge and freedom: one knows only when one *wants* to know. Thus: «Vis cognitiva non movet, nisi mediante appetitive» (*S. Th.*, q. 20, a. 1, ad 1), or «Intelligo enim quia volo» (*De Malo* 6). On this, Klaus Demmer, *Sein und Gebot: Die Bedeutsamkeit des transzendentalphilosophischen Denkansatzes in der Scholastik der Gegenwart für den formalin Aufriss der Fundamentalmoral* (Munich: Schöningh, 1971).
13 Phenomenologists speak, in this context, of *value blindness* (*Wertblindheit*). See especially Dietrich von Hildebrand, *Sittlichkeit und ethische Werterkenntnis: Eine Untersuchung über ethische Strukturprobleme*, 3rd ed. (Vallendar-Schönstatt: Patris Verlag, 1982), 47-86. Bernard Lonergan has developed a similar perspective in *Insight: A Study of Human Understanding* (New York: Harper and Row, 1957), 595-618.
14 The recent magisterium seems to look skeptically at the attempt to expand the notion of conscience beyond practical judgment. In *Veritatis Splendor* one finds a kind of alternating between the language of moral personalism and the traditional essentialism of neo-Scholastic theology: «The judgment of conscience is a *practical judgment*....It is a judgment which applies to a concrete situation the rational conviction that one must love, do good and avoid evil. The first principle of practical reason is part of the natural law...but whereas the natural law discloses the objective and universal demands of the moral good, conscience is the application of the law to a particular case....Conscience thus formulates *moral obligation* in light of the natural law (59).
15 «Idiocy» refers, in its etymological meaning, to the *idios*, i.e., the intimate. For an articulation of the «idiotic» as a potency of the ethical, I am indebted to William Desmond, *Ethics and the Between* (Albany: State University of New York Press, 2001). John F. Crosby's notion of «incommunicability» speaks to the same phenomenological reality. See his *The Selfhood of the Human Person* (Washington, DC: Catholic University of America Press, 1996), 41-81.

identify a reality in which insight and decision toward the good exist in unity. Thus, conscience is the voice of *transcendence* because it is also the reality to which all dimensions of moral agency, that is, commanding, prohibiting, inviting, and so on are *transcendentally* reduced in the identity of a moral subject who is fully himself or herself when before God (*coram Deo*).[16] In conscience, the person sees his or her ultimate destination, in a *visio* irreducible to a purely intellectualistic anticipation of one's ultimate fulfillment. The apprehension of a particular value system entails, at the same time, a fundamental option, a *decision* to the absoluteness of the good that binds knowledge to action, insight to freedom. In the sacred space of conscience. The person sees himself or herself in light of the value system to which he or she has committed. Conscience occupies a position of (transcendental) ultimacy with respect to moral action and to the norms that regulate it: It is the original source from which ethical principles and norms depend; as *derivative* functions of conscience, moral norms are grounded by the horizon of understanding and interpretation that generates them.

Theological Premises of a Renewed Doctrine on Conscience

In the second part of this chapter, I reflect on the meaning of the shift initiated by the Second Vatican Council. I said that the theological premises of such a shift do not always come to the fore unequivocally. Thus, the importance of a more systematic articulation that unpacks what is implicit or latent, and this in the context of the larger turning point of post-Vatican II theology, known in the language of Peter Eicher as the «anthropological turning point.»[17] The

---

16 For a transcendental reinterpretation of the notion of conscience and the interplay of «reduction» and «deduction» see Klaus Demmer, especially *Fundamentale Theologie des Ethischen* (Freiburg: Herder, 1999), 183-233, and Demmer, *Sein und Gebot*, 15-119. In an analogous vein, Lonergan speaks of the transcendental notion of value. See Walter E. Conn, «Conscience and Self-Transcendence in the Thought of Bernard Lonergan,» in *Conscience*, ed. Charles E. Curran, 151-62.
17 See Peter Eicher, *Die anthropologische Wende: Karle Rahners philosophischer Weg von Wesen des Menschen zur personalen Existenz* (Freiburg: Universitätsverlag, 1970).

attempt to reduce the meaning and function of moral conscience, pushing theological reflection back, behind the theoretical threshold set by the council, can be understood as a critique of the broader anthropological turning point in question. Thus, the ambiguity inherent in the *letter* of the council's documents very quickly leads conservative readers to a dismissal of the aggiornamento that unequivocally grounds their *spirit*.[18] In this systematic reconstruction, unpretentious as it may be, I start from the premise that the «anthropological turning point» in theology constitutes, at least methodologically, a point of no return, even when remaining open to the possibility of more nuanced developments.[19] Karl Rahner, one of the main exponents of such a turn, explains its meaning thus:

> Theology wants to tell man what he is, and what he still remains even if he rejects the message of Christianity in disbelief. Hence, theology itself implies a philosophical anthropology, which enables this message of grace to be accepted in a really philosophical and reasonable way, and which gives an account of it in a humanly responsible way.[20]

For Rahner, anthropology is the hermeneutical locus and the point of entry to the theological question.[21] For an anthropologically oriented moral theology, conscience designates the sacred space in which, in faith, the moral subject confronts the ultimate telos of his

---

18 Consult, as an example, the articles of German Grisez, Russel Shaw, and William E. May, in the collection previously quoted: Curran, ed., *Conscience*. The works of Janet E. Smith follows in the same direction. The contribution of Joseph Ratzinger on the topic cannot be homologized to that of the authors above, for it stands out for historical precision and systematic depth. Ratzinger's commitment to a more Augustinian anthropology has made him ambivalent about too optimistic a celebration of conscience. This can be seen already in his commentary on *Gaudium et Spes* for the 1986 edition of the *Lexikon für Theologie und Kirche* (Ergänzungsbad III, 313-54), and more recently, in his rendition of *synderesis* in terms of the Platonic notion of *anamnesis*. See Joseph Ratzinger, *On Conscience* (San Francisco: Ignatius Press, 2006), 30-37.
19 The critique of Hans Urs von Balthasar of Rahner might be taken here as symptomatic of the reaction. See his very polemical *The Moment of Christian Witness* (San Francisco: Ignatius Press, 1984).
20 Karl Rahner, *Foundations of Christian Faith: An Introduction to the Idea of Christianity* (New York: Seabury Press, 1978), 25.
21 Such an anthropological turning point must be understood in Christological terms.

or her moral experience. In conscience, the moral subject encounters, in all its clarity and urgency, the central question of Christian morality: How can the disclosure of moral truth in Christological event represent a meaningful alternative to the reality of moral conflict experienced by every human being?

1. *Conscience and Norms: The Meaning and Function of Moral Normativity*

In his 2013 Apostolic Exhortation, *Evangelii Gaudium*, Pope Francis states:

> The centrality of the kerygma calls for stressing those elements which are most need today: it has to express God's saving love which precedes any moral and religious obligation on our part; *it should not impose the truth but appeal to freedom*; it should be marked by joy, encouragement, liveliness....All this demands on the part of the evangelizer certain attitudes which foster openness to the message: approachability, readiness for dialogue, patience, a warmth and welcome which is non-judgmental. (165)

Truths about the human being are implications of truths about God's free and forgiving self-communication, reaching its eschatological measure in the Christ-event. Consequently, Christian morality represents the practical articulation of a renewed anthropological understanding, flowing from the encounter with the Christ-event and its «effectual history» in the life of the Christian tradition. To posit a structural correlation between theological anthropology and ethics entails a shift in moral discourse: First and foremost, one is «invited» to share in an anthropological ideal. Of course, the latter finds expression in models of moral conduct and norms; yet, the «good news» of Christian moral discourse consists in the fullness of meaning disclosed by a renewed human teleology, on which any prescriptive or normative functions depend. Norms are always relative to an anthropological self-understanding articulated in ideologies of human fulfillment:

> Before we ask what we must do, there is a more fundamental question: «What can I be?» This question is prompted by a self-understand-

ing that perceives ethical truth as a promise of existential possibilities. Morality is no oppressive burden but an empowerment to act, which leads to ever-greater freedom.[22]

In conscience, the person encounters the question of his or her own ultimate identity. Thus, the intentional correlate of conscience is already beyond the normative level. For sure, the communication of moral claims – be they values, ideals, or virtues – takes place through the mediation of normative language. Yet, the moral language of rules and norms is secondary to the reality of human conscience. It is *in* conscience and *through* conscience that the communication of ethical insights and notions of human flourishing takes place. Here the meaning of moral experience discloses itself to a subject willing to recognize its power of fascination and attraction. I am speaking of «experience» not just in its empirical meaning but also as the interpreted sedimentation of lived intuitions of the good, which have matured historically within a moral tradition through progressive confrontations with and solutions to situations of conflict. It is such experience that saturates the moral identity of a community and defines its *institutional* conscience. Moral norms are *grounded* in the community's experience: They express the concrete existential potential of a moral community, articulated through trial and error over time. One could say that moral norms convey a standard of «moral performance,» a consensually defined moral threshold behind which the community stands as a definition of its moral identity. Because they are an interpretation of experience, norms are also being kept in constant movement by the very experience that grounds them: Institutional conscience refers to a dynamic rather than static, reality.

The discussion of moral norms, however, needs preliminary clarifications. This is so because moral normativity is *analogical*, rather than univocal: It entails different levels of meaning, which, in turn, must be analyzed in their own specific nature. Moral principles, such as the first principle of practical reason, the categorical imperative, or the Golden Rule, cannot, by themselves, provide standards for action. They can do so when the substance of their moral nor-

---

22 Klaus Demmer, *Living the Truth: A Theory of Action* (Washington, DC: Georgetown University Press, 2010), 11.

mativity is unpacked, or «broken down,» in concrete moral norms.[23] The path to concreteness in the language of morality signals a move from the *formal* orientation of moral principles, aimed at the goodness of the moral agent, to a progressively *content-filled* determination of moral norms, directed at the rightness of the action. Whereas the former refers directly to conscience, the latter is driven by the work of practical reason.

Consider the relevance of the last observation when applied to the distinction between «transcendental» and «categorical» norms.[24] Transcendental norms appeal to moral values and, consequently, directly speak to the moral intentionality, that is, the conscience, of the agent. In categorical norms, however, the language of values is enriched with reference to particular goods, which now become the intended object of a specific action. In the passage from intention to action, from goodness to rightness, values intermediate with goods in a creative synthesis that ultimately aims not only at the pursuit of partial goals, but at the complete flourishing of the moral agent. Moral norms, now clearly understood as action norms, will be the result of a kind of «probing» of experience, in which both standards of practicability and consequences of action converge into the definition of the morally right. Far from being understood as fixed determinations, moral norms will be seen as «signposts» (Heidegger) in the historical between. Their epistemic status remains that of prima facie rules, namely, rules that have a presumption of validity, as long as their claim can be supported by existence. This might happen in two ways: first, in relation to a change of the circumstances envisioned by a norm as the concrete conditions of its own practicability, and, second, in relation to a possible shift in the standards of freedom presupposed by the norm. The practicability

---

23 We need images to articulate the meaning of this process. See Thomas R. Kopfensteiner, «The Metaphorical Structure of Normativity,» *Theological Studies* 58 (1997): 331-46. Demme privileges the notion of «sedimentation,» especially with reference to a community of moral discourse: «It is not a misunderstanding to say that action norms represent the consensus *sedimented* [*geronnene*] within a community of moral discourse.» Klaus Demmer, «Sittlicher Anspruch und Geschichtlichkeit des Verstehens,» in *Heilsgeschichte und ethische Normen*, ed. Hans Rotter (Freiburg: Herder, 1984), 75.

24 For an analysis of the distinction see Klaus Demmer, *Shaping the Moral Life: An Approach to Moral Theology* (Washington, DC: Georgetown University Press, 2000), 46-47.

of a moral norm traditionally considered true might be falsified by the evidence of changed historical circumstances or by the existential condition of the community to whom the norm is directed.²⁵ When this happens, a shift in paradigm sets in:

> Each science attempts to remember its own past. The guiding question is whether or not fundamental paradigms have been exhausted in their ability to provide adequate explanations. Only when this has been determined is it possible to responsibly speak of a shift of paradigm. One can discover new paradigms; indeed this happens when the chasm between empirical verification and theoretical explanation becomes disjointed.²⁶

Thomas Kopfensteiner has shown the importance of such development in the meaning of moral norms by reference to the change in the interpretation of the principle of totality.²⁷ Issues pertaining to the ethics of transplantation provide a context for the application of the principle.

---

25   John T. Noonan has shown the logical of moral development in relation to the cases of usury, slavery, and religious freedom. See «Development in Moral Doctrine,» in *The Context of Casuistry*, ed. James F. Keenan and Thomas A. Shannon (Washington DC: Georgetown University Press, 1995), 188-204. More broadly Albert R. Jonsen and Stephen Toulmin, *The Abuse of Casuistry: A History of Moral Reasoning* (Berkeley: University of California Press, 1988)
26   «Klaus Demmer, «Theological Argument and Hermeneutics in Bioethics,» in *Catholic Perspectives in Medical Morals: Foundational Issues*, ed. Edmund D. Pellegrino, John P. Langan, and John Collins Harvey (Dordrecht: Kluwer Academic, 1989), 103. Demmer's moral epistemological analysis of «paradigms» is inspired by Thomas S. Kuhn, *The Structure of Scientific Revolutions* (Chicago: University of Chicago Press, 1962). For an intensive confrontation with Kuhn's thesis see Klaus Demmer, *Moraltheologische Methodenlehre* (Freiburg: Herder, 1989), especially 34-52.
27   See Thomas R. Kopfensteiner, *Paradigms and Hermeneutics*, diss., Pontifical Gregorian University (Rome, 1988). Also «Historical Epistemology and Moral Progress,» *Heythrop Journal* 33 (1992): 45-60. Traditionally, the principle rests on a physicalistic and individualistic understanding, seeing the removal of an organ in the body as a «mutilation» justified by the axiom *pars propter totum*. A body part can be surgically mutilated only for the good of the whole. However, in 1944 Bert Cunningham made possible a further expansion of the principle. See Bert Cunningham, «The Morality of Organic Transplantation,» *Studies in Sacred Theology*, no. 86 (1944). For an analysis of Cunningham's thesis, see David F. Kelly, *The Emergence of Roman Catholic Medical Ethics in North America: An Historical, Methodological, Bibliographical Study* (New York: Edwin Mellen Press, 1979), 332-41.

In a different context, defined by the discovery of new medical possibilities, an expanded understanding of the notion of therapy, and the development of the field of transplantation, also the principle of totality and the norms supported by it undergo a «shift of paradigm.» In the end, moral norms safeguard the integrity of freedom by protecting *historically* defined standards of moral performance from possible erosion. Moreover, they articulate goals for action, making progressively more transparent their relation to the ideologies of human fulfillment on which they depend.[28]

2. Conscience and Moral Truth

In light of the previous reflections on the meaning and function of moral normativity, it becomes clear that the moral truth conveyed by concrete norms is always the mediation of two dimensions: one orienting the conscience of the moral agent to a value-attunement, and the other measuring such orientation with the concrete reality of moral goods to be protected for the sake of the agent's integrity and dignity:

> Human beings learn by reflecting on their own reactions in the face of challenging demands. In this way, the concept of moral value can be concretize within one's own *experienced* context. This concretization does not invalidate the universality of values. The morally true [*das sittliche Wahre*] is always also the possible [*das Mögliche*]. But what is possible is grasped through the experience of one's own freedom.[29]

A reflection on conscience needs to highlight the distinctive phenomenological character of moral truth as truth pertaining to meaning. As Klaus Demmer has repeatedly observed, moral theology hardly seems to have any interest in the theoretical discussion on truth; as a result, it rarely engages in serious confrontation with con-

---

28   Thomas R. Kopfensteiner, «Science, Metaphor, and Moral Casuistry,» in *Context of Cauistry*, ed. Keenan and Shannon, 207-20.
29   Klaus Demmer, «Wahrheit und Bedeutung: Objektive Geltung im moraltheologischen Diskurs,» *Gregorianum* 81 (2000): 59-99. The relation of freedom to truth constitutes a central theological topos of John Paul II's magisterial teaching. The emphasis is on the correspondence of freedom to truth.

temporary theories of truth, feeling rather at home in the Scholastic definition it inherits from the tradition.[30] Unfortunately, in doing so, it precludes itself from becoming aware of several difficulties affecting a naïve use of the notion of truth, especially one that fails to distinguish between the truth of empirical states of affairs, which presupposes an objective distance between subject and object, and historical, anthropological, and ethical truths, which entail a reference to a meaning recognized and interpreted by a subjectivity.

It is customary for moral theology to rely on Thomas's definition of truth as the *adequatio intellectus et rem*, the conformity of the knowing intellect to its object (*De Veritate* q. I, a. 1). This definition conveys an understanding of truth in terms of a formal notion expressing a relation; as such, it presupposes as its condition of possibility the ontological truth of things, manifesting itself to a *receptive* mind, open to the manifestation of the object. Although certainly not wrong, the above definition must be further nuanced when applied to moral truth. This can be done by looking at the analogical relation of knowledge and truth. The knowledge of an object entails an act of receptivity on the part of the mind toward the object, a receptivity which, however, is grounded in the active anticipation (*Ausgriff*) of the totality of being. Following Karl Rahner's reinterpretation of Thomas Aquinas's metaphysics of knowledge, one could say that the relation between *intellectus possibilis* and *intellectus agens* is one in which the former always presupposes the activity of the latter.[31] In an analogous way, the truth of the object is already grounded in the active anticipation of its meaning by the mind, against whose horizon the truth of this object can be recognized as such, that is, as true. What is meant with this analogy? As truth pertaining to meaning, moral truth is, by its very mature, the truth *intended* by an agent, not by a detached spectator. Moral truth is the truth that measures one's life-project, the happiness and moral fulfillment pursued by a moral subject in a delicate balance of personal expectations and disposition to sacrifice. Moral reason

---

30 Klaus Demmer, «Wahrheit und Bedeutung: Objektive Geltung im moraltheologischen Diskurs,» *Gregorianum* 81 (2000): 59-99. The relation of freedom to truth constitutes a central theological topos of John Paul II's magisterial teaching. The emphasis is on the correspondence of freedom to truth.
31 Karl Rahner, *Spirit in the World* (New York: Continuum), 1994).

is *practical* because its knowledge of the truth is always energized and sustained by freedom's desire of the good.[32] In Thomistic terms, the anticipation of the mind (*excessus mentis*) toward being as true always corresponds to an anticipation of the will (*excessus voluntatis*) toward being as good.[33] Finally, the singular character of moral truth can be recognized when seen as the result of an *intersubjective* or communal engagement. Although truth cannot be reduced to consensual agreement or social constructionism (*consensus non facit veritatem*), it is nevertheless important to recognize that moral experience intermediates the historical meaning of values and ideals relative to a specific lifeworld and its interpretive framework. Standards of moral performance can be gained only through a «process of communication» within a community of moral discourse, in which all the participants bring to the table their own contribution, in a spirit of freedom and fairness, In such an *ideal* community, echoing the Kantian «kingdom of ends,» theoretical presuppositions must be disclosed in order to exclude positions that ground their claim on the basis of privilege or other uncontrollable sources, and to ensure the best results in terms of intellectual honesty and transparency.[34]

How to understand, in the end, the relativity of moral truth to freedom?[35] More specifically, how can the «transcendental reduc-

---

[32] The practical nature of truth, as well as its beauty, is already presupposed by the unity of the transcendentals. On this aspect insists especially von Balthasar: «Because…it is a formal relation of correspondence, [the true] displays a certain correctness, but this is not yet the same as saying why anyone should care about this adequation. Thus, even the light of truth could seem cold and joyless if it did not also have the warmth of the good.» Hans Urs von Balthasar, *Theo-Logic/Volume I: Truth of the World*, trans. Adrian J. Walker (San Francisco: Ignatius Press, 2000), 221.

[33] «Dicendum quod verum et bonum in se invicem coincident, quia et verum est quoddam bonumm et bonum est quoddam verum; unde et bonum potest considerari cognition speculative, prout consideratur veritas eius tantum» (*De Veritate* III, q. 3, ad 9). Also: «Vis cognitive non mvet, nisi mediante appetitive» (*Summa Theologiae* I. q.20, a.1, ad 1).

[34] This is the basis of «communicative» or «discourse ethics,» especially in the version of Jürgen Habermas. See Maureen Junker-Kenny, *Habermas and Theology* (London: T&T Clark, 2011), 81-94; more recently, Maureen Junker-Kenny, *Religion and Public Reason: A Comparison of the Positions of John Rawls, Jürgen Habermas, and Paul Ricoeur* (Berlin: De Gruyter, 2014), 103-83.

[35] For the reflections on the relation of freedom and moral truth, I am indebted to Schockenhoff, *Das umstrittene Gewissen*, 115-33.

tion»³⁶ of moral truth to freedom avoid the pitfall of a possible *reductionism* of moral truth to a pragmatic function of self-preservation and self-interest? The problem is clearly at the heart of the magisterial preoccupation, as the following quotation from *Veritatis Splendor* shows:

> The way in which one conceives the relationship between freedom and law is thus intimately bound up with one's understanding of the moral conscience. Here the cultural tendencies – in which freedom and law are set in opposition to each other and kept apart, and freedom is exalted almost to the point of idolatry – lead to a «*creative» understanding of moral conscience*, which diverges from the teaching of the Church's tradition and her Magisterium.³⁷

One must keep in mind that the term «freedom» does not denote freedom of choice in its empirical facticity, but rather a fundamental disposition toward the good, which, in turn, constitutes the ontological condition for the possibility of each concrete and free choice. We could speak here of *transcendental* freedom, understood as the original source that grounds and sustains empirical freedom. In its dynamic actualization (*Vollzug*), however, freedom is not pitted against its own metaphysical and, ultimately, theological ground, as in the Kantian version of autonomy; rather, it is released to its own identity as love. Each moral decision can be considered good when it mediates the transcendental ground of love upon which it rests; only thus can each human decision articulate, in a historical way, the radical openness (*Vorgriff*) of human freedom to its own *true* fulfillment (*finis ultimus*).³⁸ Freedom expresses the ultimate nature of conscience, and this is why I can say, I *am* my conscience; that is, I am the act of a

---

36 The term «reduction» is, of course, technical, in that it refers to the transcendental movement of grounding.
37 *Veritatis Splendor* 54. Later on in the encyclical one reads: «In their desire to emphasize the "creative" character of conscience, certain authors no longer call its actions "judgments" but "decisions": Only by making these decisions "autonomously" would man be able to attain moral maturity» (55).
38 James F. Kennan, *Goodness and Rightness in Thomas Aquinas's Summa Theologiae* (Washington, DC: Georgetown University Press, 1992). For a reinterpretation of the notion of *finis ultimus* in contemporary moral theology, see Joseph Fuchs, *Moral Demands and Personal Obligations* (Washington, DC: Georgetown University Press, 1993).

radical openness to the source that gives to be, a source giving out of love, and whose call is, in the end, the call to love. Perhaps the Christian *form* of conscience might be properly reformulated in terms of a de-mystification of autonomy: conscience released to its own true self in love, as a freedom beyond autonomy. A conscience grounded in the Christological event turns autonomy into a free service for the good of the other, beyond the calculative self-interest of utilitarian prudence, beyond the determination of self-legislating will. It is an *agapeic* service, which serves the good of the other out of a release of freedom toward the other, a release that is an overflow of generosity.[39]

The systematic reconstruction of the notion of conscience offered above entails implications for Catholic health care. The very meaning of a Catholic normative system for health care, the interplay of norms and experience, and the risky nature of individual decision-making within the framework of a faith-based institution – all these questions will be affected, one way or the other, by the way we understand conscience.

Those who work in Catholic health care, whether professionals or patients, look at the *Ethical and Religious Directives for Catholic Health Care Services* as a basic normative statement. They guide both institutional and personal decision-making not against, but *through* the mediation of conscience. And this is so because their prescriptive character remains intimately tied to the recognition of an agency that commits to them. Unlike traffic signs, or the orders of a military commander, both of which obtain independently of the agent's appropriation of their claim, principles and norms do so because they bind as *moral*, rather than legal, norms.[40] Thus, the *Directives* ought to serve, first and foremost, as a guide in the formation of responsible decision-makers and their conscience. This holds true despite the fact that the *Directives'* reference to conscience is, to say the least, sparse.[41]

---

39   For a demystification of autonomy and the articulation of an ethics of generosity and service in a philosophical perspective, see the work of Leuven philosopher William Desmond, in particular *Ethics and the Between*.

40   The statement in question does not contradict the recognition that the *Directives* are not only morally, but also *legally,* binding within Catholic institutions.

41   At the end of the «General Introduction,» one finds allusion to «a correct conscience based on the moral norms for proper health care.» Later, at the end

A richer understanding of conscience is needed to articulate both the importance of *institutional conscience* and the respect accorded to its *individual* exercise. Trust in the maturity of responsible decision-makers need not be pitted against the clarity of normative provisions. It is precisely the positive interplay of both that nourishes, as Pope Francis reminds us, the paradoxical conviction of Christian ethics: «The centrality of the kerygma ... *should not impose the truth but appeal to freedom.*» Only in freedom can moral norms adequately convey the truth that sustains them. The grammar of love neither stifles the vitality of freedom nor hinders the intelligence needed for a faithful discernment of the good; through the latter, our hearts are open to a generosity mindful of human measure, yet ultimately guided by divine mercy.

---

of the introduction to Part 1 («The Social Responsibility of Catholic Health Care Services»), the *Directives* speak to the fact that...«within a pluralistic society, Catholic health care services will encounter requests for medical procedures contrary to the moral teachings of the Church. Catholic health care does not offend the rights of the individual conscience by refusing to provide or permit medical procedures that are judged morally wrong by the teaching authority of the Church.»

CHAPTER 19
INSIGHTS FOR A MORAL THEOLOGICAL REFLECTION ON CONSCIENCE
Pope Francis Apostolic Exhortation *Amoris Laetitia*

I introduce my topic with a rather long quotation from a document of the International Bioethics Group, dedicated to the sexual abuse crisis in the Church.¹ In it, the authors of the document ask:

> What cognitive and psychological distortions could develop in the abusers such habits of deviant sexuality, which, eventually, make them into predators? What understanding of the moral culpability of their acts? And what of their conscience, so obviously incapable of recognizing the vulnerability of the victims? We feel that a certain way of understanding the morality of actions might be partly responsible for the abusers' inability to perceive the seriousness of their misconduct, together with the devastating psychological, emotional, and spiritual consequences for their victims. We are referring here to a conception of morality that judges the ethical quality of sexual acts in objectivist terms, an objectivism not entirely alien to the traditional understanding of lack of *parvitas materiae in sexto*. If everything in the sexual sphere is equally a sin, i.e., an infraction against the sixth commandment, then, is molesting a minor necessarily more serious than an act of sexual self-gratification, or other violations of chastity? Such an objectivist, i.e. abstract, mindset encourages the perpetrator to «disengage» from his very actions, to «suspend» his own moral agency, and to forfeit full responsibility for the consequences of his acts. The abuser ends up looking at his crime from a distance, so to speak, rather than in a «first person» perspective. He judges his deeds as objectively sinful but silences his conscience, never fully exploring the condition of moral immaturity that has led him to such actions in the first place. Failure to recognize the centrality of conscience as a personal call to responsible actions, and a lack of sensitivity for the consequences of one's actions on the victims, are the inevitable shortcomings of an objectivist perspective in ethics.

---

1   The text of the document, with an introduction to its scope and intentionality, can be found in Marriage, Families & Spirituality 26 (2020): 248-258.

This chapter is an attempt to unpack the premises of such observations, which I take as a kind of heuristic starting point. *Amoris laetitia*, the apostolic exhortation of Pope Francis,[2] offers important insights for a rehabilitation of the notion of moral conscience. Its teaching, though not immediately directed at the predicament of sexual abuse within family constellations, may be of relevance in unpacking the premises of an ill-formed conscience in the abuser, and in bringing to the fore the centrality of personal responsibility for the consequences of one's action. It is such responsibility that seems most lacking in the perception of the person who takes advantage of another's vulnerability.

Furthermore, an emphasis on the education of conscience may help the potential victim to promptly recognize ideological mystifications and justificatory postures in the potential predator. Insofar as the asymmetry of power – of parents over children, priests over underage parishioners, sport coaches over trainees, etc., provides the context for the exploitation involved in sexual abuse, it becomes paramount for the victims to recognize the signs of prevarication, and to discern them in their deeper intentionality. Here the victim's conscience functions as a watchdog, a buffer against any requirement for docile obedience incompatible with the dignity and respect due to a vulnerable human being.

The chapter's emphasis will be on a patient reading of *Amoris laetitia*, rather than a creative interpretation. My intent is to help the interlocutors of the Church's magisterium, whether pastors or lay people, gain a deeper understanding of the document, without forcing upon it too strong an interpretation, driven by a pre-defined agenda. Of course, no view is «a view from nowhere»; thus, the reader will also recognize theological sensibilities and pastoral interests at work in the reflections I will develop. The essay comprises the following parts: a brief entry into the topic; an examination of the theological tradition, which, in turn, leads to the definition of conscience implied by the exhortation; finally, an analysis of what I call «a *praxis* of conscientious decision making» for moral judgment in particular situations.

---

[2] Pope Francis: Apostolic Exhortation, *Amoris Laetitia*, 19 March 2016; hereafter: AL.

The hope is that the analyses in this essay will provide a different moral compass in addressing the issue of sexual abuse within family constellations. To begin, a rehabilitation of conscience invoked by Pope Francis might offer a deeper understanding of the place of personal responsibility, the relevance of consequences in the moral judgment on actions, and the centrality of honest transparency in discerning the roots of personal evil.

Prolegomena to an Understanding of Moral Conscience

The philosophical tradition, at least since Aristotle, speaks of virtue as a mean between extremes. To be moral is to discern «the middle», to be able to respond to the call of values with an answer that stirs the depths of a person's moral capacities, equally far from either the fancy of impossible idealism or the complacency of cheap pragmatism. Both attitudes ultimately stifle the energy of resilience, driving moral imagination into a resignation unwilling to dream, because it is unable to cope.

To appreciate the concreteness of such a «virtuous middle» implies neither giving up on the absoluteness of values, nor forfeiting the seriousness of the moral call. For theological, no less than for philosophical, ethics, «the morally true is also always the morally possible», as the eminent ethicist Klaus Demmer suggests, echoing, with a different terminology, the Augustinian *Deus impossibilia non jubet*: «God does not command the impossible».[3]

Yet, only the eyes opened by grace can bear the vision that liberates our freedom into the generosity of love. The moral call, in all its concreteness and urgency, is born of a contemplative outlook. This is why *Amoris laetitia* begins with an invitation, not a

---

3   The entire quotation reads: «Non igitur Deus impossibilia iubet: sed iubendo admonet, et facere quod possis, et petere quod non possis»: «God does not command the impossible, but by his commandments he counsels to do what you can and to pray for his aid in that which you cannot do.» Augustine: *De natura et gratia*, 43 (50); English translation in Augustine: *Four Anti-Pelagian Writings*, trans. J.A. Mourant/W.J. Collinge, Washington, DC: The Catholic University of America Press, 1992, 60. Klaus Demmer's statement can be found in «Sittlich Handeln aus Erfahrung», in *Gregorianum* 59 (1978), 677.

command, gently directing our gaze to the biblical account of love and family life.[4] Commenting on the portrayal of family in Psalm 128, Pope Francis invites us to «cross the threshold of this tranquil home, with its family sitting around the festive table» (AL 9), and to reflect on the «majestic early chapters of Genesis», which «present the human couple in its deepest reality» (AL 10). Later on, the gaze will repose on Jesus's teaching, and the amazing hymn to love in Saint Paul's first letter to the Corinthians (1 Cor 13). What the contemplative outlook becomes able to see is the *mystery* at the heart of human love: its openness to the depth of God himself, its grace-given capacity to become the «symbol of God's inner life», a communion of love, of which the couple and the family are a «living reflection» (AL II).

However, to be true to reality, contemplation cannot simply indulge in the idyllic picture, when «the presence of pain, evil, and violence that breaks up families and their communion of life and love» (AL 19) run through the history of salvation, indeed through our very own history, a veritable «thread of suffering and bloodshed» (AL 20). The Christian vision is one of *realism*, because it reflects the logic of incarnation. It emerges from a more complex intermediation with the frailty of individuals, the struggles of couples and families, the disappointments that end in separation and distance from one another. Thus, the initial admission of Pope Francis, stunning for an official document of the Church, stands almost like an admission of failure, encouraging a different pastoral approach:

> We need a healthy dose of self-criticism … At times we have proposed a far too abstract and almost artificial theological ideal of marriage, far removed from the concrete situations and practical possibilities of real families. This excessive idealization, especially when we have failed to inspire trust in God's grace, has not helped to make marriage more desirable and attractive, but quite the opposite. (AL 36)

What should such an alternative approach look like? Moreover, how could it be articulated, when it means, obviously, something other than surrender to moral defeatism, often hiding behind the call to «keep up with the times»? If there is «another» route, it would

---

[4] One should not forget that the document itself is called an «exhortation»!

have to be the *only one* available to the Church, i.e., the route that «clearly reflects the preaching and attitudes of Jesus, who set forth a demanding ideal yet never failed to show compassion and closeness to the frailty of individuals, like the Samaritan woman or the woman caught in adultery» (AL 38).

Not surprisingly, in his exhortation, Pope Francis calls for a rehabilitation of the notion of conscience. I say not surprisingly, because the tension between extremes, always shadowing the «virtuous middle», can only be resolved by a mature moral conscience. Summoned to rehabilitation is, first of all, the conscience of the Church as a whole, and its ability to convey to the world the gospel's message of love. The Church is the community that exists in the *anamnesis* of the Christ event, a memory that is living rather than stifling, triggering a constant process of interpretation and actualization of the gospel's message. As it journeys through history, the Church faces the challenge of remaining faithful to its origins. Such faithfulness, however, is more than a nostalgic looking back: it is a process of creative development that, guided by the Spirit, keeps the conscience of the Church alert to the «signs of the times» and the change in historical circumstances.

Maturity of conscience is invoked also as a necessary condition for pastors to confront the existential predicament of the faithful and the concrete experiences of couples and families. In particular, «while clearly stating the Church's teaching, pastors are to avoid judgments that do not take into account the complexity of various situations, and they are to be attentive, by necessity, to how people experience and endure distress because of their condition» (AL 79). A different attitude is clearly called for, one that gives the moral conscience of lay people a more robust recognition: «We also find it hard to make room for the consciences of the faithful, who very often respond as best as they can to the Gospel amid their limitations, and are capable of carrying out their own discernment in complex situations.» (AL 37) The quotation ends with the extraordinary statement: «We have been called to form consciences, not to replace them.» (AL 37)

No one can deny that the task entrusted to conscience, whether the conscience of the Church as a whole, or the individual conscience of pastors and of the laity, is a complex one. For Christian married couples, the voice of conscience reverberates in the call to weave the *form* of their identity, defined by the sacrament of marriage they have received, into the tapestry of ordinary life, welcoming unexpected challenges no less than undeserved blessings. Thus, the Church «cannot encourage a path of fidelity and mutual self-giving without encouraging the growth, strengthening, and deepening of conjugal and family love. Indeed, the grace of the sacrament of marriage is intended before all else "to perfect the couple's love".» (AL 89)

A «work of love» involves «an intuition that can enable us to hear without sounds and to see the unseen» (AL 255), thus spiritual discernment is needed, and the recognition that Christian life unfolds in the delicate balance of giving and receiving. Only a conscience mindful of the gift received can sustain husband and wife in the fundamental option to love, building in to the courage of daily choices the life-long commitment that stands the test of time.

Defining Conscience: The Legacy of the Moral Theological Tradition

I spoke of a «rehabilitation of conscience» in *Amoris laetitia*. Does such language entail the implicit overcoming of a past amnesia, perhaps even of a willing undermining?[5] We need to wait for a complete judgment on the issue. Let us notice, to begin with, the abundance of references to the notion of conscience in the document. All of them stand against the background of the warning Pope Francis raised in his first apostolic exhortation, *Evangelii gaudium*: «The great danger in today's world, pervaded as it is by consumerism, is the desolation and anguish born of a complacent yet covetous heart, the feverish pursuit of frivolous pleasures, and a blunted conscience.»[6]

---

[5] John Paul II's encyclical *Veritatis splendor* speaks critically of more recent theological developments on the topic, as promoting, in the pope's view, a «creative» notion of conscience. «There seems to be an emphasis within *Veritatis splendor* on the suppression of conscience and a move of power toward the magisterium»; J. Hoose: «Conscience, in *Veritatis Splendor* and the *Catechism*», in C.E. Curran (ed.): *Conscience*, New York: Paulist Press, 2004 (Readings in Moral Theology; 14), 89.

[6] Pope Francis *Evangelii gaudium* (*The Joy of the Gospel*), 24 November 2013, n. 2.

But what is the opposite of a *blunted* conscience? It is a wakeful conscience, a conscience defined by vigilance and openness to the movements of the Spirit. Here a broad definition is implied. Conscience is not a «thing» in the person, a «function» among others, but the very center, the most pristine source of her decision making, in which the person is confronted with her deepest moral identity before God (*coram Deo*). This is true, first, in relation to decisions that have already been made: what we have done, if wrong, stirs in us a sense of guilt and shame, even a sense of personal defeat; on the other hand, doing what is right makes us happy, proud of ourselves, not because of what we have achieved – the language of moral achievement, in a Christian perspective, turns automatically into the language of grace – but because, through the action in question, we have grown in communion with the good, taking yet another step in the direction of what we are called to be in truth.

When looking at it, somewhat superficially, conscience seems *purely* like a «mechanism» of praise or reproach, and not without reason the moral theological tradition speaks of conscience as a *judgment*: the «ultimate practical judgment on the act that has been made, or the act to be made».[7] Though not wrong in itself, the definition in question is, nonetheless, insufficient. Its shortcomings are also consistent with the main features of the traditional paradigm of moral theology up to the Second Vatican Council: the restriction to an act-centered morality, the reduction of conscience to a purely subjective function, and the overall syllogistic character of moral reasoning, understood as a deductive conclusion from rigidly defined premises. I want to mention briefly these three characteristics of pre-Vatican II moral theology to show the relevance of the council's turning point, whose conceptual achievements are then integrated in the broader vision of *Amoris laetitia*. To begin, in the traditional paradigm of neo-scholastic moral theology, the focus is on the act considered in isolation from the contextual features of the acting person. When expressing a judgment on the moral quality of an action, attention is paid to the action «in itself», so to speak, and only secondarily

---

7   For a discussion of the definition, see the classic work of B. Schüller: *Die Begründung sittlicher Urteile*, 2nd ed., Düsseldorf: Parmos Verlag, 1980, 45.

to other dimensions, such as the intentionality of the agent, or the specific circumstances defining the context in which the action takes place. More broadly, the tendency in this paradigm is to undermine the historical character of the action, its place within the trajectory that distinguishes the life of the agent, his/her concrete existential predicament.

Because, detached from these ethically relevant markers, the object of the action, i.e., what defines its rightness or wrongness, seems to be determined purely by its congruency with the existing norm. Consider the case of sexual morality: any act that is not conducive to the goal envisioned by the norm, i.e., the end of procreation to be pursued within marriage, is deemed immoral in itself, by its very nature. Furthermore, no attention is paid to the circumstances that might influence the moral judgment on the act.

According to the traditional manuals of moral theology, when it comes to sexuality, a realm «covered» by the treatise on the sixth commandment, every deviation from the norm is a mortal sin (*de sexto non datur parvitas*).[8] Even for Thomas Aquinas, the spouses that engage in sexual relations for reasons other than to pursue the end of procreation, or to fulfill the conjugal debt, *always* commit sin.[9]

This position, further documented in the theological literature, finds expression in magisterial teaching. In his 1930 encyclical on Christian marriage, *Castii connubii*, Pius XI unequivocally states:

> Since, therefore, openly departing from the uninterrupted Christian tradition some recently have judged it possible solemnly to declare another doctrine regarding this question, the Catholic Church, to whom God has entrusted the defense of the integrity and purity of morals, standing erect in the midst of the moral ruin which surrounds her, in order that she may preserve the chastity of the nuptial union from being defiled by this foul stain, raises her voice in token of her divine ambassadorship and through Our mouth proclaims anew: any use whatsoever of matrimony exercised in such a way that the act is deliberately frustrated in its natural power to generate life is an offense against the law

---

[8] For an example, see the widely used manual of A. Koch: *Lehrbuch der Moraltheologie*, 7th ed., Freiburg: Herder, 1907, 634.

[9] «Unde, quando conjuges convenient causa procreandae, vel ut sibi invicem debitum reddant, quae ad finem pertinent, totaliter excusandem a peccato», Thomas Aquinas: *IV Sententiarum*, dist. 31, q. 2, a. 2.

of God and of nature, and those who indulge in such are branded with the guilt of a grave sin. (56)[10]

What becomes of conscience in such a paradigm? Since only the norm determines the rightness of an action with respect to its object (*finis operis*), conscience inevitably assumes a secondary importance. The motivation and intention (*finis operantis*) emerging in the conscience of the agent are now reduced in their meaning to «psychological», rather than strictly «ethical», qualifiers.[11] The result is an abstract rendition of the moral event, and the relegation of conscience to a purely subjective, rather than objective, determinant.

The reduction entailed by the traditional understanding of conscience will become clearer as my argument moves closer to a more systematic definition. For now, suffice it to say that the «anthropological turning point» in moral theology, that is, the emergence of a different focus on subjectivity, and the recognition of the centrality of the person as a moral agent, will bring about a renewed attention to conscience as a dimension of self-understanding.[12] In conscience, the moral subject comes to himself as a historical being; thus, conscience becomes relevant not only as the *locus* of moral judgment, but, more broadly, as the hermeneutical context in which values are being perceived and interpreted. The traditional understanding of sexuality conveyed by both the manuals of moral theology and the magisterial

---

10 For a reconstruction of the history of Church doctrine before *Amoris laetitia*, see S. Goertz/C. Witting (eds.): *Amoris laetitia: Wendepunkt für die Moraltheologie?* Freiburg: Herder, 2016.
11 Certainly, the traditional manuals discuss possible extenuating circumstances to the culpability of the agent, as far as defects concerning voluntariness, for example, with reference to the elements of fear, manipulation, phobia, etc. A robust analysis of the anthropological meaning of the action, as defining the reality of the moral subject in his/her existential totality is, however, absent. In this sense, the old manuals of moral theology operate with different methodological premises. Interestingly, *Amoris laetitia* speaks of the difference between «voluntariness» and «freedom» (AL 273).
12 For a recent account, see R. Dell'Oro: «Conscience after Vatican II: Theological Premises for a Discussion of Catholic Health Care», in: D. De Cosse/T. Nairn (eds.): *Conscience and Catholic Health Care: From Clinical Contexts to Government Mandates*, Maryknoll, NY: Orbis Books, 2017, 33-48. Also, R. Dell'Oro: «Moral Truth and Anthropological Mediation in the Theological Ethics of Klaus Demmer», in: *Gregorianum* 98/I (2017), 37-50.

statements become increasingly problematic not only because they are detached from a more holistic account of agency but also because they are progressively unable to integrate a vision in which sexuality is no longer seen as exclusively directed at procreation.[13]

One final observation: the neo-scholastic tradition rests on an understanding of moral reasoning in the form of a «top-down» process, that is, of a deductive application of higher principles to individual cases. The higher principles of morality, which are also the most abstract and general, such as the first principle of practical reason, «do good and avoid evil», define the realm of *synderesis*, or «original conscience». Individual cases and situations, on the other hand, open up the space in which so called «conscience in situation» operates. Original conscience and conscience in situation have different degrees of certainty. *Synderesis* is absolute and infallible; therefore, it cannot err. This «infallibility» of original conscience speaks to the fact that we are given to be into the difference of good and bad, right and wrong, a difference we cannot produce, only receive. Conscience in situation, on the other hand, carrying out the operations of practical reason through the application of the first principles to individual cases, is inevitably open to the possibility of mistakes.

When it comes to articulating the relation between *synderesis* and conscience in situation, the traditional manuals of moral theology adopt a notion of application that is virtually equivalent to the conclusion of a syllogism: the major premise expresses the moral law, the norm; the minor refers to the act to be done in the particular situation, to which the law is supposed to obtain; the conclusion states whether the action comes under the moral law. Clearly, this type of reasoning is a form of deductive syllogism.[14]

---

13 A more personalistic vision of sexuality, already anticipated before the Second Vatican Council in the work of ethicists such as Dietrich von Hildebrand and Herbert Doms, found its full recognition in the statements of *Gaudium et spes*.

14 See C.E. Curran: «Conscience in the Light of the Catholic Moral Tradition», in: C.E. Curran (ed.): *Conscience*, 3-24. The difference between Thomas and the neo-scholastic tradition should not go unnoticed. Although the latter developed out of the former, the two differ on important points, especially with regard to the character of the judgment of conscience in question. The neo-scholastic tradition develops in dialogue with a modern conception of ethics defined by a geometric model (*more geometrico demonstrata*). On this, D. Lachterman: *The Ethics of Geometry: A Geneology of Modernity*, New

Toward a Broader Definition of Conscience

We may ask whether such a paradigm is able to convey the richness entailed by the notion of conscience. To begin, I want to bring attention to the fact that the traditional definition points not only to the past («a judgment on the act that has been made») but to the future as well («a judgment on the act *to be made*»). This means that, in conscience, not only do we look back at what we have done; we also envision possible decisions to make. In so doing, conscience is already *beyond* the judgment on a particular act: it calls into question the way in which the character of the agent, her history and complexities, are implicated in the moral event.

To reduce the rightness of an action (past or future) to its conformity to the law would entail a truncated version of morality that abstracts the action from its agent, and brackets considerations that are relevant precisely as conditions to understanding the moral subject's responsibility for the action in question. In the end, an action can be considered *right* when it articulates the *goodness* of the acting person, her moral identity as a whole, one that will never be fully exhausted by simple correspondence to the law.[15] Thus, *Amoris laetitia* suggests, «it is reductive simply to consider whether or not an individual's actions correspond to a general law or rule, because that is not enough to discern and ensure full fidelity to God in the concrete life of a human being» (AL 304).

The place of *time*, and the dialectical relation of past and future in ethical decision making, is especially relevant here.[16] Consider

---

York: Routledge, 1989. Thomas Aquinas, on the other hand, distinguishes *determination* from *conclusio* in *Summa Theologiae* I-IIae, q. 95, a. 2. See E. Schockenhoff: *Das umstrittene Gewissen: Eine theologische Grundlegung*, Mainz: Grünewald, 1990, 77-82. Thomas underscores the autonomy of *practical reason*, and the singularity of practical judgment in relation to theoretical reason. On this, see for example M. Rhonheimer: *Natural Law and Practical Reason: A Thomist View of Moral Autonomy*, New York: Fordham University Press, 2000.

15   In his work, James Keenan has successfully demonstrated that this is precisely how Thomas Aquinas understands the matter in question. See J. Keenan: *Goodness and Rightness in Thomas Aquinas' Summa Theologiae*, Washington, DC: Georgetown University Press, 1992.

16   See K. Demmer: *Living the Truth: A Theory of Action*, Washington, DC: Georgetown University Press, 2010, especially 85-149. More broadly, and

the following: when we foresee our decisions, we confront not only the persons we are, but also the persons we will become. In this movement of the spirit, in the *nexus* of intentionality and motivation that triggers our choices, our *real* selves are already transcended toward the *ideal* selves ahead of us. The moral realism mentioned above consists precisely in the wisdom to measure the span of this transcending movement, the ability to maintain in us the continuity between past and future.[17]

In the tension that impels us toward a new horizon, we cannot bracket the history of decisions that has already defined us, making us, for good or bad, the persons we are. This is why conscience does more than take responsibility for the «here and now» of compliance with the law: «conscience can do more than recognize that a given situation does not correspond objectively to the overall demands of the Gospel. It can also recognize with sincerity and honesty what for now is the most generous response which can be given to God, and come to see with a certain moral security that it is what God himself is asking amid the concrete complexity of one's limits, while yet not fully the objective ideal.» (AL 303) The substance of this last statement is further buttressed by an equally relevant conclusion: «In any event, let us recall that this discernment is dynamic: it must remain ever open to new stages of growth and to new decisions which can enable the ideal to be more fully realized.» (AL 303)

A definition of conscience, implicit in the last quotation, seems to emerge. The reference to «discernment», «stages of growth», together with the recognition of the «concrete complexity of one's

---

with a thorough articulation of the implications of time consciousness for action theory, G. Höver: *Sittlich handeln in Medium der Zeit: Ansätze zur Handlungstheoretischen Neuorientiering der Moraltheologie*, Würzburg: Echter, 1988, especially 200-218 and 253-273. To my knowledge, Höver's reflections have remained without significant echo in moral theology.

17 William Desmond has beautiful observations about the relevance of memory in balancing the tension between desire and imagination: «Self-transcendence, when equivocal ... can be in flight from self, and not towards the other either. We dissipate the original energy of our being because we cannot face the selves that we are in promise. We need to remember who we are in our self-transcendence. What this means is that *memory* is needed to balance desire and imagination, and to ballast the quest for self-transcendence. Memory must mediate the metaphors of the equivocal self.» (W. Desmond: *Being and the Between*, Albany, NY: SUNY Press, 1995, 393).

limits», makes clear that, in the perspective of *Amoris laetitia*, conscience ought to be understood more as an *experience* than a judgment.[18] Although it never offers a precise definition, one could summarize the exhortation's understanding of conscience as the «experience of responding to the moral call in freedom». Let me parse out the individual elements of such definition.

1. *The Place of Experience*

A first element: experience points to a journey (see the root of the Latin word for experience, i.e., *experior*), and presupposes a notion of moral identity that is dynamic, rather than static, one that encompasses the entire life history of the person, thus framing the judgment on a particular action or decision in light of a broader trajectory. Our response to the challenge disclosed in particular moral calls, and to the moral norms that mediate such calls, progressively build in us a moral identity. Who we are morally is both what we have been and what with God's grace we intend to be; and yet, «each person's situation before God and their life in grace are mysteries which no one can fully know from without» (*Evangelii gaudium*, 172).

We all come from afar, so to speak, each of us with a different life narrative that comprises failures as well as successes, the recognition of God's providential presence, no less than the anguish at being abandoned by him. Desolation, the feeling of moral defeat, perhaps even the loss of faith, are part of this journey, almost like necessary ingredients of the testing it entails (the Latin *experior* also points to the Greek word *peira*, which means trial!)

---

18 There is here a noticeable development with respect to John Paul II's tendency to look somewhat skeptically at any attempt to expand the notion of conscience beyond practical judgment. In the encyclical *Veritatis splendor* one sees a kind of alternating between the language of traditional neo-scholastic theology, and the one that finds more obvious recognition in the magisterium of Pope Francis: «The judgment of conscience is a *practical judgment* ... It is a judgment which applies to a concrete situation the rational conviction that one must love, do good and avoid evil. The first principle of practical reason is part of the natural law ... but whereas the natural law discloses the objective and universal demands of the moral good, conscience is the application of the law to a particular case ... Conscience thus formulates *moral obligation* in the light of the natural law» (*Veritatis splendor*, 59).

In this journey, conscience grows *with* us, or better, *in* us, never ceasing to alert us to the signs of hope God mysteriously disposes on our path, warning of the dangers, and gently moving us past our failures, a voice of reproach at times, of consolation always. In conscience we truly are «at home with ourselves»,[19] and come to know ourselves in the deepest sense. Perhaps we might even say that we do not simply *have* a conscience, we *are* one.

2. The Dialogical Character of Conscience

A second element: conscience is the experience of *responding to the moral call*. Knowing about ourselves is always a «knowing with» (*conscientia* comes from *cum scire*). Thus, conscience is never just a monologue; in conscience we are summoned to the call of the good, which, ultimately, elicits our responsibility for the other, not only to the autonomy of the self.[20]

Conscience entails a «responsorial» quality, a coming to oneself out of a dialogue with another. I would qualify the otherness in question as twofold. First, we come to grasp the deepest roots of our moral identity in dialogue with God; secondly, we do so out of a dialogue with others.

We speak of the «voice» of conscience as the voice of God himself. With a beautiful text of the Second Vatican Council, we can say that «conscience is the most secret core and sanctuary of a person. There each one is alone with God, whose voice echoes in the depths of the heart.» (*Gaudium et spes*, 16) Such is «the dignity of conscience» (*Gaudium et spes*, 17), expressing itself in the right of the individual «not to be forced to act against conscience, nor be prevented from acting according to conscience» (*Dignitatis humanae*, 3).

Amoris laetitia's broad vision of conscience is deeply rooted in the lessons of the Second Vatican Council. In particular, the latter brought out a more anthropologically sophisticated understanding of human selfhood, and the full theological implications of a notion

---

19   Thomas Aquinas speaks of a «return to onself», a *reditio ad seipsum*.
20   On this, see the reflection of Jewish philosopher E. Levinas, especially in *Totality and Infinity: An Essay in Exteriority*, Pittsburgh, PA: Duquesne University Press, 1969.

of conscience that transcends the judgment on individual actions in relation to the norm. Retrieving the biblical notion of «heart», the council alludes to a complex set of dimensions that synergistically bring together knowledge, decision, and power of ratification by a moral agent.[21] Furthermore, according to authoritative commentators, the heart metaphor reconnects the theology of conscience to the mystical tradition, especially Meister Eckhart's reference to the «ground of the soul» (*Seelengrund*).[22]

The second dimension of the dialogue in question alludes to the intersubjective character of conscience. Already in the texts of the council one finds reference to the dialogical quality of conscience, for an example, when *Gaudium et spes* 16 encourages a more inductive, rather than deductive, path in the discovery of moral truth, and calls for the collaboration of Christians with others in society in the search for common values and criteria for action.

I think *Amoris laetitia* goes even further, stressing the dialogical character of moral truth in the very process that leads the Church to the formulation of its own doctrine. The following quotation, at the beginning of *Amoris laetitia*, is most telling: «I would make it clear that not all discussions of doctrinal, moral or pastoral issues need to be settled by interventions of the magisterium. Unity of teaching and practice is certainly necessary in the Church, but this does not preclude various ways of interpreting some aspects of that teaching, or drawing certain consequences from it.» Later in the same paragraph, Pope Francis speaks of different experiences that might be gathered from local churches, which, in turn, can translate into a pluralized spectrum of solutions: «Each country or region, moreover, can seek solutions better suited to its culture and sensitive to its traditions and local needs.» (AL 3)

The moral doctrine of the Church is not an abstract system of principles to be applied univocally across the board, once and for all, by means of a top-down approach. Moral norms must be understood as *interpretive* in nature, that is, as an attempt to articulate the

---

21 For a philosophical anthropology inspired by the «heart tradition» see A. Tallon: *Head and Heart: Affection, Cognition, Volition as Triune Consciousness*, New York: Fordham University Press, 1997.
22 K. Golser: «Das Gewissen als verborgenste Mitte im Menschen», in W. Ernst (ed.): *Grundlagen und Probleme der heutigen Moraltheologie*, Würzburg: Echter Verlag, 1989, 113-137, at 116.

main insights of the Christian gospel – for example, the indissolubility of marriage, the centrality of love at the heart of the couple, and the fruitful nature of such love – with the lived experience of the faithful in the journey of history, and the particularity of circumstances and contexts entailed by such a pluralized experience.

Ultimately, the doctrine of the Church witnesses to a moral truth that is symphonic in character; in fact, the more rich and original the more it remains hospitable, like in a musical score, to the individuality of different instruments, the particularity of their contribution to the whole, and the synergistic character of their interaction with one another.[23]

3. The Mediating Function of Freedom

A final element in the definition of conscience points to the experience of responding to the moral call *in freedom*. Morality is typically an exercise in freedom. Think of the difference between consenting to the appeal of a moral value and the response to a legal injunction. Whether I want it or not, I have to stop at a red sign, file my tax return on time, repay my debts, etc. In principle, except in the case of extenuating circumstances, I am not free to choose my options in regard to the law. The latter imposes itself upon me quite independently of the way I feel about it. We say that the law is *heteronomous* in nature.[24]

With moral values it is an entirely different situation. My response to them, though felt as an obligation, i.e., as something I *ought* to do, energizes in me the freedom to consent to their appeal or, conversely, to close myself to it. In a way, my freedom remains latent in me, up to the point when, stirred by the call of values, it is released into the possibility of its full actualization. It is because

---

23 On this, see the classic work of H.U. von Balthasar: *Truth is Symphonic: Aspects of Christian Pluralism*, trans. Graham Harrison, San Francisco: Ignatius Press, 1987.

24 The point should not be absolutized for, after all, there is also such a thing as a «conscientious objection» to an unjust law. But this, in a way, further stresses the logic of the point in question, for it shows that the relation to freedom is inescapable. Perhaps it is more correct to say that law and morality appeal to the freedom of the individual differently, although the latter in a more perfect way than the former.

moral values appear on the horizon of my conscience that I discover myself to be free, growing in that freedom through the act of my responding. Any «value response» (the phenomenologists speak of *Wertantwort*) can emerge only as an act of «correspondence» with the values disclosed by experience, a transcending of the person toward their importance and meaning out of an act of *autonomous* recognition.[25]

The ultimate identity of the person, what or who a person wants to be, is «the very theme», that is, what is most profoundly at stake, in the moral situation: although some moral obligations are also legal in nature, and thus carry judicial consequences, it is not the sanction entailed by the latter that identifies their central phenomenological character. Rather, it is the fact that, *qua* obligations, they appeal to the core of our freedom.[26] I ask: is freedom the equivalent of unbridled subjectivity, or a form of idiosyncratic license; is it, in the end, only the subtle pretext for doing whatever one wants? If not, what does freedom mean in ethics?

Generally, we think of freedom as the ability to choose this or that, based on a retreat from the immediacy of what is offered to us as an object of choice. In regaining something like a condition of neutrality or distance from all the available options, we hope to better envisage their discrete goodness, and to commit to one of them in the choice we make. Think about the selection of food in a cafeteria: one takes a step back from the array of available choices, thus gaining, in the act of withdrawing from the proximity of pluralized offerings, an «objective» outlook, i.e., the ability to measure

---

25 The entire phenomenology of moral experience articulated by the Christian philosopher D. von Hildebrand goes in this direction. See, most systematically, his *Ethik* [GW II] (Stuttgart: Kohlhammer, 1973), 2nd edition. On «value response», J. Seifert: «Dietrich von Hildebrands philosophische Entdeckung der "Wertantwort" und die Grundlegung der Ethik», in *Truth and Value: The Philosophy of Dietrich von Hildebrand*, Bern: Peter Lang, 1992, 34-58. For a critical assessment of von Hildebrand's phenomenological account, see R. Dell'Oro: *Esperienze morale persona: Per una reinterpretazione dell'etica fenomenologica di Dietrich von Hildebrand*, Roma: Editrice Pontifica Università Gregoriana, 1996.
26 In his *Grounding to the Metaphysics of Morals* Immanuel Kant distinguishes between the law as the condition whereby we come to know of a moral obligation (*ratio cognoscendi moralitatis*) and freedom as the ontological condition of morality itself (*ratio essendi moralitatis*).

the correspondence between what is being offered and what one really desires. A certain distance from reality, absence of external influences, and neutrality vis-à-vis constraining preferences are the marks of freedom in this definition. Freedom thus understood looks more like a kind of indeterminacy, a negative separation from the object of choice, rather than a response to its value and meaning. Yet, why would anything be worth choosing, if not because it presents itself as an object of importance, a source of fulfillment and happiness we recognize? What if the search for the good had already taken place in us, almost behind our back, before we set out to choose anything?

To understand the way in which *Amoris laetitia* speaks of freedom, one has to recognize the constitutive relation of freedom to the good as such, a relation that precedes and grounds our determining this or that thing as a *good for us*.[27] The point is made in a seemingly unrelated passage of the document, one that speaks of life in the family, yet it touches prominently on the issue at stake: «We are all sons and daughters. And this always brings us back to the fact that we did not give ourselves life but that we received it. The great gift of life is the first gift that we received.» (AL 188) Yes, we are already born into the goodness of creation, often unaware of the grace that grounds its always enigmatic presence, a presence that might be taken for granted, if being is, yet need not be. To recognize something as good is to allow ourselves to be swept up in the mystery of goodness in which we dwell, and to which the particular object of our choice refers. We inhabit a good that is *given* to us, graced by the glory of the morning sky, the splendor of the summer colors, the beauty of a caring smile, before we set out to choose a particular good in an act of self-determination.

Freedom, when considered in its deepest meaning, must be more than freedom of choice, because in our power of self-determination we actualize an ultimate decisionality *toward* and *for* the good; better, we are *freed* into the goodness of the whole that calls us to itself, released into something, rather than separated for the sake of a

---

27 The point is metaphysical, in that it concerns the grounding character of the good with respect to being as such. On the «metaphysical» character of the good, see especially W. Desmond: *Ethics and the Between*, Albany, NY: SUNY Press, 2001.

better option contingent upon our own determining. Through good choices, we grow into who we truly are, who we are meant to be.

The Second Vatican Council, especially in *Dignitatis humanae*, the *Declaration on Religious Freedom*, leads the Church to better understand that it is the mystery of freedom that defines the depth of conscience: it is a freedom that calls for respect for the person choosing, and reverence that abstains from manipulation and coercion: «in all his activities a man is bound to follow his conscience faithfully ... It follows that he is not to be forced to act in a manner contrary to his conscience.» (*Dignitatis humanae*, 3)

Following in the same spirit, Pope Francis reminds us that freedom is the condition for our relation to truth.[28] Consider the way he speaks of the ethical education of children: «Freedom is something magnificent, yet it can also be dissipated and lost. Moral education has to do with cultivating freedom ... Human dignity itself demands that each of us "act out of conscious and free choice, as moved and drawn in a personal way from within".» (AL 267)

In his first exhortation, Pope Francis also spoke eloquently about evangelization as an appeal to the freedom of conscience: «The centrality of the kerygma calls for stressing those elements which are most needed today: it has to express God's saving love which precedes any moral and religious obligation on our part; *it should not impose the truth but appeal to freedom*; it should be marked by joy, encouragement, liveliness ... All this demands on the part of the evangelizer certain attitudes which foster openness to the message: approachability, readiness for dialogue, patience, a warmth and welcome which is non-judgmental.» (*Evangelii gaudium*, 165).[29]

If freedom is nourished by a more original relation to the good, it is not a question of thinking of the moral call as oppressive, as thwarting one's expectations for happiness. Ethical truth is always the promise of existential possibilities; rather than an oppressive burden, it is an empowerment to act, leading to ever greater freedom.[30] To see the truth as an appeal to freedom is the condition for

---

28 The relation of freedom to truth constitutes a central theological *topos* of John Paul II's magisterial teaching. The emphasis here is on the correspondence of freedom to truth. For an excellent systematic analysis of the relation, see E. Schockenhoff: *Das umstrittene Gewissen*, 115-133.
29 The emphases in the text are mine.
30 K. Demmer: *Living the Truth*, II.

recognizing that what may come across as an obligation, as the call of duty, is in fact an offer of fulfilment, the response to a seeking that dwells already in us like an indeterminate cry,[31] a promise rather than a demand.

Think of the meaning of fidelity, and the promise of total self-giving spouses vow to one another in the sacrament of marriage. Is such promise burdensome? Certainly, but it is also fulfilling, in that through it, two persons become what they are bound to be in love, one in the flesh. Thus, in speaking of the indissolubility of marriage, *Amoris laetitia* says: «The indissolubility of marriage ... should not be viewed as a "yoke" imposed on humanity, but as a "gift" granted to those who are joined in marriage ... God's indulgent love always accompanies our human journey; though grace, it heals and transforms hardened hearts, leading them back to the beginning through the way of the cross.» (AL 62)

Toward A Praxis of Conscientious Decision Making

So far, I have sketched out something like a «pre-understanding», a set of premises that are relevant to unpacking the notion of conscience in *Amoris laetitia*. To begin, such a pre-understanding is important in itself, for purely theoretical reasons: it points to a sophisticated and articulate conceptual framework by a magisterial document, which seems to integrate the lessons of the tradition, especially of the Second Vatican Council, and of the «effectual history» the latter has triggered in both theological reflection and the life of the Church.

A pre-understanding is indeed a way of looking at things, but it is also a criterion for dealing *practically* with reality. In this final part of the article, I look at the way in which the notion of conscience emerging in the document guides the Church's approach to the complexity of moral life in the realm of marriage and family. I ask: how can the Church recast what is essential in the gospel's message of love? How is it to do so, when «the message we preach runs a greater risk of being distorted, or reduced to some of its secondary aspects. In this way certain issues which are part of the Church's

---

[31] Remember Saint Augustine's statement that our heart is restless until it rests in God (*Confessions*, I.I).

moral teaching are taken out of the context which gives them their meaning. The biggest problem is when the message we preach then, seems identified with those secondary aspects which, important as they are, do not in and of themselves convey the heart of Christ's message.» (*Evangelii gaudium*, 34).

Building on the premises of his previous exhortation, Pope Francis further asks in *Amoris laetitia* how the Church can unpack, once more, her treasure of hope and joy not only for those who already witness with their lives to its timeless credibility but also for those who live, so to speak, on the margins, struggling to find their way to the truth, when burdened by feelings of exclusion, and the perception of being shunned or cast out of the community. For many, in and outside the Church, the path taken leads nowhere, the road back appears closed, even faith seems finally lost.

The response to these difficult questions depends, among other things, upon a certain understanding of the Church. Pope Francis reminds us of what we might have forgotten, namely, «that the Church's task is often like that of a field hospital» (AL 291). This is certainly not inconsequential: images yield insights, which, in turn, offer food for thought. Thus, it is perfectly logical for the image of the Church as a field hospital, or as a mother «who always does what she can, even if in the process, her shoes get soiled in the mud of the street» (*Evangelii gaudium*, 45) to generate an ecclesiology of inclusion and mercy.

Two alternative ecclesiological models seem to be available in this situation, and *Amoris laetitia* is very clear about the path the Church must take:

> Here I would like to reiterate something I sought to make clear to the whole Church, lest we take the wrong path: «There are two ways of thinking which recur throughout the Church's history: casting off and reinstating. The Church's way, from the time of the Council of Jerusalem, has always been the way of Jesus, the way of mercy and reinstatement ... The Way of the Church is not to condemn anyone forever; it is to pour out the balm of God's mercy on all those who ask for it with a sincere heart ... For true charity is always unmerited, unconditional and gratuitous.» Consequently, there is a need «to avoid judgments which do not take into account the complexity of various situations» and «to be attentive, by necessity, to how people experience distress because of their condition» (AL 296).

The «logic of integration» (AL 299) and the «invitation to pursue the *via caritatis*» (AL 306) are more than rhetoric, or worse, an attempt by the Church to stop the hemorrhage of people leaving its tents, increasingly dissatisfied with a doctrinal system they feel unable to reconcile with their own existential predicament. Furthermore, to say that «such persons need to feel not as excommunicated members of the Church, but instead as living members» (AL 299) cannot be seen as just another strategic move dictated by pastoral necessity either; rather, it is a way to remain faithful to the truth of the Christ event, understood as the final and unequivocal disclosure, indeed, the epiphany of God's love in the midst of history: «We put so many conditions on mercy that we empty it of its concrete meaning and real significance. That is the worst way of watering down the Gospel. It is true, for example, that mercy does not exclude justice and truth, but first and foremost we have to say that mercy is the fullness of justice and the most radiant manifestation of God's truth.» (AL 311)[32]

When it tries to formulate a «*praxis* of conscientious decision making» based on such ecclesiology, *Amoris laetitia* relies upon three basic principles: the gradualness of the law, the distinction between objective morality and subjective culpability, and, finally, the relation between the general norm and the particular situation of individuals.

What I call «praxis of conscientious decision making» gives an illustration in *actu exercito*, i.e., in the concrete, of what, early on, I called the «Christian realism» of *Amoris laetitia*. It also shows how the definition of conscience implicitly entailed by the document becomes relevant in addressing the most difficult pastoral questions of the day.

The Law of Gradualness

The Church encourages the faithful to live according to the Gospel's ideal, yet it does not close her eyes to the reality of those situations in which the fullness of the sacramental marriage is not

---

32 On mercy, see W. Kasper: *Mercy: The Essence of the Gospel and the Key to Christian Life*, trans. William Madges, New York: Paulist Press, 2013.

yet realized. For example, what do we make of those who have contracted a merely civil marriage, who live in «simple cohabitation» (AL 294), or who have entered a new union after divorce (AL 298)? I am not addressing, in this paper, the difficult questions of canonical status and sacramental access. The paper reflects on the notion of conscience, and thus focuses more explicitly on ethical decision making. At stake here is the *praxis* of pastors, the wisdom of their discernment, and the forms of their accompaniment, but also at stake is the identity of concrete individuals, their self-understanding as moral agents for whom «the grace of God works also in their lives» (AL 291). They should, therefore, not only *not* feel abandoned by the Church, but, as still living members, «be encouraged to participate in the life of the community» (AL 243).

The «law of gradualness», that is, the idea that the human being «knows, loves and accomplishes moral good by different stages of growth» (*Familiaris consortio*, 34), does not provide the person with a space of normative exceptionalism, an excuse to forfeit the demands of the law; rather, it throws into relief the fact that to act with conscience always entails responding to the moral call as a moral *subject*, that is, as a concrete human being defined by a personal narrative that honestly recognizes the presence of failures, no less than moral accomplishments.

For sure the «law of gradualness» is not the «gradualness of the law», i.e., the reduction of the law to the measure of a person's whim; yet it cannot be other than the only *existentially* possible way to heed the moral call, if the moral norm mediates in freedom the invitation of the good, not the yoke of an impossible demand. Thus the «law of gradualness» provides a path for the formation of conscience, that is, for a journey of growth that cannot be improvised, but calls for cultivation, caring, and the adventure of ever deeper understanding.[33]

Consider the way *Amoris laetitia* speaks of the much-debated issue of responsible parenthood and birth control. Although the traditional teachings of the encyclical *Humanae vitae* and the apostolic exhortation *Familiaris consortio* are still offered as a point of reference for the conscience of the faithful, the allusion to so

---

33 Only an *informed* conscience can articulate the *praxis* of «conscientious decision making» presupposed by these reflections.

called «natural methods» is made as a measure «to be promoted» (AL 222), rather than imposed as the only available option.

Furthermore, in the same context, a clear reference to conscience frames the normative indication. Following the quotation of the well-known passage of *Gaudium et spes* 16, which speaks of conscience as «the most secret core and sanctuary of a person», the place where «each one is alone with God, whose voice echoes in the depths of the heart», the exhortation adds: «The more the couple tries to listen in conscience to God and his commandments, and is accompanied spiritually, the more their decision will be profoundly free of subjective caprice and accommodation to prevailing social mores.» (AL 222)

What is being said here of birth control applies, analogically, to other situations. Given the fairly rich casuistry disseminated throughout the document, I take the predicament of divorced and remarried couples as a case in point. The analysis is striking for its sense of nuances, and the recognition that the identification of the *right* moral path calls for a weighing of circumstances, and careful discernment of different situations. Any rigid «application» of abstract principles, supposedly fitting for every single case, is to be avoided: «The divorced who have entered a new union ... can find themselves in a variety of situations, which should not be pigeonholed or fit into overly rigid classifications leaving no room for a suitable personal and pastoral discernment.» (AL 298)

When thinking of the 1930 «solemn» proclamation of Pius XI quoted earlier in this chapter, one cannot fail to observe that a different language also reflects a different appreciation of the complexity of the reality of which the language speaks, a more humble attitude toward the predicament of personal situations, and, most especially, a refraining from an approach that talks *down* to people, rather than *with* them.

Moral discernment, as Pascal suggests, is an exercise in *esprit de finesse*, rather than *esprit de géometrie*. This explains the recurring emphasis on the need to carefully distinguish the particularity of each situation: «One thing is a second union consolidated over time, with new children, proven fidelity, generous self-giving, Christian commitment, a consciousness of its irregularity and of the great difficulty of going back without feeling *in conscience* that one would fall into new sins ... *There are other cases* of those who made every effort to save their first marriage and were unjustly abandoned ... *Another thing* is a new union arising from a recent divorce.» (AL 298)[34] Pope Francis concludes, with statements borrowed from the *Relatio* of the 2014 Synod, that «the discernment of pastors must always take place "by adequately distinguishing", with an approach which "carefully discerns situations". We know that no "easy recipes" exist.» (AL 298)

As a way to articulate the relation between general principles and concrete situations, I find note 329 in the paragraph just quoted (AL 298) especially pertinent. When acknowledging situations in which «for serious reasons, such as the children's upbringing, a man and a woman cannot satisfy the obligation to separate» (AL 298), the question of the feasibility of their living together «as brothers and sisters», as the norm of Church requires, is implicitly left open to the judgment of the spouses' conscience. Thus, the reference to *Familiaris consortio* 84 is juxtaposed to *Gaudium et spes* 51: «In such situations, many people... point out that if certain expressions of intimacy are lacking, it often happens that faithfulness is endangered and the good of the children suffers.» Of course, the norm remains valid in its generality, but, in this case, the principle of «lesser evil» seems to insinuate the moral plausibility of a different course of action.

Objective Morality and Subjective Culpability

The distinction between objective morality and subjective culpability is nothing new for moral theology. Only, it does receive in this context renewed importance.[35] The justification for the distinction lies in the recognition that the discipline of the Church cannot by itself exhaust the mystery of a person's conscience, if her situation before God (*coram Deo*) is one of ultimate radicalness. Indeed, «each person's situation before God and their life in grace are mysteries which no one can fully know from without» (*Evangelii gaudium*, 172).

---

34 The emphases in the text are mine.
35 On this see the insightful piece of C.M. Kelly: «The Role of the Moral Theologian in the Church: A Proposal in Light of *Amoris Laetitia*», in: *Theological Studies* 77 (2016), 922-948.

Though a being-in-relation, the person is ultimately given over to an «idiocy»[36] inaccessible to complete communicability, even within the communion of the Church. The singularity of each human being can be fully transparent, in its ultimate truth, to God alone, who, as Augustine suggests, exceeds the intimacy each individual experiences with himself, since He is «interior intimo meo et superior summo meo» («higher than my highest and more inward than my innermost self»).[37]

Two quotations from *Amoris laetitia* follow in the same direction, articulating the depth of the spiritual insight into a moral principle. The first: «For this reason, a negative judgment about an objective situation does not imply a judgment about the imputability or culpability of the person involved» (AL 302). Again, the task of pastoral discernment, «while taking into account a person's properly formed conscience» is to «take responsibility for these situations» (AL 302). The second quotation is even more striking in its rhetorical power: «It is possible that in an objective situation of sin – which may not be subjectively culpable, or fully such – a person can be living in the life of grace and charity, while receiving the Church's help to this end.» (AL 306)

As I mentioned earlier, the mystery of conscience is grounded in the mystery of freedom itself, whose unfathomable depth cannot be purely measured by correspondence to the norm. Thinking that «everything is black and while, we sometimes close off the way of grace and growth, and discourage paths of sanctification which give glory to God» (AL 305). The letter of the law cannot, in the end, choke the life of the Spirit, even when it expresses itself in «small steps, in the midst of great human limitations» (AL 305).

General Norms and Particular Situations

The final strategy proposed by *Amoris laetitia* concerns the interpretation of moral norms: what does it mean to discern the moral

---

36 I use the term idiocy to indicate, as the Greek root *idios* suggests, the intimate in the person. On this, W. Desmond: *Ethics and the Between*. For the ultimate incommunicability of the person, emerging in a phenomenological analysis of selfhood, see J.F. Crosby: *The Selfhood of the Human Person*, Washington, DC: The Catholic University of America Press, 1996, especially 41-81.

37 *Confessions*, III, 6, II.

meaning of an individual's actions, when recognizing the inevitable gap that exists between the generality of the law, and the concrete predicament of a person? Can the latter throw into relief an *existential* singularity, irreducible to the *essential* capacity presupposed by the norm?[38] If so, what a discerning conscience does is to bridge the gap between «essential» and «existential» morality, recognizing that, if a moral subject cannot have an ethical direction without the former, she ultimately makes decisions based on an interpretive judgment of the latter.

The principle in question is perfectly traditional, as it echoes the distinction known to the manuals between the law as the «remote» norm of morality (*norma remota moralitatis*), and conscience as the «proximate» norm (*norma proxima moralitatis*).[39] A more philosophically astute justification, however, might seek a different framework. *Amoris laetitia*, quoting Thomas Aquinas's descending order of certitude in the articulation of the precepts of natural law (AL 304), fails, in my opinion, to provide such framework. More hopeful is the reference to the statement of the International Theological Commission: «Natural law could not be presented as an already established set of rules that impose themselves *a priori* on the moral subject; rather, it is a source of objective inspiration for the deeply personal process of decision making.» (AL 305)[40]

Its methodological shortcomings notwithstanding, it seems clear that the spirit of *Amoris laetitia* hints at something like a hermeneutic model of interpretation. It argues from the premise that it is impossible to fully grasp a concrete decision in its singularity, and thus a certain degree of abstractness inevitably remains. On the

---

38 Consider the observations of Klaus Demmer on this: «norms must respect the concrete possibilities of the individual they address. Thus, norms cannot be imposed on individuals beyond their moral capacities. The principle of reasonableness must be respected: what surpasses the limits of reasonableness can neither be true nor obligatory» (K. Demmer: *Shaping the Moral Life: An Approach to Moral Theology*, trans. Roberto Dell'Oro, Washington, DC: Georgetown University Press, 2000, 47).
39 So, for an example, J.P. Gury: *Compendium Theologiae Moralis*, Tomus I, 17th ed., Rome: Typis Civilitatis Catholicae, 1866, 19-20.
40 The quotation is from International Theological Commission: *In Search of a Universal Ethic: A New Look at Natural Law* (2009), 59; available at http://www.vatican.ca/roman_curia/congregations/cfaith/cti_documents/rc_con_cfaith_doc_20090520_legge-naturale_en.html

one hand, such abstractness, expressing the generality of the norm, is necessary for the sake of moral communication. The alternative would be a situational ethics deprived of universal validity. On the other hand, something like a space of discretion in the application of the norm cannot be excluded, if the model of conscience presupposed by the exhortation must have any bearing on concrete decision making.

Concluding Reflections

Amoris laetitia has brought fresh insights to the attention of the Church, and something new is obviously at work in *Amoris laetitia*, its continuity with previous doctrine notwithstanding. Still, because continuity is not a rigid uniformity, the doctrine of the Church remains always hospitable to developments, if not radical shifts in paradigms. In the end, *Amoris laetitia* throws into relief three important elements: the stress on the dialogical nature of the Church; the openness to a more nuanced sexual anthropology, which, in turns, grounds a richer understanding of both marital sacramentality and moral subjectivity; and, finally, a model of moral normativity defined by the centrality of discernment in the integration of contextual dimensions.

First, the dialogical nature of the Church is reflected in Pope Francis's frequent appeal to the work of the Bishops' synods, which, in turn, nourishes the content of the exhortation, its constant reference to the experiences of local churches, and its attempt to articulate the gospel of love for our times by way of a careful interpretation of the contemporary situation of marriage and family. The gospel of love does not come down from heaven, a ready-made textbook simply available for judicious application; rather, it is the honest articulation of a communal discernment, requiring the work of empirical reason no less than the insight of faith-inspired understanding. The approach of *Amoris laetitia* is clearly inductive, rather than deductive; integrative, rather than exclusivist; sensitive to the plurality of voices, rather than fixed on the clarity of a univocal refrain.

Secondly, the portrayal of sexual anthropology entailed by the document is worth noting: building on phenomenological attentiveness to the nuances of human experience, it conveys a picture of the person mindful of the complexity of self-appropriation, the centrality of growth, and the importance of development in the definition of sexual identity. Thus, the emphasis on the normativity of sexual ideals is tempered by the recognition that the journey toward a perfect love cannot be embarked upon without adequate attendance to the fallibility of human beings, a realistic assessment of their moral capacity, and the encouragement to persevere in the adventure of constant learning. Only a realist account of the human condition (*conditio humana*) can open up the space for a deeper recognition of the work of grace, the gift of the Spirit effecting, in the sacramental economy, the glorious destiny promised to the poverty of the flesh.

In the end, a renewed attention to the centrality of conscience, and a more concrete picture of the moral subject, emerge from such anthropology: the moral subject is not an abstract agent, but an incarnate actor, not an appendix in the description of the moral event, but a central point of reference in the formulation of moral judgment. The importance of conscience formation, itself premised on mindfulness of the Church's teachings, calls spouses to the silence of prayer, and the honesty of transparent dialogue; in turn, it reminds pastors of the need to listen, and the *finesse* required by careful discernment.

Finally, the question of whether the exhortation signals a change in doctrine was raised, in the wake of its publication, by the media, and by commentators less attuned to the nuances of magisterial developments. Clearly, no changes have occurred, as far as the Church's commitment to fundamental teachings: indissolubility of sacramental marriage, the centrality of love and procreation in marriage, the relevance of gender difference, the inability of homosexual unions to express the reality of a sacramental bond, etc. Certainly, Pope Francis seems to convey a different attitude toward so called «irregular situations» of those who live outside the space envisioned by the canonical law: but to be outside the norm is not to be outside the Church. Whether changes *might* ensue in the future, at the level of a concrete articulation of moral norms, remains to be seen.

The strategies for «conscientious decision making» articulated by *Amoris laetitia*, the law of gradualness, the distinction between objective morality and subjective culpability, and the relation between general norms and concrete situations, point in the direction of something like a *praxis* of pastoral accommodation. Is such a strategy sufficient? Or will the very premises upon which the strategy is based call for further developments in the formulation of the normative system itself? Let the answer be left open for now, entrusted to hope for dialogue within the Church, and to the «work of love» that «enables us to hear without sounds and to see the unseen» (AL 255).

CHAPTER 20
SHOULD WE HOPE FOR A CURE?
Raising a Child with Autism

The adventure of parenting a child with autism has forced my family to reckon with a world previously unknown, leading us all through an experience both puzzling, at times disconcerting, yet also full of grace. We have been in this journey for the past twenty five years, since our son was diagnosed, at the age of two, on the autistic spectrum disorder. He is now a young adult man. Through intensive interventions during all these years, especially applied behavior analysis, speech therapy, etc., our son has made immense strides, but remains almost non-communicative, at least from a verbal point of view. Of course, there is a deeper communion one can establish with him, as I will mention later: to the language of love, affection, and care he is most sensitive and responsive. One may ask: What to expect in a marriage, when confronted with the challenge of disability? How does a child with autism affect the intimacy of the spouses, the balance of relations with other siblings, the flow of everyday life, now filled with unforeseen preoccupations and fears, at times even the disruption of chaos?

All these are very good questions, but, for the sake of this essay, I want to entertain another, perhaps more pressing query: should a family hope for a cure? More specifically, should a family explore all avenues made available by medicine and clinical research in order to overcome the predicament in which they find themselves, as they see their child grow into adulthood with minimal improvements? Or should they just learn to grieve for the child they once had (my child was born and developed perfectly normal until the age of two), and face reality as it is, suppressing any dreams for the adult person they had expected to raise?

Like other parents of children with autism, I have followed the research in this field with great expectations and hope; yet, as someone who works in bioethics, I am aware of the need to submit such effort to a more rational scrutiny. It is a difficult balancing act, which might be described most appropriately as a tension between resistance and surrender, to echo a linguistic pair made famous by Dietrich Bonhoeffer. Resistance: to the deterministic aspects of a condition parents would want to see overcome in their child; and this through research and experimentation for the sake of a cure. Surrender: to the reality of an autistic person, whose world, with the passing of time and the harsh realization that little truly works, becomes increasingly distance and unreachable, yet suggestive of something like an integrity of being, unique for sure, yet no less real. Of course, such a world is different from the world of non-autistic persons; nevertheless it is meaningful in its own mysterious terms.

This essay addresses the question of what justifies biomedical research on persons with autism, given their inability, as it is in the case of our son, to account for decision making capacity. More specifically, I will ask whether a cure for autism is, in the case of adults affected by the condition, in their best interest. My argument will emerge slowly. Having touched upon issues of informed consent and standards for proxy decision making in research, I will eventually reflect, somewhat dialectically, on the notion of autistic integrity, which poses the question of whether research on a cure for autism can be ethically justified at all. For my part, I do not believe something like an autistic integrity exists; what is there, rather, is the plea of parents and guardians to care for persons with autism. The search for a cure is an expression of such care, indeed, an expression of the responsibility we owe the most vulnerable in our midst.

What Justifies Biomedical Research on Persons with Autism?

Individuals affected by autism are regularly offered the opportunity to enter clinical trials as research subjects. One type of research, especially in the case of children, is behavioral research[1]: it addresses sensory difficulties described by persons with autism, not aiming at a cure, per se, but at the alleviation of the symptoms experienced. A more invasive biomedical research, on the other

---

1 U. Frith/F. Happé: «Theory of Mind and Self-Consciousness: What Is it Like to Be Autistic?», in: *Mind & Language* 14/1 (1999), 1-22

hand, aims at curing autism, as opposed to merely relieving symptoms. In this case, the central ethical question focuses on the issue of informed consent, for the problem of decision capacity of persons with autism applies not only to the obvious case of research on children, but also to adults. Standards for guiding research with less than autonomous subjects have been developed historically, and can certainly serve as a point of reference for our particular issue. Still, questions about the specific nature of the deficits faced by persons with autism have forced investigators to consider the issue of research in its most radical version: should we use persons with autism in biomedical research at all?

It is well known that the principle of informed consent represents the cornerstone in the ethics of medical experimentation with human subjects.[2] It is found in all major ethical codes, from Nuremberg to the Belmont Report. Such requirement is necessary because experimentation risks functionalizing the person experimented upon to the ends pursued by science for the wider benefits of society. The plausibility of such means-end relation is less than problematic for someone who argues on utilitarian premises. It requires, on the other hand, a higher level of justification for those committed to the rights of individuals. The utilitarian justification assumes as a starting point the impartiality of a rational calculus of consequences, in which the benefit of the «greatest good for the greatest number» warrants the sacrifice of individual interest for the sake of social utility. An ethics committed to individual rights, on the other hand, strives to see others as others, endowed with a dignity that deserves a respect not subsumable under the finalities defined by personal interest, or the interests of society. For both these justifications, the presence of a «theory of mind» in the research subjects is presumed as a condition for either recognizing the ends of social utility, or for understanding the requirements of individual rights. Now, the core deficits found in autism can be explained precisely by the lack of a theory of mind: persons with autism are not able to recognize that

---

2   R.R. Faden/T.L. Beauchamp. *A History and Theory of Informed Consent*, New York: Oxford University Press, 1986; A Buchanan/D.W. Brock: *Deciding for Others: The Ethics of Surrogate Decision Making*. New York: Cambridge University Press, 1990; T.L. Beauchamp/J.F. Childress: *Principles of Biomedical Ethics* 7th edition, New York: Oxford University Press, 2013.

other persons have minds; they don't see other persons as possessing mental life independent of their own, with beliefs, preferences, desires, and the whole range of intentional attitudes. A lack of theory of mind, as defining the condition of the person with autism, entails, of necessity, failure to appreciate the demands of social utility, as well as those that ground the rights of others as an object of absolute respect.

The conclusion to be drawn here is that, since none of the familiar arguments for the use of subjects in research can be convincing for persons with autism, the latter cannot be presumed to be competent agents, possessing the very mental premises, such as theory of mind, necessary for informed consent. Should we then categorically exclude that all persons with autism can be competent to offer informed consent? If we do not come to such radical conclusion, what precautions ought to be in place for research to take place?

Reflections on the ethics of research with vulnerable population, such as subjects with mental retardation, might offer insights to the predicament in question.[3] It has been suggested that, rather than excluding cognitively disabled entirely, a presumption obviously at odds with requirements of justice, researchers focus on minimal competencies they might possess, so as to, at least, make some research possible. In her work, Celia Fisher pleads for what she calls a «goodness-of-fit» ethic of informed consent, which, rather than concentrating on the deficiencies of potential research subjects, looks instead to «an examination of those aspects of consent setting that are creating or exacerbating consent vulnerability,» as well as «considerations of how the setting can be modified to produce a consent process that best reflects and protects the consumer's hopes, values, concerns and welfare».[4] She describes her «goodness-of-fit» ethic as relational. The term, alluding to «responsiveness to the abilities, values and concerns of research participants» presupposes what I would call a «relational concept of autonomy,» in which the agency of the subject is understood beyond individualistic parameters, as an

---

[3] C.B. Fisher: «Goodness-of-Fit Ethic for Informed Consent for Research Involving Adults with Mental Retardation and Developmental Disabilities», in: *Mental Retardation and Developmental Disabilities Research Review* 9 (2003), 27-31.

[4] *Ibid.* 29.

expression of «connectedness to others» on the part of persons with mental deficiency: «Adults with mental retardation, like all persons, are linked to others in a relationship of reciprocity and dependency. A relational ethic calls for scientists to construct informed consent procedures based upon moral principles of respect, care, and justice guided by the responsiveness to the abilities, values, and concerns of research participants and awareness of the scientists' own competencies and obligations.»[5]

The emphasis on relationality is certainly a positive step in the direction of a less individualistic rendition of autonomy. The question of whether it should be assumed in persons with a theory of mind deficit is, however, equally pressing. A possible answer might come from a careful look at the necessary conditions for competency. Bioethicists Allan Buchanan and Dan Brock define competency in terms of the «capacity for understanding and communication,» and as «the capacity for reasoning and deliberation».[6] Even if not all autistic persons are incapable of the latter, the former seems to present special problems for them on account of their theory of mind's deficits: «Understanding is not merely a formal or abstract process, but also requires the ability to appreciate the nature and meaning of potential alternatives – what it would be like and «feel» like to be in possible future states and to undergo various experiences – and to integrate this appreciation in to one's decision making...In light of these considerations, one has to conclude that to understand is to appreciate alternatives. More specifically to understand what it is to consent to a particular form of action is to understand what would happen if the action were not performed».[7]

Two difficulties emerge, when applying this notion of understanding to persons with autism. First, persons with autism might have difficulties with the abstraction that is required to employ counterfactuals that capture a kind of experience different from their own. Secondly, the significance of self-awareness to understanding presents a problem. In their work, Frith and Happé have highlighted the unique nature of self-consciousness in persons with autism, one in which the communicating character of one's expe-

---

5 Ibid.
6 A. Buchanan/D.W. Brock: Deciding for Others, 23.
7 Ibid. 24.

riences to others becomes problematic for them. More specifically, the problem with persons with autism, is not communicating per-se, but is being aware, i.e., self-conscious of what is being communicated: «...Simply put, they lack the cognitive machinery to represent their thoughts and feelings as thoughts and feelings.»[8] The conclusion to draw from these considerations is that, given the unique difficulties they face, criteria for competent decision making cannot be ascribed to persons with autism.

What Standards for Proxy Decision Making?

If persons with autism are not competent to make decisions, then a surrogate or proxy decision maker will be asked to make decisions on behalf of the person. Two standards are normally invoked in this situation.[9] The first is the substituted judgment standard. According to this standard, surrogates make decisions for another agent based upon the agent's prior beliefs and preferences. If an agent has consented to a particular course of action, then that is what the surrogate should consent to on behalf of the agent. Thus the surrogate «substitutes his judgment» for that of the agent on whose behalf he offers consent, according to the now incapacitated agent's own prior beliefs. Seen in this way, the surrogate is someone who, given his knowledge of another's beliefs and preferences, has moral authority to speak on that person's behalf. Clearly, that substituted judgment standard presupposes that the person in question was at some point competent to make decisions. Thus the standard is invoked in the case of progressively worsening conditions that lead to incompetence, such as Alzheimer's. In the words of Devettere, «...proxies can use substituted judgment only when they know what the patient would have wanted».[10] But this is not the case with autism, for here there was never a time during which the person was in fact competent. Thus the substituted judgment standard is not an appropriate

---

8   U. Frith/Happé: «Theory of Mind and Self-Consciousness», 7.
9   R.J. Devettere: *Practical Decision Making in Health Care Ethics: Cases and Concepts,* 3rd edition, Washington, DC: Georgetown University Press, 2010.
10 *Ibid.* 102.

means by which decisions can be made to participate in research for incompetent autistic subjects. We need to explore another standard.

A second standard is the best interest standard. According to this standard, an intervention is ethical, if it is in the best interest of the incompetent person. The fact that the intervention is expected to be in that person's best interest has justificatory power over the lack of the person's consent. Typically, decisions made on behalf of very young children, or persons born with cognitive disabilities that affect decision making capacity, can be made utilizing this standard. It would appear that such is the case also for persons with autism and yet it has been argued that participation in some biomedical experiences of adults with autism cannot be justified on the basis of a best interest standard. It must be noted that the focus of the thesis in question does not affect research studies designated to investigate the nature of autism in noninvasive ways. In addition to genetic research, I mentioned before that behavioral research, or minimally invasive research is ethically unproblematic. The focus, rather, is on biomedical studies that are more invasive, and are done with the aim of investigating potential cures for autism, such as remedies for theory of mind deficits. The unique ways in which theory of mind deficits shape a person with autism has stunning repercussions for the application of the best interest standard. Here are some of the arguments. To being with, the best interest standard is only applicable when it is genuinely in the best interest of the person on whose behalf consent is offered to participate in research. Any non-therapeutic research, that is, research that might benefit others, but not autistic subjects themselves, is not in the latter's best interest. Could then research on adults, aiming at a potential cure for autism, be of benefit to autistic subjects at all? Some have rejected such a research as beneficial.[11] To live a full human life certainly entails the ability to enter into relationships of the kind that are not available to persons without a theory mind. For sure it would be better for future children to be born without autism. It is not clear that adults who have always lacked theory of mind would be benefitted by gaining theory of mind in mid-life: who and what they are would be compromised tremendously by gaining a theory of mind. Therapeutic

---

11   D.R. Barnbaum: *The Ethics of Autism: Among Them, But Not of Them*, Bloomington, IN: Indiana University Press, 2008.

research designed to benefit them may not be a benefit to autistic adults at all. What to make of the shift from a world in which full intentionality is not ascribed to other humans to one in which others are rendered, all of a sudden, more complex? Can individuals who are non-autistic imagine what it would be like to move from the autistic into the non-autistic world? The complexity of the adjustment, in this argument, pertains to the sudden realization of the complexity of the presence of others, of their «coming-into-being», so to speak, for the first time. But this also entails a second dimension, a coming-into-being of oneself as well, something like the discovery of oneself in the form of a new presence to oneself. Being cured of autism requires a person to undergo a radical change in the understanding not only of other persons, but of oneself; it requires adjusting oneself to a new way of being-in-the-world. If this is not a benefit to the person with autism, there is no reason to assume there will be aspirational benefits to be gained from this research either, allowing future adults with autism to be cured. It is clear that, if one were to abide by this argument, the best interest standard cannot be used to offer ethical justification for the use of adults with autism for research studies that have the therapeutic benefit of possibly restoring the subject's theory of mind. It is a stunning conclusion, which warrants digging more deeply into some of the premises of the argument. I will highlight three of them.

First, the argument stresses the uniqueness of autism relative to other conditions. Such an emphasis seems necessary, unless one comes to the untenable position that no therapeutic research on any condition can be considered morally acceptable, insofar as it might entail a dramatic change in the person's being-in-the-world. For an example, if one were to substitute «cancer», or «blindness» for autism, one could see how the argument runs into difficulties. But precisely here is the difference: autism does not merely change the way individuals interact with the world – it changes the very nature of the self, and the very nature of the other inhabitants in the world with whom the autistic person interacts. Blindness restricts the nature of communication with other persons, but it does not challenge a blind person's ability to interact with other persons qua persons. Consider the following observations of Barnbaum: «Narratives of illness are replete with descriptions of the ways in which cancer changes the relationship that people have with themselves, and with others. But the poignancy that characterizes these narratives results from the pain that can emerge from loneliness, isolation, or loss of relationships with others. The uniqueness of autism is that the pain in the loneliness, isolation, or loss of relationships is not necessarily as great and in some cases, is not there at all.»[12]

A second consideration: what if research on sensory difficulties were to lead, in the long run, to restoring theory of mind? It is well known that, in persons with autism, the connections between the amygdala, which determines appropriate emotional responses to sensory stimuli, and the sensors themselves, is disrupted. The result is that the autistic person's perception of the landscape that surrounds him is not what it should be, rendering bright lights, high-pitched sounds, or scratchy clothing almost unbearable. Especially in cases where sensory overload is causing a person with autism pain, it is clear that such research is of therapeutic benefit to the participant himself, as well as to future persons with autism. Thus, such research is morally permissible, especially if it is done with children. In the case of young children, it is possible that their view of other persons as well as their own self-concept might not be solidified, such that the acquisition of theory of mind would not prove to be harmful. Instead, the early acquisition of theory of mind would allow a child to ultimately experience the full range of human capabilities. Children with autism may yet be able to acquire theory of mind without forcing them to radically re-think their notions of others and of self. For them, participation in research study that may hold out the possibility of a cure would be ethically justifiable. On the other hand, such participation might not be justifiable for adults with autism. Rather than involving adults with autism in biomedical experiments that might drastically change their relationship with others and their understanding of self, shouldn't adults with autism be allowed to live out the lives the way they are?

An Ethic of Autistic Integrity

The notion of autistic integrity questions the condition of autism as a disease in need of a cure, pointing to the fact that being

---

12   *Ibid.* 199.

autistic is just that, a way of «being in the world».[13] The call for autistic integrity concerns those persons with autism who cannot recognize others in the fullest sense; second, it resigns to the notion that changing autistic persons into persons endowed with the theory of mind would require them to undergo a fundamental shift in the way they interact with others and understands themselves. For the most part, people without autism do not know what it means to have autism, that is, what it would be like to have autism. Similarly, those who have autism do not know what it would be like not to be autistic. If parents are justified in making sure that their future children will not have autism, for adults with autism, this expectation comes too late. To have certain capabilities or potentialities is, *objectively* speaking, better than not having them. Autistic integrity though, calls for the recognition of persons with autism as individuals with personalities and preferences of their own. Thus, to foist a cure on persons with autism is failing to recognize him as a person in his own right, because that cure assumes that the person would be better off cured. As I said above, whereas curing cancer or restoring sight to a person who was blind does not fundamentally change that individual *qua* person, restoring theory of mind would. For sure, an ethic that requires the non-autistic population to respect the differences of the autistic population places a burden on non-autistic society: it calls for a notion of distributive justice that provides autistic persons with the services they need. Distributive justice demands the integration of the persons with autism into society, not unlike any other effort to integrate persons with disabilities into a society that often does not do enough to promote accessibility. According to Francesca Happé, «…the central coherence account of autism…predicts skills as well as failures, and can be best characterized not as a deficient account, but in terms of *cognitive style*».[14] If it is so, then the lack of theory of mind is simply the way some adults are. It would have been better for autistic

---

13 D.R. Barnbaum: *The Ethics of Autism;* R.E. Barnes/H. McCabe: «Should We Welcome a Cure for Autism? A Survey of the Arguments», in: *Medicine, Health Care, and Philosophy* 15(2012), 255-269.
14 F. Happé: «Parts and Wholes, Meaning and Minds: Central Coherence and Its Relation to Theory of Mind», in: S. Baron-Cohn et al. (eds.): *Understanding Other Minds: Perspectives from Developmental Cognitive Neuroscience*, 2nd edition, New York: Oxford University Press, 2000, 205.

individuals if, from birth, they had an intact theory of mind. Such persons would have been able to enjoy all the human capabilities, enter into full relationships, and speak the same moral language as non-autistic persons. But as adults, each person has to be appreciated for what he/she is.

Autistic Integrity: A Tentative Rejoinder

What to make of such arguments? Let me conclude my essay with a brief rejoinder to the autistic integrity account. I do so with full awareness of the value of such account, which insofar as it stresses the singularity of persons with autism, their unique value and dignity, might even distinguish itself as morally superior. Autistic integrity speaks, on premises hard to understand at first, of what it means «to-be-on-the-world» differently. Thus, it also entails a call for acceptance and care beyond the obvious effort to return persons with autism to a rigid definition of *normalcy*. To fail to accept the otherness of autism makes us all, the *others* to them, poorer, because closed to the richness of diversity, ourselves autistic in a sense, because disengaged from the vast realm of alterity, of which autism is just another dimension. And yet, I find the argument of autistic integrity ambiguous, if it entails the barring of a search for cure *for the sake* of persons with autism. This seems, at least on the face of it, paternalistic and pretentious. I ask: how to discern, in the recognition of an autistic integrity without further qualifications, the fine line between commitment to the good of the autistic person, and capitulation to the complacency of resignation? More positively, how to galvanize the efforts of scientists, the generosity of parents, the solidarity of civic institutions, the call of churches to service of the vulnerable, if it all ends with a simple attestation of difference in «cognitive style»? My argument is twofold. I look, first, at a cure for autism as a requirement of commutative justice, beyond the insufficient warrants of informed consent standards highlighted so far. Second, I question on a philosophical basis the notion of an absolute discontinuity between autistic and non-autistic worlds; consequently, I also question the radical incommensurability of the life-worlds of persons transitioning from autistic self-enclosure to other-orientation and relationality.

First, to cure autism is a requirement of *commutative* justice. The search for a cure is a moral imperative because we owe persons with autism what is *due to them*, on account of the fact that they share in our human condition. To be human, for them as for us, is to be «given to be,» a predicament that is neither conditional upon a choice on the part of the autistic person – so that one would be free to accept or refuse such predicament from a position of neutral distance, nor depends upon the recognition of non-autistic others, as if the conditions for human belonging were to be defined by criteria of social acceptability. Thus, we cannot say for sure whether the person with autism would be better or worse off, were he to transition to a non-autistic reality, because such judgment would entail an all-encompassing viewpoint that, standing above two worlds, it includes them both. But the truth is that we live and make sense of the world around us, and of other human beings, from perspectives that can only be partial and limited, never completely wise, nor entirely ignorant. In our human condition, we are given to be in the enigmatic expectation of a community of being, of all beings made possible by their openness to each other. Such anticipation of otherness at the heart of all new life (think of the child's expectation for the mother's embrace captured by the artist!) is not chosen, but given. We do have to reckon with the painful realization that the gift of such openness is not actualized by everyone, and perhaps will never be. And yet the therapeutic restoration of such predisposition to be filled by the other's embrace, and to reach out to the alterity that awaits human beings, is to be pursued not only because it is deemed a good in a list of selected capabilities, but because it is *given* to use as a dimension of our integrity of being, before our recognition of it as a functional value, or as a pleasure-producing quality.

I come to a second rejoinder. The autistic integrity argument assumes the incommensurability of autistic and non-autistic worlds, thus postulating an absolute discontinuity between the two. At the same time, it fails to take into account the fact that discontinuity of mental states, such as the passage from a lack to a possession of theory of mind, is grounded by the *continuity* of an embodied, temporal individuality. How does such an underlying continuity affect the process of mental re-adjustment? If the acquisition of a theory of mind represents a process of *coming to mindfulness* that involves not only the «mind», but the very «being» of the person, how can we predict a priori the results of such transition? Could it not be that «traces» of presence to oneself, rooted in the embodied and temporalized flesh of the person, provide the passage way, literally, the «metaphor» (*meta-pherein*) for a still to be fully articulated release into mindfulness? If that were the case then, to acquire a theory of mind is less akin to passing from not having to *having something* – as in buying a house or winning cash in the lottery. Also, it cannot be like becoming someone else either, for, if the embodied and temporal preconditions of one's identity are to be taken seriously as conditions for becoming mindful, we can only become *more fully* what we already are. I think of the communication with my autistic son, the exchange of caresses, and kisses, and touches that has built, over the years, something like a memory of my presence in him, inarticulate for sure, perhaps even shrouded in a lack of awareness (does he know I am his father?), yet mysteriously recognized by him. It is *me* he feels in the abandonment of spontaneous trust, or the search for proximity at the startled realization of a foreign presence, a person, an animal, a scary thing. Isn't such «living in the flesh» of his already sparked with stray flashes of a more mindful intimation? Isn't this indeterminacy of *being* as elementally embodied a raw anticipation of a more fully coming to mindfulness, as in a fully developed theory of mind? And how could he gain the latter without the former?

All this seems to suggest that the recognition of the otherness of autism, of its mysterious integrity, can still coexist with the effort to bring such unexpressed intimacy into a gift for others, a gift for us. This is why I say, let the search for a cure continue. Let it continue for the benefit of persons with autism, indeed, for their best interest.

CHAPTER 21
LIFE TAKEN, LIFE GIVEN
Meta-Ethical Reflections on Covid-19 Pandemic

I write from the United States of America. Here, as everywhere else in the world, the pandemic has brought desolation and much social havoc. After a rather long period of «lock down,» and the short-lived hope for things returning to «normalcy,» we have undergone a resurgence of concerns, first, in light of the so-called «delta variant,» and, more recently, of the new «omicron mutation» of the virus. It seems like the story is not likely to end soon, though it has lasted long enough to spread dread and fear. We wonder what to make of it.

The pandemic escapes definition. And this not just because we remain unclear about its etiology and the conditions that led to its rapid spreading. The very nature of the «thing itself» is in question, for its *phenomenic* appearance bespeaks an indeterminacy hard to define.

Following the lead elicited by the realization in question, I offer a number of considerations, confronting (1) the epistemological question, (2) the existential crisis generated by the universal experience of uncertainty under COVID, and (3) the challenges on the responsibility of individuals and the duties of society in terms of public health policy measures.

What Are We Undergoing? The Epistemological Challenge

A *caveat*, to begin, about the scope of my contribution. In the United States, the reflection on the pandemic has been mostly about ethics. This is understandable. Given the pressing issues faced by health care institutions and civic society, such focus is not only necessary; it is even praiseworthy.

Indeed, the United States has produced very quickly a substantive body of ethical statements, from those concerning criteria for resource rationing in intensive care units, to guidelines on public health restrictions and sound research protocols, to the articulation of global strategies in the distribution of vaccines.[1]

In a very significant way, the *pragmatic* character of the contributions in question responds to the dramatic quality of the pandemic, with all its facets, some of them unquestionably tragic. I think of health care professionals' struggle to find adequate resources for their patients; of the elderly (and not so old) dying in the loneliness of isolation; of families unable to reach out to their loved ones affected by the virus, deprived even of the most basic expressions of human *pietas*. Since ethics is a conduit to action, «doing something» to address the current predicament, if not doing everything possible, gives witness to the courage of resistance.

In this contribution, however, I intend to go beyond a purely *ethical* analysis, pleading for a mindfulness beyond the pressure of «doing,» the latter collapsing *all* reflection into moralistic statements.

One might object: such *theoretical* pausing cannot keep us from falling into inertia of complacency or, worse, connivance in resignation. I ask for thoughtful «stepping back» that is other to inaction, a thinking that un-makes what brought us here in the first place. In the face of life compromised by sickness or taken by death, is there a *thinking* that might mutate into *thanking* for life initially given? If yes, such thinking could turn out to be a passageway to life's rebirth.

COVID-19 is the name of a global crisis (*pan-demic*). It has come to the world-scene as a strange predicament, long-since predicted yet never seriously addressed. The strangeness pertains to the nature of the phenomenon itself. On the one hand, there is the effort of the scientific community to delimit as clearly as possible the dimensions of the crisis, to reduce it to the univocity of a controllable «object.» The search for the infection's causes, the algorithmic descriptions of its quick spread, the projections into the future impact of the pandemic on a number of social variables –

---

[1] See the *COVID-19 Ethics Resource Center* offered by the Hastings Center and available at https://www.thehastingscenter.org/category/COVID-19/. One finds a wealth of information in this repository of resources, spanning all the ethical issues impacted by the pandemic. More systematically, the volume of Gregory E. Pence, *Pandemic Bioethics* (Peterborough: Broadview Press, 2021) offers an analysis of the ethical issues triggered by COVID within the broader historic perspective of previous pandemics.

from the economical, to the psychological, and the political, are all expressions of an attempt to control, contain, and, potentially, solve the problem.

In all these attempts, however, the scientific approach turns out to be only limited, if not entirely unsuccessful. It bumps at every step into further variables, new and unpredictable scenarios occasioned by what appears like an endless series of mutations, so called «variants,» on the original virus. The shortcomings of previously held statistical projections soon become patent, what seemed established theories must face the trial of new experimental verifications.

There is nothing surprising about this, of course. Philosophers of science have known for a long time that, intrinsic to the very *objectivity* of the scientific method is the testing of hypotheses and, subsequently, the potential falsifiability of every established theory.[2] The indeterminacy of scientific theories, the fact that their validity is never *absolute*, because relative to the broader validity of the paradigm within which they function, only attests to the definitive overcoming of a scientific approach based on positivistic premises. It does not invalidate the truth of the scientific approach altogether.[3]

What is obvious to the scientific community, however, is less clear to the average person, who is still prone to judge the indeterminacy of empirical results as a sign of their unreliability, rather than expression of their only *relative* validity within a process of scientific determining. Those results are indeed *in the truth*, but

---

[2] Justification of the thesis in question goes well beyond the scope of this contribution. Also, the documentation in philosophy of science would take us far afield. For the sake of historical clarity, one might recall the progressive shift from verifiability to falsification in the trajectory that leads to the «critical rationalism» of Karl Popper, most famously in his *The Logic of Scientific Discovery* (London: Routledge, 2002).

[3] However, the positivistic outcome, as Michel Henry suggests, is not without connection to the fundamental premises of modern scientific methodology, insofar as the latter progressively abstracts from life. See Michel Henry, *La Barbarie* (Presses Universitaires de France, 2004). The notion of «shift of paradigm» was popularized by Thomas Kuhn in *The Structure of Scientific Revolutions* (London: Continuum, 2008) and has found wide applicability in the humanities, including ethics, philosophy, and theology. As an example, see Klaus Demmer, *Moraltheologische Methodenlehre* (Freiburg: Herder, 1989).

only as hitherto established. They are not *the truth* tout court, if one sees the latter for what it is, i.e., a moment in the longer trajectory toward *final* scientific validation.

The impact of such indeterminacy on the public opinion in the U.S., however, has been enormous, with dire consequences on the lack of confidence toward the political class, and the generalized perception that public policy decisions have no firm grounding or justification in science.[4] The generalized skepticism toward the political class has had also a rebound effect on the credibility of the scientific community itself, based on the positivistic premise, implicitly shared by the public, that science must be able to provide us with unequivocal findings. When this expectation falters, *any* scientific claim seems to become dubious as well, and the distinction between provisions backed up by *proven* data and counter-proposals based in fancy unleashes an uncontrollable conflict of interpretations that turns, soon enough, into a clash of mere opinions.[5]

Is there a *tertium quid* to the contrast between what appear like intractable alternatives, between absolute univocity and relativistic equivocity? What would that be?

I suggest looking in the direction of a notion of *phenomenality* that transcends the measures of control imposed by positivistic premises. I stand, with contemporary phenomenology, for something like a «reversal of intentionality,» an opening toward «saturated phenomena» in which the fullness of intuition exceeds

---

4   Of course, expressions of the so-called «anti-vaccination movement» can be found outside the U.S. as well. The emphasis on «freedom of choice» and personal defense of bodily integrity, key dimensions of «expressive individualism,» are unequivocally central to the American ethos. On this, see Robert Bellah et al., in *Habits of the Heart: Individualism and Commitment in American Life* (Berkeley: University of California Press, 1996). A useful collection on the ethical and public policy issues of vaccination is Jason Schwartz and Arthur L. Caplan, *Vaccination Ethics and Policy: An Introduction with Readings* (Cambridge: MIT Press, 2017).

5   This is indeed the case when one looks at the (pseudo) arguments of the anti-vaccination movement. The latter presents strange features, not least because it espouses elements of the already mentioned «expressive individualistic» ideology with a vaguely Foucaultian critique of power in matters of health care and public health policy. For the latter, see especially *The Birth of the Clinic: An Archeology of Medical Perception* (New York: Pantheon Books, 1973) and *Power and Knowledge: Selected Interviews and Other Writings, 1972-1977* (New York: Vintage Books, 1980).

intentionality, whether defined by scientific categories or conceptual prejudices. I side with an epistemology that lets reality be in its event-like character, already beyond the stricture of a subject-object relation; a «trans-objectivity,» one might say, which is over-determinate, because unpredictable in its capacity for self-disclosure.[6]

I take the current predicament generated by the COVID pandemic to be an example of such an event of disclosure, calling for the patience of attention and for discerning *finesse*, irreducible to both the objective gazing of science and the trivial reductionism of common sense.[7]

Science will continue to approach the phenomenon in its objectivity, hopefully coming to an acceptable measure of control that minimizes risks and restores reality to the ordinariness of daily life. Still, it will not exhaust the phenomenon in its exceeding fullness, for the latter reverses the categories with which we approach it, suggesting something more, something unexpected, an *advent* (*adventus*) of meaning whose eschatological tonality (for everything can come to nothing) might be suggestive of a more radical presence that resists the annihilating force of any pandemic. In the contingency of our suffering, worse, in the experience of

---

6   The reference is to Jean Luc Marion and his notion of «saturated phenomenon.» See his *Studies in Saturated Phenomena* (New York: Fordham University Press, 2002), chap.1, «On Phenomenology of Givenness and First Philosophy.» Also *The Visible and the Revealed* (New York: Fordham University Press, 2008), chap. 3: «Metaphysics and Phenomenology: A Relief for Theology.» For the idea of «over-determinacy,» see William Desmond, *Being and Between* (Albany: SUNY, 1995). The proximity (and distance) of Desmond to Marion is articulated by the former in «Saturated Phenomena and the Hyperboles of Being: On Marion's Postmetaphysical Thought,» in *The Voiding of Being: The Doing and Undoing of Metaphysics in Modernity* (Washington, D.C.: The Catholic University of America Press, 2020), 193-225.

7   I draw the distinction between *event* and *object* with respect to COVID from Marion. See his conference for the Diocese of Besancon from July 2020, available at https://www.youtube.com/watch?v=s5xMAEpfOSs. (As of now, I am not aware of whether the text of such intervention is available in a published form.) The notion of «event,» drawing on Heidegger's phenomenology of *Ereignis*, has become a *topos* of post-modern theological discourse as well. On the «unconditional affirmation» of the event, see John D. Caputo, *The Weakness of God: A Theology of the Event* (Bloomington: Indiana University Press, 2006).

death and desolation, we might be awaken to the goodness of life given to us in the first place, and now, only too late, potentially poised to be lost in the assault of disease.

Such realization, though in too many cases endowed with the quality of regret (think of those on their deathbed cursing the decision to forfeit vaccination!), might open up a different appreciation for being, an awakening from the slumber of indifference towards the miracle of life's beauty. Perhaps the very experience of *contingency* can provide an access to the simple, yet unexplainable mystery of life's givenness.[8] Such experience has nothing esoteric to it. As an experience available in principle to everyone, it touches all (*cum-tangere*) in the intimacy of one's consciousness. Life should be welcome as the «intimate universal.»[9]

A Tale of Desolation? The Existential Challenge

Like so many processes in our contemporary world, COVID-19 is the most recent manifestation of globalization. From a purely empirical perspective, globalization has effected many benefits to humankind: it has disseminated scientific knowledge, medical technologies, and health practices, all potentially available for everyone's benefit. More than anything, it has elicited interconnectedness at the widest possible range: global information networks link researchers and health professionals in the search for solutions and remedies for international challenges, and new communication technologies facilitate interaction among people who are at physical distance.[10]

---

8 The argument is not so new. I find myself thinking of Thomas Aquinas famous «third way» on contingency and necessity as having something to do with it. Of course, the point of arrival of my considerations is not God as such, but the affirmation of life's goodness.
9 For this category and its systematic implications see William Desmond, *The Intimate Universal* (New York: Colombia University Press, 2016).
10 The impact of globalization on the ethics of health care and medicine is carefully analyzed by Henk Ten Have, *Global Bioethics: An Introduction* (London: Routledge, 2016). More broadly, from a sociological perspective, Ulrich Beck, *The Metamorphosis of the World* (Cambridge: Polity Press, 2016).

At the same time, with COVID-19, we have found ourselves differently linked, sharing in a common experience of contingency. Sparing no one, the pandemic has made us all equally vulnerable, all equally exposed. Such a realization has come at a high cost. The pandemic has offered the spectacle of empty streets and ghostly cities, of human proximity wounded, of physical distancing. It has deprived people of the exuberance of embraces, the kindness of hand shakings, the affection of kisses, and turned relations into fearful interactions among strangers, the neutral exchange of faceless individualities shrouded in the anonymity of protective gears. For elderly people in the last stages of life the suffering has been even more pronounced, for the physical distress can be coupled by diminished quality of life and lack of visiting family and friends.[11]

The way the media in particular speak of our current predicament points to metaphors that emphasize hostility and a pervasive sense of menace. We undergo repeated encouragements to «fight» the virus, read press releases that sound like «bulletins of war,» with daily updates on the number of infected turning quickly into grief for the «fallen victims.»[12] The world has become somber, and the silence bequeathed on our lives, when the day is over and the night comes, feeds our fear of the unknown.

In the suffering and death of so many, we have learned the lesson of *fragility*. Hospitals still struggle with overwhelming demands, facing the agony of resource rationing and the exhaustion of health care personnel. Immense, unspeakable misery, and the struggle for basic survival needs, has brought into evidence the condition of prisoners,[13] those living in extreme poverty at the margins of society, especially in developing countries, the abandoned destined to oblivion in refugee camps from hell.

---

11 See, among others, Larry R. Churchill, «On Being an Elder in a Pandemic,» *Hastings Bioethics Forum*, 13 April 2020.
12 It is as if the pandemic has emphasized the Hobbesian foundational tale of *bellum omnium contra omnes* as constitutive of modern society. On this Charles Taylor, *Sources of the Self: The Making of Modern Identity* (Cambridge: Harvard University Press, 1989).
13 On the predicament of prisoners during COVID, see Gregory E. Pence, «Prisoner's Dilemma and Vaccination Uptake,» in *Pandemic Bioethics, op. cit.*, 138-141.

We have witnessed the most tragic face of death, the loneliness of separation both physical and spiritual, life coming to its end, without heed for age, social status, or health conditions. The virus reaps indiscriminately: the young, no less than the middle-aged, the strong, no less than the frail.

But «frail» is what we *all* are: radically marked by the experience of finitude at the core of our existence, not just occasionally there, visiting us with the gentle touch of a passing presence, leaving us undeterred in the confidence that everything will go according to plan.[14] We emerge from a night of mysterious origins: called into being beyond choice, we come soon to presumption and complaint, asserting as ours what was only vouchsafed to us. Too late do we learn consent to the darkness from which we came, and to which we finally return.

Some say this is all a tale of absurdity, for it all comes to nothing. But how could this nothing-ness be the final word? If so, why the fighting? Why do we encourage each other to the hope of better days, when all that we are experiencing in this pandemic will be over, when resistance will be given its due, and the solicitation to hold on, in spite of everything, reap its reward?

Life comes and goes, says the custodian of cynical prudence. Yet its rising and ebbing, now made more evident by the fragility of our human condition, might open us to a different wisdom, a different realization. For the sorrowful evidence of life's frailty may also renew our mindfulness of its *given* nature. Coming back to life, after savoring the bitter fruit of its contingency, we might even be wiser, more grateful and less arrogant.

With the pandemic, our claims to autonomous self-determination and control have come to a sobering halt, a moment of crisis that elicits deeper discernment. It had to happen, sooner or later, for the bewitchment had lasted long enough. Since the beginning of modernity, we have bought into the Cartesian dream to be «masters and possessors of nature.»[15] Now the dream has turned into night-

---

14 The issue of finitude, though hardly a major *topos* in American bioethics, defines the philosophical reflection, at least since Heidegger's *Sein und Zeit*. A thorough investigation, moving from an «eidetic» to an «empirical» analysis of fragility and finitude as the condition of possibility of fault can be found in Paul Ricoeur, *Fallible Man* (New York: Fordham University Press, 1986).

15 On the story of modernity, with respect to the dialectic of objectification of being and subjectification of value, see William Desmond, *Ethics and the*

mare. The environmental crisis in all its gravity and the glory of creation reduced to inhospitable wasteland have finally given the lie to all our empty pretenses.[16]

The COVID-19 epidemic has everything to do with our depredation of the earth and the despoiling of its intrinsic value. It is a symptom of our earth's malaise and our failure to care. More it is a sign of *our own* spiritual malaise, as Pope Francis reminds us in *Laudato Si*.[17] Will we be able to remedy the fracture that has separated us from our natural world, turning our assertive subjectivities into a menace to creation, a menace to one another?

The emergence of COVID is not without connection to a chain of disastrous events, all link to one another. I do not intend to offer an etiology of COVID, yet a number of phenomena give us pause. Consider how increasing deforestation pushes wild animals in the proximity of human habitat, how virus hosted by animals are then passed on to humans, exacerbating the reality of zoonosis, a phenomenon well known to scientists as a vehicle of diseases. There is also the demand for meat in first world countries, itself the result of a compromised diet, poor in healthier choices of vegetables, grains, and fruit, giving rise to enormous industrial complexes, like pig and chicken farms. It is easy to see how infected meat might ultimately bring about devastation of global proportions, occasioning the spread of a virus through international transportation, mass mobility of people, business travelling, tourism, etc.[18]

Between (Albany: SUNY, 2001), 17-45. In the same vein, already Henri De Lubac, *Le drame de l'humanisme athée* (Paris: Cerf, 1999) and Romano Guardini, *Das Ende der Neuzeit. Ein Versuch zur Orientierung* (Würzburg: Echter Verlag, 1951). More recently also Remi Brague, *The Kingdom of Man: Genesis and Failure of the Modern Project* (Notre Dame: University of Notre Dame Press, 2018).

16 On this Pope Francis, Encyclical Letter *Laudato Si': On Care for Our Common Home*(2015), at https://www.vatican.va/content/dam/francesco/pdf/encyclicals/documents/papa-francesco_20150524_enciclica-laudato-si_en.pdf.

17 «Nor must the critique of a misguided anthropocentrism underestimate the importance of interpersonal relations. If the present ecological crisis is one small sign of the ethical, cultural and spiritual crisis of modernity, we cannot presume to heal our relationship with nature and the environment without healing all fundamental human relationships,» *Laudato Si'*, no. 119.

18 On the conditions that facilitate COVID's spread, see the statements of the Center for Disease Control (CDC). COVID-19, at https://www.cdc.gov/coronavirus/2019-ncov/index.html. Also Adam Kucharski, *The Rules*

In light of all this, it becomes clear that the phenomenon of COVID-19 is not just the result of *natural* occurrences, in the face of which we stand, powerless, as if in the hands of an impersonal fate. What happens in nature is already the result of a complex intermediation with the *human* world of economical choices and models of development, themselves infected with a *different virus* of our own creation. It is the result, more than the cause, of financial greed, the self-indulgence of life styles defined by consumption and excess.

We have built for ourselves an ethos of prevarication and disregard for what is given to us, in the elemental promise of creation.[19] This is why we are called to reconsider our relation to the natural habitat, to recognize that we dwell on this earth as guests, not as masters and lords. We are welcome to a feast of abundance and generosity from life's own sources, summoning us to care and responsibility. Those sources nourish our incarnate condition: the air we breathe, the water we drink, the bodies we feed.[20]

We have been given everything, but ours is only an endowed, not an absolute, sovereignty. Mindful of its origin, it carries the burden of finitude and the mark of vulnerability. Our destiny is a *wounded freedom*. We might reject it as a curse, a provisional condition to be soon overcome. Or we can learn a different patience: capable of consent to finitude, of renewed porosity to neighborly proximity and distant otherness.[21]

When compared to the predicament of poor countries, especially in the so-called Global South, the plight of the «developed»

---

of Contagion: Why Things Spread – and Why they Stop (New York: Basic Books, 2020).

19 Such promise is the primal ethos of life, which we endangers with the construction of a second ethos, often inimical to the first. See Roberto Dell'Oro, «On the Ultimate that Is the First: Thinking Beyond (Bio)ethics,» in *Gregorianum* 100, no. 3 (2019): 621-647.

20 This is why the premises of a dualistic anthropology of person against nature, person against person, and the disregard for our condition of *carnal singularity*, show their dreadful consequences. See, Le Corps in *Cahiers d'Études Lévinassiennes* XVIII (2021). In this issue, I develop the notion of the person as «incarnate singularity, coming to itself, in relation to others.» See Roberto Dell'Oro, «Thinking about the Person with Levinas,» 156-180.

21 The notion of «porous self» stands in contrast to the «buffered self» of an atomistic social philosophy in which individuals are opposed, rather than related, to one another. See Charles Taylor, *A Secular Age* (Cambridge: Harvard University Press, 2007).

world, such as the United States, looks more like a luxury. Only in rich countries people can afford the safety of separation. In those not so fortunate, on the other hand, «social distancing» is just an impossibility due to necessity and the weight of dire circumstances. Crowded settings and the lack of affordable distancing confront entire populations as an insurmountable fact. The contrast between the two situations throws into relief a strident paradox, recounting, once more, the tale of disproportion in wealth between poor and rich countries.

To learn finitude and to consent to the limits of our own freedom is more than a sober exercise in philosophical realism. It entails opening our eyes to the reality of human beings who experience such limits *in their own flesh*, so to speak: in the daily challenge to survive, to secure minimal conditions for subsistence, to feed children and family members, to overcome the threat of diseases in spite of the availability of cures too expensive to afford. I think of the immense loss of lives in the Global South: malaria, tuberculosis, lack of drinkable water and basic resources still sow the destruction of millions of lives per year, a situation known for decades.[22] All these predicaments could be overcome by committed international efforts and policies less obtuse and cynical: how many lives saved, how many diseases eradicated, how much suffering avoided!

Maybe only a real conversion of mind and spirit can effect changes. A conversion rooted in the final realization that the other side of finitude, visible on the faces of entire populations in the Global South, is morally culpable: its roots extend into the coldness of our hearts, its shortsightedness is imputable to our unwillingness to look, not to our inability to see what is so obviously plain. Can a different openness expand the horizon of our moral imagination, to finally include what for too long has been left unattended, what has been blatantly passed over in silence?

Our pretentions to monadic solitude have feet of clay. With them, there crumbles the edifice of individualism, the false hopes for an atomistic social philosophy built on egoistic suspicion toward what is different and new, an ethics of calculative rationality bent toward

---

22 The point is forcefully made by Henk Ten Have in *Global Bioethics, op. cit.*

self-fulfillment, impervious to the responsibility of the common good on a global, and not only national, scale.[23]

Our interconnectedness is a matter of fact. It makes us all strong or, conversely, vulnerable, depending on our own attitude toward it.[24] Consider its relevance at a national level, to begin. While COVID-19 may affect everyone, it is especially harmful for particular populations, such as the elderly, people with associated diseases and compromised immune systems. Policy measures are taken for all citizens equally. They ask for the solidarity of the young and healthy with vulnerable populations.[25] They ask for sacrifices from many people who depend from public interaction and economic activity for their living. In richer countries these sacrifices can be temporarily compensated, but in the majority of countries such protective policies are simply impossible.

For sure, in all countries the common good of public health needs to be balanced against economic interests. Here in the United States, during the early stages of the pandemic the focus was on saving the most lives.[26] Hospitals, and especially intensive care

---

23   I am aware that the issue touches upon complicated debates in political philosophy concerning liberty and privacy. Libertarians in general expressed strong disagreement toward any limiting measures imposed by various states in order to control the spread of the pandemic. In their thinking, libertarians rely upon Robert Nozick, *Anarchy, State, and Utopia* (New York: Basic Books, 2013), but in the U.S., they have also appropriated positions expressed by Michel Foucault and Giorgio Agamben's critique of «techno-medical dispotism.» On the American «appropriation» of Agamben, see Christopher Caldwell, «The Coronavirus Philosopher,» *New York Times*, 21 August 2020.

24   The notion of vulnerability has received little attention in American bioethics. On the other hand, together with solidarity, it forms one of the basic principles of European bioethics and biolaw. See Jacob Dahl Rendtorff and Peter Kemp, *Basic Ethical Principles in European Bioethics and Biolaw. Vol. 1: Autonomy, Dignity, Integrity and Vulnerability* (2000). More recently, and with a systematic intention covering different spheres of bioethics, see Henk Ten Have, *Vulnerability: Challenging Bioethics* (London: Routledge, 2016).

25   The point can be best articulated in light of solidaristic and communitarian political philosophies, but is not without merit according to a Rawlsian «theory of justice.»

26   This approach is documented by Jenny Jarvie, «Ethical Dilemmas in the Age of Coronavirus: Whose We Save?» *Los Angeles Times*, 19 March 2020. Also the analysis of Gregory E. Pence, *Pandemic Bioethics*, *op. cit.*, especially

services became progressively insufficient to meet the demands of the pandemic, and were only expanded after enormous struggles. Remarkably, care services survived because of impressive sacrifices of doctors, nurses, and other care professionals, more than technological investment.

The focus on hospital care, however, diverted attention from other care institutions. Nursing homes, for an example, were severely affected by the pandemic, and sufficient protective equipment and testing only became available in a late stage. Ethical discussions of resource allocation were primarily based on utilitarian considerations, without paying attention to vulnerable populations. The role of general practitioners, though for many people the first contact in the care system, was largely ignored. The result has been an increase in deaths and disabilities from causes other than COVID-19.

Common vulnerability calls for international cooperation as well, and the realization that a pandemic cannot be withstood without adequate medical infrastructure, accessible to everyone at the global level. Nor can the plight of individual countries be dealt with in isolation, without forging international agreements, and with a multitude of different stakeholders. The sharing of information, the provision of help, the allocation of scarce resources, will all have to be addressed in a synergy of efforts.

We still need to assimilate the lessons learned during COVID. For sure, the seeds of hope have been sown in the obscurity of small gestures, in acts of solidarity too many to count, too precious to broadcast. Communities in the U.S. have struggled honorably, in spite of everything, sometimes against the ineptitude of their political leadership, to re-imagine a life based on ideals of solidarity and reciprocal solicitude.

Still, we have not payed sufficient attention, especially at the global level, to human interdependence and common vulnerability. While the virus does not recognize borders, countries have sealed their frontiers. In contrast to other disasters, the pandemic does not impact all countries at the same time. This might offer the opportunity to learn from experiences and policies of other countries, yet

---

chap. 4: «Policies for Containment.» Pence lays out four different models of fighting pandemics (55-60).

learning processes at the global level have been minimal. In fact, some countries have sometimes engaged in cynical games of reciprocal blame.[27]

The same lack of interconnectedness can be observed in efforts to develop remedies and vaccines. Absence of coordination and cooperation is now increasingly recognized as an obstacle to address COVID-19. The awareness that we are in this disaster together, and that we can only overcome it through cooperative efforts of the human community as a whole, is stimulating shared endeavors. The articulation of cross-border scientific projects is an effort going in that direction. It should also be demonstrated in policies, through strengthening of international institutions. This is particularly important since the pandemic is enhancing already existing inequalities and injustices, and many countries lacking the resources and facilities to cope adequately with COVID-19 are dependent on the international community for assistance.[28]

Toward Life's Rebirth: The Normative Challenge

The lessons of fragility, finitude, and vulnerability bring us to the threshold of a new vision: they foster an ethos of life that calls for the engagement of intelligence and the courage of moral conversion. To learn a lesson is to become humble; it entails change and the search for resources of meaning hitherto untapped, perhaps disavowed. To learn a lesson is to become mindful, once more, of the goodness of life that offers itself, releasing an energy that runs deep, deeper than our own foolishness. Under what conditions can this occasion the promise of a new beginning, the promise of life's rebirth? I offer a few final considerations with an eye to the global situation.

---

[27] The Trump administration in the United States has not refrained from such games of blame. On the American failure of leadership see Gregory E. Pence, *Pandemic Bioethics, op. cit.*, 182-188.

[28] Paul Farmer, a physician, author and activist, has insisted on this point for years. See already his «Rethinking Medical Ethics: A View from Below,» *Developing World Bioethics* 4, no. 1 (May 2004), long before the break out of COVID.

We must come, first, to a renewed appreciation of the existential reality of *risk*.[29] All of us may succumb to the wounds of disease, the killing of wars, the overwhelming threats of disasters. In light of this, there emerge very specific ethical and political responsibilities toward the vulnerability of individuals and groups who are at greater risk for their health, their life, their dignity.

COVID-19 might be seen, at first glance, as only a *natural*, if certainly unprecedented, determinant of global risk. The pandemic, however, forces us to look at a number of additional factors, all of which involve a multifaceted *ethical* challenge. To focus on the natural genesis of the pandemic, without paying heed to the economic, social, and political inequalities among countries in the world, is to miss the point about the conditions that make its spread faster and more difficult to address. A disaster, whatever its origin, is an ethical challenge because it is a catastrophe that affects human life and harms human existence in multiple dimensions.

Though we can now rely on the availability of a number of vaccines, we cannot count on the ability to defeat permanently the virus that caused the pandemic. The virus could be with us for still a very long time. Immunity against COVID-19, especially in light of its variants, remains something of a hope for the future. If so, we need to flesh out a concept of *solidarity* that extends beyond generic commitment to helping those who are suffering. A pandemic urges the global community to address and reshape structural dimensions that are oppressive and unjust.

The basic contours of an ethics of risk, grounded in a broader concept of solidarity, entail a *definition of community* that rejects provincialism, the false distinction between insiders, i.e., those who can exhibit a claim to fully belonging to the human family, and outsiders, i.e., those that can hope, at best, in a putative participation to it. The dark side of such separation must be unmasked as a conceptual impossibility and a discriminatory practice. No one can be seen as simply standing «in waiting» for full status recognition, as if at the doors of the *humana communitas*. The *Universal Declaration*

---

29   On the notion of «risk society,» already Ulrick Beck, *Risikogesellschaft. Auf dem Weg in eine andere Moderne* (Frankfurt: Suhrkamp, 1986).

on Bioethics and Human Rights recognizes as a universal human right access to quality health and to essential medicines.[30]

Two conclusions follow logically in the wake of such premise. The first concerns *universal access* to the best preventive, diagnostic, and treatment opportunities, beyond their restriction to a few. The distribution of vaccines is a point in case. The only acceptable rule, consistent with a just allocation of the vaccine, is access for all without exceptions.[31]

The second conclusion touches upon the need to renew the commitment to clinical research and pertains to the definition of responsible scientific investigation. The stakes here are very high and the issues complex.[32] The first concerns respect to the integrity of science and the awareness that the advancement of science though not entirely «objective,» as was argued above, cannot eschews dimensions of objective control.

Also the ideal of freedom of investigation must be safeguarded in terms a freedom from conflicts of interests. At stake is the very nature of scientific knowledge as social practice, defined, in a democratic context, by rules of equality, liberty, and fairness. In particular, scientific freedom of inquiry and policy decision making should remain distinguished, if not separate, from one another. Insofar as it voices the reality of public consensus and the will of responsible citizenry, policy decision making and the realm of politics as a whole should maintain their autonomy from the encroachment of scientific power, especially when the latter turns into manipulation of public opinion.

Finally, what is in question here is the essentially «fiduciary» character of scientific knowledge in its pursuit of socially beneficial results, especially when knowledge is gained through experimentation on human subjects and the promise of treatment tested in

---

30 *Universal Declaration on Bioethics and Human Rights*, no. 14. At https://www.aaas.org/sites/default/files/SRHRL/PDF/IHRDArticle15/Universal%20Declaration%20on%20Bioethics%20and%20Human%20Rights_Eng.pdf.
31 The issue is tackled by Gregory E. Pence, *Pandemic Bioethics, op. cit.*, chap. 6: «Developing Vaccines,» with relevant literature.
32 Among others, see Daniel M. Hausman, «Challenge Trials: What Are the Ethical Problems?» *Journal of Medicine and Philosophy* 46(1) (2021): 137-145.

clinical trials.[33] The good of society and the demands of common good in the area of health care come before any concern for profit. The *public* dimensions of research cannot be sacrificed on the altar of *private* gain. When life and the well-being of a community are at stake, profit must take the back seat.

Solidarity extends also to efforts in international cooperation. In this context, a privileged place belongs to the World Health Organization (WHO). Deeply rooted in its mission to lead international health work is the notion that only the commitment of governments in a global synergy can protect, foster, and, ultimately, make effective a universal right to the highest attainable standard of health. The role of WHO must be reaffirmed, securing its vision and action, enhancing the timeliness of its interventions, and strengthening its scientific leadership.

At the same time, shortcomings in the approach taken by the W.H.O. in confronting the COVID pandemic cannot be denied. First, leaders of W.H.O. downplayed the seriousness of the virus. They took, rather naively, China's word as to the virus' origin in a Wuhan market that sold live animals. Since China funds a good portion of the organization's budget, one cannot exclude that certain conflicts of interests were at stake. Also the W.H.O. gave credit to the notion that no community spread had occurred. Second, the W.H.O. failed to recognize and inform the public that masks reduce both the spread of COVID and a person's own chance to contract the virus. Finally, the W.H.O.'s leaders initially claimed that transmission of COVID through asymptomatic individuals was rare.

Still, the WHO is the only international organization with a global outreach, including specifically the needs and concerns of less developed countries coping with an unprecedented catastrophe. The foolishness of provincialism and the narrow mindedness of national self-interests have led many countries to vindicate for themselves a policy of independence and isolation from the rest of the world, as if a pandemic could be faced without a coordinated global strategy. Such an attitude might pay lip service to the idea of *subsidiarity*,

---

33 This point is dear to the philosopher of medicine Edmund Pellegrino in many of his writings. Programmatically in Edmund Pellegrino and David Thomasma, *A Philosophical Basis of Medical Practice* (New York: Oxford University Press, 1981).

and the importance of a strategic intervention based on the claim of a lower authority taking precedence over any higher one, more distant from the local situation. In reality, the attitude in question feeds into a logic of separation that is, to begin, less effective against the pandemic. The disadvantage, furthermore, is not only *de facto* short sighted. It does result in the widening of inequalities and the exacerbation of resource imbalances among different countries. Though all, rich and poor, are vulnerable to the virus, the latter are bound to pay the highest price, and to bear the long-term consequences of lack of cooperation. It is clear that the pandemic is worsening the inequalities that already are associated with processes of globalization, making more people vulnerable and marginalized without health care, employment, and social safety nets.

Ultimately, the moral, and not just strategic, meaning of solidarity is the real issue in the current predicament faced by the human family. Solidarity entails responsibility toward the other in need, itself grounded in the recognition that, as a subject endowed with dignity, every human being is a *person*, an end in itself, not a mean. The articulation of solidarity as a principle of social ethics rests on the concrete reality of a *personal* presence in need, crying for recognition. Thus, the response required of us is not just a reaction based on sentimental notions of sympathy; it is the only *adequate* response to the dignity of the other summoning our attention, an ethical disposition premised on the apprehension of the intrinsic value of every human being.[34]

As a duty, solidarity does not come free, without cost and the readiness of the rich countries to pay the price required by the call for the survival of the poor and the sustainability of the entire planet. This holds true both synchronically, with respect to the different sectors of the economy, and diachronically, that is, in relation to our responsibility for the well-being of future generations and the gauging of available resources.[35]

---

34 With a bow to Emmanuel Levinas, to act morally is to transcend oneself, better to embark in a movement of *trans-ascendence*. See *Totality and Infinity: An Essay on Exteriority* (Pittsburgh: Duquesne University Press, 1969), 35.
35 The preoccupation in question lies at the heart of the birth of bioethics, which Van Rensselaer Potter conceived as a science of survival in his *Bioethics: Bridge to the Future* (Englewood Cliffs: Prentice-Hall, 1971). On the notion of responsibility toward future generations, the classic work remains Hans

Almost a Coda...

There is nothing pretentious or definitive about the suggestions just made. More importantly, there is no contradiction between the initial epistemological claim of over-determinacy, as to the phenomenic nature of the pandemic, and the rather univocal character of the recommendations put forth in the ethical call for global solidarity and justice. After all, something *excessive* define such ethical claims as well. Their urgency only speaks to the overwhelming force with which they summon the conscience of those who are willing to listen and to see.

For sure, no ethical claim, especially when it cries for justice and fairness to strangers, can be entirely reduced to the univocity of norms, if not through the mediation of conscience. Even love can become a commandment, yet one that can be met only by a *loving* freedom. As a moral norm, love, no less than justice, can only be an invitation, not an imposition.

In the end, there is a disclosure of *giftedness* revealed in the strange phenomenality of this pandemic, for under the veil of ignorance that covers our eyes in the darkness of suffering and death there appears the light of a life that is given all along. The gift is too precious to be sacrificed on the altar of unwise freedom, heedless to the call of solidarity and the solicitude of care. What a vulnerable and fragile gift life is! We can only hope for the pandemic to be over soon. And may a «posthumous mindfulness» find us still alive.

Jonas, *The Principle of Responsibility: In Search for an Ethics for the Technological Age* (Chicago: Chicago University Press, 1984).

MIMESIS GROUP
www.mimesis-group.com

MIMESIS INTERNATIONAL
www.mimesisinternational.com
info@mimesisinternational.com

MIMESIS EDIZIONI
www.mimesisedizioni.it
mimesis@mimesisedizioni.it

ÉDITIONS MIMÉSIS
www.editionsmimesis.fr
info@editionsmimesis.fr

MIMESIS COMMUNICATION
www.mim-c.net

MIMESIS EU
www.mim-eu.com

Printed by
Rotomail Italia S.p.A.
October 2024

www.ingramcontent.com/pod-product-compliance
Lightning Source LLC
Chambersburg PA
CBHW031844220426
43663CB00006B/492